Plant Signal Transduction

Frontiers in Molecular Biology

SERIES EDITORS

B. D. Hames

*Department of Biochemistry
and Molecular Biology
University of Leeds, Leeds LS2 9JT, UK*

D. M. Glover

*Department of Genetics,
University of Cambridge, UK*

TITLES IN THE SERIES

Plant Signal Transduction

EDITED BY

Dierk Scheel

Institute of Plant Biochemistry,
Halle,
Germany

and

Claus Wasternack

Institute of Plant Biochemistry,
Halle,
Germany

OXFORD
UNIVERSITY PRESS

This book has been printed digitally and produced in a standard specification
in order to ensure its continuing availability

OXFORD
UNIVERSITY PRESS

Great Clarendon Street, Oxford OX2 6DP
Oxford University Press is a department of the University of Oxford.
It furthers the University's objective of excellence in research, scholarship,
and education by publishing worldwide in
Oxford New York
Auckland Cape Town Dar es Salaam Hong Kong Karachi
Kuala Lumpur Madrid Melbourne Mexico City Nairobi
New Delhi Shanghai Taipei Toronto
With offices in
Argentina Austria Brazil Chile Czech Republic France Greece
Guatemala Hungary Italy Japan South Korea Poland Portugal
Singapore Switzerland Thailand Turkey Ukraine Vietnam

Oxford is a registered trade mark of Oxford University Press
in the UK and in certain other countries

Published in the United States
by Oxford University Press Inc., New York

ISBN 978-0-19-963879-6

Printed and bound in Great Britain by CPI Antony Rowe,
Chippenham and Eastbourne

Contents

4 The role of active oxygen species in plant signal transduction   45

EVA VRANOVÁ, FRANK VAN BREUSEGEM, JAMES DAT, ENRIC BELLES-BOIX AND DIRK INZÉ

9 Recognition and defence signalling in plant/bacterial and fungal interactions

JONG HYUN HAM AND ANDREW BENT

Contributors

DOROTHEA BARTELS
Botanisches Institut, University of Bonn, Kirschallee 1, D-53115 Bonn, Germany

ENRIC BELLES-BOIX
Laboratoire de Biologie Cellulaire, Institut National de la Recherche Agronomique, Route de St-Cyr, F-78026 Versailles Cedex, France

ANDREW BENT
Department of Plant Pathology, University of Wisconsin-Madison, 1630 Linden Drive, Madison, WI 53706-1598, USA

KAPIL BHARTI
Department of Molecular Cell Biology, Biocenter N200, Goethe University, Marie-Curie-Strasse 9, D-60439 Frankfurt/Main, Germany

TON BISSELING
Laboratory of Molecular Biology, Department of Plant Sciences, Wageningen University, Dreijenlaan 3, NL-6703 HA Wageningen, The Netherlands

JAMES DAT
Département de Biologie et Ecophysiologie, Université de Franche-Comté, F-25030 Besançon Cedex, France

MARCEL DICKE
Laboratory of Entomology, Wageningen University, PO Box 8031, NL-6700 EH Wageningen, The Netherlands

S. P. DINESH-KUMAR
MCDB Department, Yale University, OML 451, New Haven, CT 06520-8104, USA

F. A. DITENGOU
Equipe de Microbiologie Forestière, Centre de Recherches, INRA de Nancy, F-54280 Champenoux, France

S. DUPLESSIS
Equipe de Microbiologie Forestière, Centre de Recherches, INRA de Nancy, F-54280 Champenoux, France

HENK FRANSSEN
Laboratory of Molecular Biology, Department of Plant Sciences, Wageningen University, Dreijenlaan 3, NL-6703 HA Wageningen, The Netherlands

JONG HYUN HAM
Department of Plant Pathology, University of Wisconsin-Madison, 1630 Linden Drive, Madison, WI 53706-1598, USA

DIRK INZÉ
Vakgroep Moleculaire Genetica and Departement Plantengenetica, Vlaams
Interuniversitair Instituut voor Biotechnologie (IB), Universiteit Gent, K. L.
Ledeganckstraat 35, B-9000 Gent, Belgium

HANS-HUBERT KIRCH
Botanisches Institut, University of Bonn, Kirschallee 1, D-53115 Bonn, Germany

H. LAGRANGE
Equipe de Microbiologie Forestière, Centre de Recherches, INRA de Nancy, F-54280
Champenoux, France

F. LAPEYRIE
Equipe de Microbiologie Forestière, Centre de Recherches, INRA de Nancy, F-54280
Champenoux, France

F. MARTIN
Equipe de Microbiologie Forestière, Centre de Recherches, INRA de Nancy, F-54280
Champenoux, France

ROSSANA MIRABELLA
Laboratory of Molecular Biology, Department of Plant Sciences, Wageningen
University, Dreijenlaan 3, NL-6703 HA Wageningen, The Netherlands

FERENC NAGY
Institute of Plant Biology, Biological Research Center, PO Box 521, H-6701 Szeged,
Hungary

LUTZ NOVER
Department of Molecular Cell Biology, Biocenter N200, Goethe University, Marie-Curie-
Strasse 9, D-60439 Frankfurt/Main, Germany

JONATHAN PHILLIPS
Max Planck Institut für Züchtungsforschung, Carl von Linne Weg 1, D-50829 Köln,
Germany

JULIO SALINAS
Departamento de Mejora Genética y Biotecnología, INIA, Carretera de la Coruña Km.7,
ES-28040 Madrid, Spain

EBERHARD SCHÄFER
Albert-Ludwigs-Universität Freiburg, Institut für Biologie II, Schänzlestrasse 1, D-79104
Freiburg, Germany

FLORIAN SCHALLER
Ruhr-Universität Bochum, Fakultät für Biologie, Universitätsstrasse 150, Geb. ND3/30,
D-44801 Bochum, Germany

DIERK SCHEEL
The Institute of Plant Biochemistry, Weinberg 3, D-06120 Halle (Saalle), Germany

FRANK VAN BREUSEGEM
Vakgroep Moleculaire Genetica and Departement Plantengenetica, Vlaams Interuniversitair Instituut voor Biotechnologie (VIB), Universiteit Gent, K. L. Ledeganckstraat 35, B-9000 Gent, Belgium

REMCO M. P. VAN POECKE
Laboratory of Entomology, Wageningen University, PO Box 8031, NL-6700 EH Wageningen, The Netherlands

C. VOIBLET
Equipe de Microbiologie Forestière, Centre de Recherches, INRA de Nancy, F-54280 Champenoux, France

EVA VRANOVÁ
Vakgroep Moleculaire Genetica and Departement Plantengenetica, Vlaams Interuniversitair Instituut voor Biotechnologie (VIB), Universiteit Gent, K. L. Ledeganckstraat 35, B-9000 Gent, Belgium

CLAUS WASTERNACK
The Institute of Plant Biochemistry, Weinberg 3, D-06120 Halle (Saalle), Germany

ELMAR W. WEILER
Ruhr-Universität Bochum, Fakultät für Biologie, Universitätsstrasse 150, Geb. ND3/30, D-44801 Bochum, Germany

STEVEN A. WHITHAM
Department of Plant Pathology, Iowa State University, Ames, IA 50011, USA

LIMING XIONG
Department of Plant Sciences, University of Arizona, Tucson, AZ 85721, USA

JIAN-KANG ZHU
Department of Plant Sciences, University of Arizona, Tucson, AZ 85721, USA

Abbreviations

ABA	abscisic acid
ABI	ABA-insensitive
ABA1	*ABA DEFICIENT 1*
ABC	ATP-binding cassette
ABP	actin-binding protein
ABRE	ABA-responsive element
ACC	1-aminocyclopropane-1-carboxylase
ACO	1-aminocyclopropane-1-carboxylic acid oxidase
ACS	1-aminocyclopropane-1-carboxylic acid synthase
ADF	actin-depolymerizing factor
AE	alternative exon
AKAP	A kinase anchoring protein
AM	arbuscular mycorrhiza
AOC	allene oxide cyclase
AOS	active oxygen species: allene oxide synthase
APx	ascorbate peroxidase
ARF	ADP ribosylation factor
AtNHX1	*Arabidopsis thaliana* Na^+/H^+ exchanger 1
Avr	avirulence
BCTV	beet curly top geminivirus
BTH	benzothiazole; benzothiodiazole-7-carbothioic acid *S*-methyl ester
bZIP	basic region-leucine zipper protein
cADPR	cyclic ADP-ribose
CaM	calmodulin
CaMV	cauliflower mosaic virus
CaN	calcineurin
CBF	C-repeat binding factor
Cdk	cyclin-dependent protein kinase
CDPK	calcium-dependent protein kinase
cFR	continuous far-red
CHS	chalcome synthase
Cln	transiently formed cyclin
CMT	cortical microtubule
CMV	cucumber mosaic virus
CNB	calcineurin B subunit
Cp	*Craterostigma plantagineum*
CP	coat protein
cR	continuous red

CRT	C-repeat
CTAD	C-terminal activator domain
CTD	C-terminal domain
DAG	diacylglycerol
DBD	DNA-binding domain
DD-PCR	differential display polymerase chain reaction
DG	1,2-diacylglycerol
DGPP	diacylglycerol pyrophosphate
DMNT	4,8-dimethyl-1,3(*E*),7-nonatriene
DN-OPDA	dinor-oxo-phytodienoic acid
DPI	diphenylene iodonium
DRE	dehydration-responsive element; drought-responsive element
DREB	DRE-binding factor
12,13-EOT	12,13(*S*)-epoxy-9(*Z*),11,15(*Z*)-octadecatrienoic acid
ER	endoplasmic reticulum
EREBP	ethylene-responsive element-binding protein
ERF	ethylene-responsive factor
FB	fine bundle
GFP	green fluorescent protein
GPCR	G-protein-coupled receptor
GPx	glutathione peroxidase
GR	glutathione reductase
GSH	reduced glutathione
GSSG	oxidized glutathione
GST	glutathione *S*-transferase
HD-ZIP	homeodomain-leucine zipper
HIR	high irradiation response
HMGR	3-hydroxy-3-methylglutaryl coenzyme A reductase
H_2O_2	hydrogen peroxide
HOG1	high-osmolarity glycerol 1
HPL	hydroperoxide lyase
13(*S*)-HPOT	(9Z,11E,15Z,13S)-13-hydroperoxy-9,11,15-octadecatrienoic acid
HPRG	hydroxyproline-rich glycoproteins
HR	hypersensitive response
Hrp	hypersensitive response and pathogenicity
hs	heat stress
HSE	heat-stress promoter element
Hsf	heat-stress/shock transcription factor
HSG	heat-stress granule
Hsp	heat-stress/shock protein
IAA	indole acetic acid
INA	2,6-dichloroisonicotinic acid
IN-Leu	1-oxo-indanoyl-isoleucine
IP_3	inositol 1,4,5-trisphosphate

IRES	internal ribosomal entry sites
ISR	induced systemic resistance
JA	jasmonic acid
JIP 60	jasmonic acid-induced protein (ribosome-inactivating protein);
JNK	c-Jun N-terminal kinase
LA	α-linolenic acid
LCO	lipo-chito-oligosaccharide
LEA	late embryogenesis abundant
LFR	low far-red
LOX	lipoxygenase
LRR	leucine-rich repeat
LT_{50}	low temperature that results in 50% plant survival
LTRE	low-temperature-response element
LZ	leucine zipper
MAP	mitogen-activated protein
MAPK	mitogen-activated protein kinase
MAPKK	mitogen-activated protein kinase kinase
MAPKKK	mitogen-activated protein kinase kinase kinase
MBP	myelin basic protein
MeJA	methyl jasmonate
MeSA	methyl salicylate
MP	membrane potential
NAADP	nicotinic acid adenine dinucleotide phosphate
NBS	nucleotide-binding site
NF-κB	nuclear factor κB
NES	nuclear export signal
NHE	Na^+/H^+ exchanger
NHERF	Na^+/H^+ exchanger regulatory factor
NLS	nuclear localization signal
NMDA	N-methyl-D-aspartate
NO	nitric oxide
$^{\wedge}O_2$	singlet oxygen
$O_2^{\bullet-}$	superoxide radical
$OH^{\bullet}$	hydroxyl radical
OPC-4:0	3-oxo-2-(2′-pentenyl)cyclopentane-1-butanoic acid
OPC-6:0	3-oxo-2-(2′-pentenyl)cyclopentane-1-hexanoic acid
OPC-8:0	3-oxo-2(2′(Z)-pentenyl)-cyclopentane-1-octanoic acid
OPDA	12-oxo-10,15(Z)-phytodienoic acid
OPR	12-oxo-10,15(Z)-phytodienoic acid reductase
ORCAs	octadecanoid-responsive *Catharanthus* AP2-domain proteins
ORF	open reading frame
os	oxidative stress
PA	phosphatidic acid
PAL	phenylalanine ammonia-lyase

PCD	programmed cell death
PCNA	proliferating cell nuclear antigen
P5CS	Δ^1-pyrroline-5-carboxylate synthase
PEBV	pea early-browning tobravirus
PI	phosphatidylinositol; proteinase inhibitor
PI5K	phosphatidylinositol 5-kinase
PIP_2	phosphatidylinositol 4,5-bisphosphonate
PI-PLC	phosphatidylinositol-specific phospholipase C
PKC	protein kinase C
PLA_2	phospholipases A_2
PLC	phospholipase C
PLD	phospholipase D
PP2C	protein phosphatase 2C
PR	pathogenesis-related
PSbMV	pea seed-borne mosaic virus
PSI	photosystem I
PSII	photosystem II
PTGS	post-transcriptional gene silencing
PTP	protein tyrosine phosphatase
PVX	potato virus X
R gene	disease-resistance gene
RAB genes	genes responsive to ABA
RLK	receptor-like kinases
RMD	RNA-mediated defence
ROI	reactive oxygen intermediate
ROS	reactive oxygen species
RyR	ryanodine receptor
SA	salicylic acid
SABP	salicylic-acid-binding protein
SAP	sequestered area of phytochrome
SAR	systemic acquired resistance
SHR	systemic hypersensitive response
SIPK	SA-induced protein kinase
SOD	superoxide dismutase
SOS	salt overly sensitive
STRE	stress-response element
SV	simian virus
TBSV	tomato bushy stunt virus
TCV	turnip crinkle virus
TEV	tobacco etch potyvirus
TGMV	tomato golden mosaic geminivirus
TIR	toll or interleukin-1 receptor
7TM	seven transmembrane
TMTT	4,8,12-trimethyl-1,3(E),7(E),11-tridecatetraene

TOR	target of rapamycin
VIC	voltage-independent cation channel
VIGS	virus-induced gene silencing
VLFR	very low far-red
WClMV	white clover mosaic potexvirus
WDV	wheat dwarf mastrevirus
WIPK	wound-induced protein kinase

1 | Signal transduction in plants: cross-talk with the environment

DIERK SCHEEL AND CLAUS WASTERNACK

The growth and differentiation of living organisms are continuously adjusted to a multitude of environmental factors, each of which underlies a perpetual variation. The sessile life form of plants further emphasizes the requirement of efficient adaptation and defence mechanisms. The changes in environmental factors may range from moderate to dramatic and concern between one and many components at the same time. They may be abiotic or biotic in nature and range from essential to toxic in their effects. Among the numerous abiotic factors are nutrients, light, oxygen, water, temperature, gravity, wind, touch and chemicals (Fig. 1). Biotic factors are represented by other organisms involved in symbiotic, pathogenic or herbivorous interactions with plants. All of these environmental factors are independently and specifically recognized by plants. Because of the absence of an equivalent of the circulating bloodstream in plants, all organs, and in many cases every cell, can perceive these signals and respond.

Perception and overall response are linked by signal transduction pathways at cellular, systemic, and interorganismic levels. The basic signalling elements and their sequence of linkage within signal transduction networks appear to be conserved between plants, animals, and lower eukaryotes to a surprisingly large extent. The coordinate action of receptors, G-proteins, ion channels, Ca^{2+}, inositol phosphates, phosphatidic acid, protein kinases, protein phosphatases, hormones, peptides, and transcription factors results in specifically altered gene expression patterns, forming the essential basis of the adaptive or defensive multicomponent response.

In order the guarantee proper adaptation to the environment, signals generated by individual perception of the multitude of environmental factors need to be integrated and evaluated according to their importance. This assessment is part of the signal transduction process, which itself is not a matter of linear chains, as usually suggested by our simplified schemes, but rather occurs within networks. Cross-talk between different signalling pathways within such networks appears to be the basis for the evaluation of the importance of incoming signals. Several cases of cross-talk

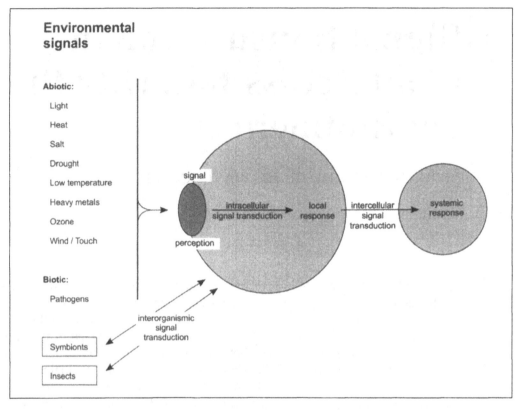

Fig. 1 Abiotic and biotic signals of the environment are specifically perceived by plant cells, and intracellular, intercellular as well as interorganismic signal transduction lead to numerous specific responses.

have been well documented, such as wounding and pathogen attack, irradiation and pathogen attack, ozone and pathogen attack, or salt and desiccation stresses, to name just a few examples.

The connection between signalling pathways occurs at all levels, since identical signalling elements are employed. One of the earliest and most common second messengers is Ca^{2+}. Changes in cytosolic Ca^{2+} levels are observed in diverse signalling processes. However, Ca^{2+} signatures and the origin of Ca^{2+} may differ from case to case, possibly explaining both, the basis of signal specificity and cross-talk. For example, moderate, sustained alterations in cytosolic Ca^{2+} levels have been observed upon attack of plants by fungal pathogens, whereas characteristic oscillations appear to occur during the early stages of plant mycorrhizal and rhizobacterial interactions. Furthermore, the Ca^{2+} source may differ, indicating the involvement of different ion channels. After pathogen attack, for example, rapid influx of Ca^{2+} from the apoplast appears to be the primary source of Ca^{2+} ions, whereas upon cold stress the release of Ca^{2+} from internal stores predominates.

A more complex picture may be expected at the level of mitogen-activated protein (MAP) kinases and transcription factors. *Arabidopsis* harbours 20 different MAP kinases that may be involved in at least as many MAP kinase cascades. In positive regulation, different cascades appear to be co-activated by distinct stresses. Since the same MAP kinases were found to be involved in different stress signalling cascades, cross-talk is very probably occurring at this level. Furthermore, negative regulation of signalling pathways by distinct MAP kinases has been described. Interestingly, those MAP kinases were found to simultaneously positively affect other signalling pathways. Therefore, MAP kinases and probably protein kinase cascades in general have the potential to guarantee both signal specificity and cross-talk.

Transcription factors involved in plant stress responses display a modular principle of interaction that maintains signal specificity and at the same time enables assessment and integration at the terminal part of signal transduction. Besides the enormous amount of information available for heat-shock transcription factors (Chapter 5), common elements have been identified in transcription factors that appear to play a role in cold acclimation (Chapter 6), dehydration stress (Chapter 7), salt stress (Chapter 8), wound response (Chapter 3), and pathogen defence (Chapter 9) (Fig. 2). Among those, the AP2/ERF domain with its single AP2 DNA-binding domain plays a central role. AP2/ERF domain-containing transcription factors are ethylene-responsive (ERF, ethylene-response factor, formerly EREBPs, ethylene-responsive element binding proteins), and regulate ethylene-responsive defence genes via the GCC-box. Interestingly, both Pti4 and Pti5 of tomato which are activated by Pto kinase (Chapter 9) belong to the family of AP2/ERF domain proteins. Furthermore, transcription factors involved in cold-acclimation and dehydration-stress response, such as DREB1, DREB2 and CBF1 (Chapters 6 and 7) are ERF domain

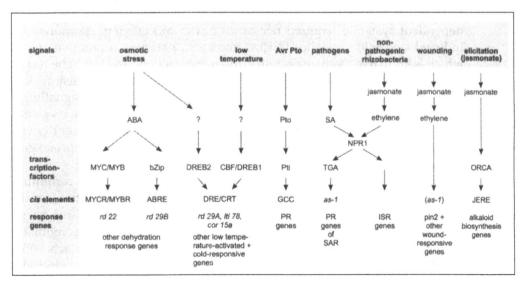

Fig. 2 Specific transcription factors and *cis* regulatory elements are known as key regulators of different stress-responsive signal transduction pathways. Upstream elements of ethylene-responsive defence genes exhibit convergent or divergent functions.

proteins. Their over-expression results in constitutive expression of stress-responsive genes and increased cold and drought tolerance. This might result from cross-talk between both signal transduction pathways at the level of a common *cis* element, the dehydration-responsive element (DRE). DRE (TACCGACAT) contains the so-called C-repeat (CRT), CCGAC, and low temperature-response elements (LTREs). Transcription factors like DREB2 (dehydration) and CBF/DREB1 (low temperature) bind to promoters harbouring this DRE/CTR/LTRE element and might therefore be responsible for the convergence of both pathways (Fig. 2). In contrast to this abscisic acid (ABA)-independent signalling, bZIP-type and MYC/MYB-type transcription factors of the dehydration-stress pathway are members of two other large families of transcription factors and act in an ABA-dependent manner (Fig. 2). The modular construction principle of transcription factors and their *cis* elements is further indicated by the fact that ERFs and DREB/CBF differ only in two amino acids of their AP2 domain which may be responsible for their different DNA-binding specificity.

A family of AP2-domain-containing transcription factors are the jasmonate-specific factors, octadecanoid-responsive *Catharanthus* AP2-domain proteins (ORCAs). They are involved in metabolic switches from primary to secondary metabolism. Another large family of transcription factors known to act in defence gene expression includes proteins binding to activation sequence-1 (*as-1*) cognate promoter elements, such as *nos-1*. Accumulating evidence suggests that many biotic and abiotic stresses converge via *as-1*-type promoter elements. *as-1* elements are composed of two tandem repeats of a TGACG motif, which occur in numerous pathogenesis-related (PR) gene promoters and bind TGA factors. Seven members of the TGA family have been described for *Arabidopsis*. Therefore, 28 different types of dimers are theoretically possible, which could explain the great diversity of transcriptional regulation of stress-related target genes. The NPR1 protein, an essential element of the salicylate-dependent systemic acquired resistance (SAR) and ethylene/jasmonate-dependent induced systemic resistance (ISR) (Chapter 9), harbours a putative ankyrin repeat which is known to mediate protein–protein interactions and might be responsible for the interaction of NPR1 with various TGA factors. In conclusion, it might be hypothesized that *as-1*-like promoter elements connect different signalling pathways of abiotic and biotic stresses by combinatorial interaction between *cis* elements, TGA factors, and NPR1. Through this knot of cross-talk, many different environmental signals may be integrated, assessed, and translated into an appropriate gene expression pattern.

The cells of plants and other multicellular organisms need to communicate with each other in order to coordinate their growth, differentiation, and adaptive measures in response to environmental variation. The mechanisms for such intercellular or systemic communication include the generation and secretion of systemic signals as a result of intracellular signalling, as well as direct contacts between cells. Extracellular signalling molecules interact with receptors of target cells (which can be every cell in the organism) and thereby initiate distinct intracellular signal transduction pathways of systemic responses. Among systemic signals of plants known to be involved in environmental signalling are hormones (including salicylate), peptides, such as the wound signal, systemin, sugars, and probably also electrical signals. Systemic signalling and responses play most important roles in wound (Chapter 3), pathogen (Chapters 9 and 10), and insect responses of plants (Chapter 13).

As elements of ecosystems, plants take part in numerous mutualistic and parasitic interactions that require interorganismic communication (Chapters 9–13). Among the more than 800 000 species living on earth, approximately 60 % represent insects of which about 50 % are herbivores. Therefore, a complex interorganismic communication system exists between plants, herbivorous insects, and their predators, which is described in Chapter 13.

The purpose of this book is to give an up-to-date summary of the enormous amount of information that is now available on signal transduction processes involved in the communication of plants with abiotic (Chapters 2–8) and biotic elements (Chapters 9–13) of their environment. Because of this specific structure of the book, many important aspects of plant signal transduction, such as hormone signalling, could not be given full treatment. World authorities in their respective fields have written the 12 chapters of this book. Each of the chapters summarizes the current knowledge of a specific stress response and emphasizes the mechanisms of signal transduction involved.

2 | Light perception and signal transduction

FERENC NAGY AND EBERHARD SCHÄFER

1. Introduction

For sessile organisms like plants, adaptations to changes in the natural environment are essential. Besides nutrient supply, light is obviously the most important environmental factor. To optimize their adaptive capacity, plants have evolved a relatively large number of photoreceptors to monitor light quality – ranging from UV to the far-red part of the spectrum – light quantity and temporal as well as spatial patterns of light. The UVA/blue-light absorbing photoreceptor, phototropin, encoded by the *NPH1* gene (1), is the most important photoreceptor for monitoring the direction of light, although cryptochromes and phytochromes are also involved in the regulation of this response (2,3). The regulation of other light-dependent physiological reactions is mediated by the UVB photoreceptors characterized primarily by action spectroscopy (4), the UVA/blue photoreceptors *CRY1* and *CRY2* (5,6) and the red/far-red reversible phytochromes (7). The physiological functions of the various photoreceptors are tightly coupled to and harmonized with those of the endogenous circadian clock, which itself can be entrained and reset by light under natural light conditions (8). To analyse and determine the biological roles of the individual photoreceptors and to study the signal transduction process underlying light-regulated gene expression and development, biochemical, cell biological, and genetic approaches have been used in recent years.

It is our goal to provide an up-to-date, critical review of the molecular events required for light-regulated signal transduction. Therefore, we summarize and discuss data derived from studying the photoreceptor families, photoreceptor mutants, which is followed by our attempt to interpret, in this context, results obtained by biochemical, genetic, and cell biological studies.

2. The cryptochrome (CRY1 and CRY2) and phytochrome (PHYA–E) photoreceptor families

The cryptochromes CRY1 and CRY2 of *Arabidopsis* are encoded by the *HY4* and *FHA* gene, respectively (5,6). Each protein consists of a photolyase homologue domain and a C-terminal extension, their functional chromophore is FAD; as autemna chromophore either a deazaflavin or a pterin is used (9).

In *Arabidopsis*, phytochromes are encoded by five genes (*PHYA–E*) (10,11). The phyA photoreceptor has an open chain tetrapyrol (phytochromobilin), which is covalently linked to the apoprotein by a thioether linkage. Expression of the phy apoproteins in yeast allowed to demonstrate that phyA (12), phyB (13), phyC, and phyE (14) can autocatalytically link the chromophore *in vitro* (12,13). Interestingly, the positions of the maxima of the difference spectra for phyA and phyB are almost identical, whereas those of phyC and phyE are shifted by 10 nm (14). These data indicate that different phytochromes may have distinct spectral quality detection capacities *in vivo*. In addition, these studies imply that all phytochrome genes encode photoreversible pigments and that these pigments exhibit specific red/far-red reversibility between their red-light and far-red absorbing P_R-P_{FR} forms, respectively. These properties render P_{FR}/P_{tot} ratio light quality dependent, thereby establishing the phytochromes as excellent light quality detectors in the red/far-red range of the spectrum.

PhyA, the dominant phytochrome of aetiolated seedlings, is rapidly degraded by proteolysis in its P_{FR} form (15). This feature of the phyA photoreceptor seems to be unique and clearly distinguishes phyA from the other phytochromes.

3. Photoreceptor mutants and over-expression studies

By now, single point or deletion mutations in all genes encoding photoreceptors, except *PHYC*, have been described. The availability of these and other double/triple and quadruple mutants, together with that of transgenic plants containing over-expressed photoreceptors, made feasible the analysing of the roles of individual photoreceptors and the characterization of the importance of their interactions in regulating light-dependent physiological responses (16). It has became evident that cryptochromes are the dominant UVA blue-light photoreceptors that control physiological responses ranging from the inhibition of hypocotyl growth and gene expression to flowering (9). CRY2 seems to be more effective at low blue-light fluence rates whereas CRY1 functions at higher fluence rates.

The members of the phytochrome photoreceptor family have not only different modes of action but also different, yet partially overlapping functions. PhyA is the most highly specialized phytochrome. It is responsible for the very low fluence rate and the far-red high irradiance responses (17). The extraordinary responsiveness (sensitivity to light) of this photoreceptor allows phyA to control germination of seeds buried in the soil and induce germination when seeds are exposed to a brief light treatment, even star light (17). The other phytochromes control, to different extents, the classical red/far-red reversible induction (low fluence rate) responses and the responses to continuous red light (18).

In the case of phyA, the proteolytical degradation of the physiologically active P_{FR} form is believed to terminate signalling. As for the other phytochromes, since their P_{FR} form is relatively stable, the switch-off mechanism is still not understood but the

P_{FR} to P_R dark reversion is a possible tool to stop signalling (14,19). PhyB seems to be the most prominent light-stable phytochrome. It is involved in most phytochrome-mediated responses, ranging from germination through inhibition of hypocotyl growth and light quality adaptation (shade avoidance syndrome), to flowering (20–22). PhyE has a major role in controlling internode elongation and shade avoidance responses (23), whereas phyD has only a marginal role in controlling red-light-mediated responses (24).

Network(s) mediating interactions between these photoreceptors seem to be extremely complex. Experimental data, indicating the existence of such network(s), known as responsiveness amplification of certain physiological reactions by different light treatments, have been obtained long before the complexity of the phytochrome and cryptochrome families became evident (25). Recently, genetic and physiological studies have demonstrated that the function of CRY1 is strongly dependent on that of phyA and phyB (26–28) and a specific interaction between the individual pathways controlled by phyD and CRY1 has also been detected (29). Furthermore, it has been shown that phyA positively controls the function of phyB (29), whereas phyB negatively effects phyA-mediated signalling (30).

4. Biochemical approaches to the elucidation of the mechanism of light-induced signal transduction

The primary molecular function of phytochromes in mediating light signal transduction remains to be elucidated. Not long after the discovery of phytochrome, a general debate started about how phytochrome regulates light responses, i.e. is phytochrome a membrane receptor, or does it function as a light-activated enzyme or does it control gene transcription directly (31)? As quite often in science, all three hypotheses may – in the end – be correct.

In lower plants like mosses, ferns, and algae, the observed action dichroism for phytochrome-controlled orientation responses strongly favours a membrane function (32). Recently, it was shown that the *Synechocystis* genome encodes for a phytochrome-like photoreceptor that has all the characteristics of a prokaryotic two-component histidine kinase phosphorelay system (33). In higher plants like oat coleoptiles or parsley cell suspension cultures, a very rapid, phytochrome-controlled, protein phosphorylation was described. This observation indicates that phosphorylation cascade(s) might play a role in mediating the early steps of phytochrome-controlled signalling (34). However, in contrast to cyanobacteria, phytochromes of higher plants are light-regulated serine/threonine protein kinases rather than constituents of a two-component histidine kinase-like receptor complex (35).

Another approach to the analysis of the mode of phytochrome-dependent signalling was microinjection of hypocotyl cells of chromophore biosynthesis-deficient tomato mutant (36–38). Results obtained by this approach indicated that the phytochrome-controlled signalling cascade includes steps that affect levels of some of the well-known second messengers identified in other eukaryotic cells: namely, light absorption by phytochromes triggers activation of a trimeric GTP-binding protein that is followed by a bifurcated signal transduction pathway. One branch activated by phyA but not by phyB modulates cGMP levels and leads to the expression of chalcone synthase (*CHS*) and to the induction of anthocyanin biosynthesis. The other branch activated by both phyA and phyB regulates chloroplast development and the expression of *CAB* genes. Additional experiments indicated a cross-talk between these branches. Although these observations have been supported by pharmacological studies, the demonstration of the modulation of internal Ca^{2+} and cGMP levels as well as the identification of the putative target proteins have remained elusive.

5. Genetic analysis of signal transduction

The combination of molecular and genetic approaches have proved highly efficient in analysing the role of different photoreceptors controlling photomorphogenesis as well as in identifying candidate genes involved in signal transduction. In a pioneering study, Koorneef screened in white light for mutants resembling a dark grown phenotype (39). This and other similar screens resulted, mainly, in the isolation and characterization of genes that either control chromophore biosynthesis or encode the photoreceptors themselves. The other type of screening, pioneered by Chory and later by Deng and their co-workers, searched for plants exhibiting light-grown phenotypes, although the plants were grown in darkness (for reviews see 7, 40, 41). This approach led to the isolation of the so-called *COP* (constitutive photomorphogenic) and *DET* (de-deetiolated) mutants. Characterization of the *COP/DET* and later the *FUS* mutants revealed that the switch between photomorphogenesis and deetiolation is regulated by a complex suppressor system that, in contrast to the photoreceptors, promotes the deetiolation pathway by repressing photomorphogenesis in darkness. These two types of screens led to the isolation of dozens of mutants displaying aberrant photomorphogenic phenotypes. However, irrespective of whether or not the genes underlying these aberrant phenotypes were isolated, it turned out to be unexpectedly difficult to unravel the function of these mutations at molecular level.

The first mutants belonging to the *COP/DET/FUS* group, namely *DET1* and *COP1* were isolated a decade ago. Genetic analysis and mapping of these and other mutants identified 11 COP/DET/FUS loci in the *Arabidopsis* genome (72). The mutants in all 11 loci are recessive and it has been shown that mutations in *COP1*, *COP10* and *DET1* loci do not affect, whereas mutations in the other eight *COP/DET* loci prevent, formation of the COP9 signalosome (73) and are lethal at seedling stage. This latter observation indicated that the COP9 signalosome might have a more general role in *Arabidopsis* development. The pleiotropic nature of the *COP/DET* mutants, however, made it difficult to decipher how the various *COP/DET* mutants interfere with the photomorphogenesis in *Arabidopsis*. Considering genetic, physiological, and molecular aspects, the most comprehensive information is available about COP1. In recent papers it has been shown that COP1 is excluded from the nucleus in the dark and interacts physically with HY5, a bZIP-type transcription factor (74, 75), a positive regulator of photomorphogenesis. These observations indicated that COP1 and HY5 interaction could only take place in darkness and should negatively regulate HY5 activity. In addition, it has been reported (76) that abundance of HY5 is directly correlated with the degree of photomorphogenic development of the seedling and that proteosome inhibitors block degradation of HY5. Therefore, it was concluded that:

- the level of the HY5 protein is primarily controlled by degradation;
- HY5 degradation is mediated most likely by the proteosome pathway.

According to this scenario, COP1 acts like an E3 ubiquitin-protein ligase by recruiting the ubiquitin-conjugating enzyme E2 and mediating transfer of the polyubiquitin from E2 to HY5 and by these steps targeting HY5 to subsequent degradation by the proteosome (76). There is evidence that all other pleiotropic COP/DET/FUS proteins are required for specific degradation of HY5 and, with the exception of COP10 and DET1, are part of the COP9 signalosome (72). The exact role of the COP9 signalosome and that of the COP10 and DET1 in regulating photomorphogenesis is not yet understood. However, it seems to be certain that all COP/DET proteins act as downstream regulators of phototransduction by controlling either directly or indirectly proteosome-mediated degradation of some of the essential signalling molecules (77).

Over the last few years, several laboratories have performed screens to isolate mutants, especially in *Arabidopsis*, which in contrast to the COP/DET mutants exhibit only specific light-dependent phenotypes (for review see 40–42). These signal transduction mutations are expected to display unique phenotypes under specific light conditions as the result of modulating specific steps of signalling rather than affecting the photoreceptors themselves. Accordingly, these mutations affect:

- only the phyA signalling pathway [either very low far-red (VLFR) or continuous far-red (cFR)];
- only the phyB signalling [low far-red (LFR) or continuous red (cR)]; or
- both phyA and phyB signalling [Fig. 1, modified after reference 42].

For example, *RED1*, *PEF2* and *PEF3* are members of a class that affects only phyB signalling. These mutations lead to reduced sensitivity to red light (43,44) and thus share some features with phyB mutants (45). These mutants flower early in short days, have elongated petioles, show decreased sensitivity to red light and defects in the shade-avoidance response. The exact molecular nature of these mutations, however, remains to be elucidated.

In a very specific screen to obtain hypersensitive rather than hyposensitive mutants for phyB signalling, only one mutant was recovered, even though more than 2 million seedlings were tested, (46). This mutant carried a point mutation within the *PHYB* gene itself and produced phyB that had the same photochemical properties as the wild-type phyB molecule. However, the mutant phyB showed three orders of magnitude higher sensitivity to continuous red light than its native counterpart. We

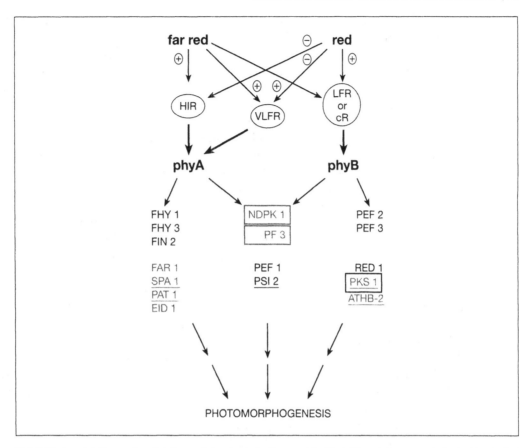

Fig. 1 Photoregulation and early intermediates in phytochrome signalling. Cloned genes are indicated in shaded boxes and negative regulators are underlined. We have only included phyA and phyB pathways although some of the components – especially those in the red signal cascade – may also be involved in phyC, phyD, and/or phyE signalling. Far-red light acts positively via HIR and VLFR on phyA and negatively via LFR on phyB. By contrast, red light acts positively via LFR or continuous red (cR) on phyB and via VLFR on phyA, but negatively via HIR on phyA.

speculate that the strong responsiveness enhancement is probably due to a defect in the cytosolic retention of the mutated phyB protein.

FHY1, FHY3, FIN2, SPA1, FAR1 and *EID1*, and *PAT1* specifically affect phyA signalling (47–52). Genetic analysis suggests that except for *SPA1* and *EID1*, they all act as positive elements in the pathway, i.e. these mutations lead to a reduced sensitivity to far-red light. Pat1 was identified in a screen for mutants with long hypocotyls in far-red light and survival in white light after far-red pre-irradiation (52). *PAT1* is a new member of the plant-specific GRAS (or VHIID) gene family (53). In contrast to the other members of the GRAS family, *PAT1*, tested as PAT1-GFP fusion protein in transgenic *Arabidopsis* seedlings, is localized in the cytosol (52). The spa1 mutant was identified as a suppressor of a weak phyA mutation (48). The *SPA1* gene codes for a putative protein that shows homology to COP1, a known negative regulator of photomorphogenesis (49). Far1 was identified as a suppressor mutation in transgenic lines overexpressing phyA. *FAR1* is a member of a plant-specific gene family (50). Both *SPA1* and *FAR1* proteins contain nuclear localization signal (NLS) motifs and are localized in the nucleus when tested in transient expression studies using GUS fusion constructs. The eid mutants were identified in a screen developed to obtain hypersensitive, phytochrome destruction mutants. It was known that wild-type *Arabidopsis* seedlings show a loss of far-red HIR if continuous far-red light is interrupted every 20 min by 20 min of red light. The aim of the screen was to identify mutants that still show a strong HIR under these conditions. The following four different, new loci could have been identified. Eid4 is a point mutation in phyA, leading to altered destruction and a hypersensitive response. Eid6 is a point mutation in cop1, having no dark phenotype but a hypersensitivity to all light qualities. Eid1 has an extremely strong phenotype. In this mutation, continuous far-red light can be substituted by short light pulses every 30 min, whereas in the wild-type, pulses must be given every 5 min (51). The mutation even shifts the action peak from far-red to red and thereby converts the native photoreceptor to a novel system that exhibits maximal sensitivity to continuous far-red light. Pef1 and psi2 mutants are affected in both phyA and phyB signalling. Psi2 is hypersensitive to both red and far-red light, whereas pef1 shows reduced sensitivity to both light treatments.

Recently, it became possible to express different phytochrome cDNAs in yeast cells and to demonstrate *in vivo* assembly of phytochrome apoproteins with chromophores (12–14). Unfortunately, when full-length phytochromes were fused with the GAL4 binding domain this fusion protein was transactivating and therefore not suitable for use in a yeast two-hybrid screen (Kretsch, Eichenberg, and Schäfer, unpublished data). Nevertheless, two-hybrid screens were successfully used to isolate phytochrome-interacting partners. Three candidates have been identified so far: PIF3 (54), PKS1 (55), and NPDK2 (56). All three proteins interact with the C-terminal part of both phyA and phyB. The function of these tentative phytochrome-interacting partners has been analysed by using knockout mutants or antisense and over-expressor lines. Over-expression of PKS1 leads to slightly elongated hypocotyls under red light, but no effect was detectable under far-red light and in antisense plants. NDPK2 loss-of-function mutants show an enhanced cotyledon opening and hook unfolding under far-red light and a reduced sensitivity in red light. PIF3 antisense lines show a decreased sensitivity to red light and only a very slight effect in far-red light. *In vitro* studies showed that the pif3 protein interacts, in a red/far-red reversible manner, with phyB (57). Furthermore, recently it was also shown that pif3, a basic helix–loop–helix protein, binds specifically to a G-box DNA-sequence motive present in promoters of various light regulated genes, whereas phyB binds reversibly to G-box bound PIF3 only in its physiologically active P_{FR} form (58). Since in PIF3 antisense plants the red-light-induced activation of several light regulated genes is reduced, it is plausible to assume that phyB-controlled transcription of various genes may be mediated by an extremely short signal transduction cascade (59).

It was demonstrated in domain swap experiments that the C-terminal halves of phyA and phyB are interchangeable and involved in signalling whereas the N-terminal halves are responsible for the light absorption and light quality specificity

(60). Nevertheless, no specific N-terminal interaction partner has been described so far. In prokaryotes, phytochrome is part of a phosphorelay system that, besides other constituents, also contains the so-called response regulatory proteins. In the *Arabidopsis* genome several genes encoding response regulator-type proteins have been identified; however, the physiological role of these plant proteins is largely unclear (61). Although several of them have been tested in a yeast two-hybrid assay, so far, only ARR4 has been shown to specifically interact with the N-terminal part of phyB (62). In addition, it was shown that ARR4 specifically and strongly inhibits dark reversion ($P_{FR} \rightarrow P_R$) of phyB in yeast cells and thus stabilizes phyB in its active form *in vitro*. Over-expression of ARR4 in transgenic plants resulted in a strong enhancement of sensitivity of responses to red light but only affected sensitivity to far-red light (62).

6. Cell biological approaches

Physiological, biochemical, and cell biological analysis as well as predictions based on phytochrome sequences indicated that phytochromes are soluble cytosolic proteins (see reference 63). In transgenic plants expressing either phyA:GFP or phyB:GFP fusion proteins under the control of the 35s promoter of cauliflower mosaic virus or that of the endogenous PHYA or PHYB promoters, it was demonstrated that the fusion proteins could complement phyA or phyB mutants, respectively (64–67). Thus these transgenic lines were suitable to analyse light-dependent intracellular localization of phytochromes. In dark-grown seedlings or dark-adapted plants, both phyA and phyB:GFP were localized in the cytosol. However, in the case of phyB, repeated red-light pulses induced accumulation of the phyB:GFP fusion protein in the nuclei, a process that was reversible by subsequent far-red pulses (64,65). Continuous red-light was also effective: accumulation of phyB:GFP showed a fluence rate dependence and phyB:GFP localized in the nuclei always formed characteristic spots (67). Furthermore, it was found that light sensitivity of this reaction is enhanced by pre-irradiation with either red or blue light but not by far-red light. Thus, accumulation of the phyB:GFP fusion protein in the nuclei exhibited a similar responsiveness amplification as previously described for many phyB-mediated responses (67). In addition, it was also shown that the wavelength dependence of the light-induced import of phyB:GFP into the nuclei closely resembles the reported, well-known action spectra of phyB-mediated responses. Therefore, it was concluded that light-controlled import of phyB into the nuclei exhibits the major characteristics of responses mediated generally by phyB.

In contrast to phyB, some immunocytochemical and *in vitro* data (see 63 for review) had previously indicated that phyA might be localized not only in the cytosol but also in the nucleus, yet the importance of these findings had been generally overlooked. However, by using transgenic tobacco and *Arabidopsis* seedlings expressing phyA:GFP, it was unambiguously demonstrated that the phyA:GFP fusion protein is also imported into the nuclei but its import, in contrast to that of phyB, is induced by short far-red or very weak red light pulses and by continuous far-red light treatment (HIR, 65,66). The import of phyA:GFP into the nuclei is more than an order of magnitude faster than that of phyB:GFP and it is preceded by the formation of cytosolic spots (65–67). These cytosolic structures closely resemble the previously described light-induced SAPs (sequestered areas of phytochrome) which probably represent an intermediate of phyA degradation mediated by the ubiquitin system. Thus, like phyB nuclear transport, phyA nuclear transport also shows the previously described light-quality and light-quantity dependence for phyA-mediated responses. Our studies also indicate that, similarly to phyA and phyB:GFP fusion proteins, phyC, phyD, and phyE:GFP are also imported into the nuclei. Under natural, diurnal light conditions, import of all phy:GFP fusion proteins show a diurnal oscillation at the level of spot formation in the nuclei. In the case of phyA:GFP, only far-red/dark cycles are inductive, whereas for the other phyto-chromes either white light/dark or red/dark cycles can be used. The reappearance of nuclear spot formation after light/dark entrainment is, in all cases, detectable prior to the onset of the light phase. These findings indicate that accumulation of phy in the nucleus is regulated by the circadian clock, which indeed could be demonstrated for phyA:GFP and phyB:GFP in continuous darkness after light/dark diurnal entrainment.

7. Conclusions

Due to the very recent progress, we are now able to develop models for the early steps of signal transduction initiated by phyA and phyB. In the case of cry1, cry2 and other phys, however, sufficient data are not yet available to do so. Based on the results of mutant screens (42), analysis of the import of phyB into the nuclei and PIF3/PHYB interactions, it is conceivable to predict that at least some of phyB responses are mediated by a very short signal transduction chain. Thus, light absorption by phyB leads to P_{FR} formation and facilitates import into the nuclei and subsequently, interaction with PIF3 bound to G-box containing promoters. The questions that await answers are what mechanisms, in addition to protein import, regulate the light-induced accumulation of phyB in the nuclei? Is the import process itself light induced or is the light-induced accumulation of phyB regulated by active retention of PhyB in the cytosol and by differential turnover in light and dark and in different compartments? Possibly, light-induced protein phosphorylation – as for phyA – may be important to regulate retention or degradation. Is PIF3 the only target in the nucleus and, if so, how is specificity achieved and the signalling terminated after a light to dark transition?

Interestingly, a relatively large number of mutants has been isolated which affect phyA specific signalling. This can be due either to the screening methods applied or to the fact that, although phyA is a red/far-red reversible pigment with photo-chemical properties almost identical to those of phyB, phyA must maintain its capability to respond maximally to continuous far-red light. It is accepted that one of the first steps after photoconversion of phyA to P_{FR} is the formation of SAPs in the cytosol and transport into the nuclei. It is tempting to predict that autophosphorylation of phyA and its kinase activity are required for formation of SAPs and possibly for eliminating retention. It seems that P_{FR} is the primary form of phyA and this is transported into the nuclei under continuous far-red light. Yet even under these conditions only a minor part of phyA will exist in its P_{FR} form in the nuclei. It follows that either both P_{FR} and P_R forms of phyA will be active in the nucleus or an unknown mechanism has to produce an HIR function not only for transport into but also for function within the nucleus. Probably, the EID1 and SPA1 gene products, localized in the nuclei, together with other components such as FAR1 will form the controlling network in the nucleus. The other candidates, namely FHY1, FHY3, and FIN2 are not yet cloned; therefore it is rather difficult to predict their functions at the molecular level.

PAT1 is so far the only known gene product that is localized in the cytosol, yet it specifically affects phyA signalling. At present, cytosolic functions of phyA and other phytochromes and cryptochromes symbolize the Achilles' heel of photoreceptor-mediated signalling. Clearly, in lower plants one can expect a dominant cytosolic function and it is quite certain that in higher plants, cytosolic functions like light-induced protein phosphorylation of protein localized in the cytosol (pKS1), induction of nuclear transport of transcription factors such as CPRF2 (68,69) are also important. In addition, data predicting involvement of G-proteins, Ca^{2+} and calmodulin, cGMP-dependent pathways still await incorporation into the network phytochrome-mediated signal transduction.

Acknowledgements

The work in Freiburg was supported by grants from the Graduiertenkolleg, SFB 388, DFG, and HFSP (Human Frontier Science Program) to E. S. and by the Humboldt Research Award to F. N. The work in Hungary was supported by a Howard Hughes International Scholar Fellowship, HFSP, DFG and OTKA grants T-025804 and T-032565 to F. N.

References

1. Christie, J. M., Reymond, P., Powell, G. K., Bernasconi, P., Raibekas, A. A., Liscum, E. and Briggs, W. R. (1998) *Arabidopsis* NPH1: a flavoprotein with the properties of a photo-receptor for phototropism. *Science*, **282**, 1698.

2. Ahmad, M., Jarillo, J. A., Smirnova, O. and Cashmore, A. R. (1998) The CRY1 blue light photoreceptor of *Arabidopsis* interacts with phytochrome A *in vitro*. *Mol. Cell*, **1**, 939.

3. Lasceve, G., Leymurie, J., Olney, A., Liscum, E., Christie, J., Vavasseur, A. and Briggs, W. R. (1999) *Arabidopsis* contains at least four independent blue-light activated signal transduction pathways. *Plant Phys.*, **120**, 605.

4. Wellmann, E. (1983) UV irradiation in photomorphogenesis. In Mohr, H. and Shropshire, W. (eds), *Encyclopedia of Plant Physiology*. NS 16B. Springer-Verlag, Berlin, Germany, p. 745.

5. Ahmad, M. and Cashmore, A. R. (1993) HY4 gene of *A. thaliana* encodes a protein with characteristics of a blue-light photoreceptor. *Nature*, **366**, 162.

6. Lin, C. T., Yang, H.Y., Guo, H. W., Mockler, T., Chen, J. and Cashmore, A. R. (1998)

Enhancement of blue-light sensitivity of Arabidopsis seedlings by a blue light receptor cryptochrome 2. *Proc. Natl Acad. Sci. USA*, **95**, 2686.

7. Fankhauser, C. and Chory, J. (1997) Light control of plant development. *Annu. Rev. Cell Dev. Biol.*, **13**, 203.

8. Somers, D. E., Devlin, P. F. and Kay, S. A. (1998) Phytochromes and cryptochromes in the entrainment of the *Arabidopsis* circadian clock. *Science*, **282**, 1488.

9. Cashmore, A. R., Jarillo, J. A., Wu, Y. J, Liu, D. (1999) Cryptochromes: blue light receptors for plants and animals. *Science*, **284**, 760.

10. Sharrock, R. A. and Quail, P. H. (1989) Novel phytochrome sequences in *Arabidopsis thaliana*: structure, evolution and differential expression of a plant regulatory phote-receptor family. *Genes Dev.*, **3**, 1745.

11. Clack, T., Mathews, S. and Sharrock, R. A. (1994) The phytochrome apoprotein family in *Arabidopsis* is encoded by five genes – the sequences and expression of PHYD and PHYE. *Plant Mol. Biol.*, **25**, 413.

12. Lagarias, J. C. and Lagarias, D. M. (1989) Self-assembly of synthetic phytochrome holoprotein *in vitro*. *Proc. Natl Acad. Sci. USA*, **86**, 5778.

13. Kunkel, T., Speth, V., Büche, C. and Schäfer, E. (1995) *In vivo* characterization of phyto-chrome-phycocyanobilin adducts in yeast. *J. Biol. Chem.*, **270**, 20193.

14. Eichenberg, K., Bäurle, I., Paulo, N., Sharrock, R. A., Rüdiger, W. and Schäfer, E. (2000) *Arabidopsis* phytochromes C and E have different spectral characteristics from those of phytochromes A and B. *FEBS Lett*, **470**, 107.

15. Hennig, L., Büche, C., Eichenberg, K. and Schäfer, E. (1999) Dynamic properties of endogenous phytochrome A in *Arabidopsis* seedlings. *Plant Physiol.*, **121**, 571.

16. Whitelam, G. C. and Devlin, P. F. (1997) Roles of different phytochromes in *Arabidopsis* photomorpogenesis. *Plant Cell Environ.*, **20**, 752.

17. Furuya, M. and Schäfer, E. (1996) Photoperception and signaling of induction reactions by different phytochromes. *Trends Plant Sci.*, **1**, 301.

18. Hartmann, K. M., Mollwo, A. and Tebbe, A. (1998) Photocontrol of germination by moon- and starlight. *Z. Pfl. Krankh. Pfl. Schutz Sonderh.*, **XVI**, 119.

19. Eichenberg, K., Kunkel, T., Kretsch, T. Speth, V. and Schäfer, E. (1999) *In vivo* characteriza-tion of chimeric phytochromes in yeast. *J. Biol. Chem.*, **274** (1), 354.

20. Reed, J. W., Nagpal, P., Poole, D. S., Furuya, M. and Chory, J. (1993) Mutations in the gene for red/far-red light receptor phytochromeB alter cell elongation and physiological responses throughout *Arabidopsis* development. *Plant Cell*, **5**, 147.

21. Smith, H. (1994) Sensing the light environment: the function of the phytochrome family. In Kendrick, R. E. and Kronenberg, G. H. M. (eds), *Photomorphogenesis in Plants*, 2nd edn. Kluwer Academic Publishers, Dordrecht, The Netherlands, p. 377.

22. Smith, H. (1995) Physiological and ecological function within the phytochrome family, *Annu. Rev. Plant. Physiol. Plant Mol. Biol.*, **46**, 289

23. Auckermann, M. J., Hirschfeld, M., Wester, L., Weaver, M., Clark, T., Okada, K., Batschauer, A. and Sharrock R. A. (1997) A deletion in the PHYD gene of the *Arabidopsis* Wssilewkija ecotype denies a role for phytochrome D in red/far-red light sensing. *Plant Cell*, **9**, 1317.

24. Devlin, P. F., Robson, P. R., Patel, S. R., Goosey, L., Sharrock, R. A. and Whitelam, G. C. (1999) Phytochrome D acts in the shade-avoidance syndrome in *Arabidopsis* by controlling elongation and flowering time. *Plant Phys.*, **119**, 909.

25. Mohr, H. (1994) Coaction between pigment systems. In Kendrick, R. E. and Kronenberg, G. M. H. (eds.), *Photomorphogenesis in Plants*, 2nd edn. Kluwer Academic Publishers, Dordrecht, The Netherlands, p. 353.

26. Ahmad, M. and Cashmore, A. R. (1993) HY4 gene of *Arabidopsis thaliana* encodes a protein with characteristics of a blue-light photoreceptor. *Nature*, **366**, 162.
27. Neff, M. M. and Chory, J. (1998) Genetic interactions between phytochrome A, phytochrome B, and cryptochrome 1 during *Arabidopsis* development. *Plant Physiol.*, **118**, 27.
28. Hennig, L., Poppe, C., Unger, S. and Schäfer, E. (1999) Control of hypocotyl elongation in *Arabidopsis thaliana* by photoreceptor interaction. *Planta*, **208**, 257.
29. Hennig, L., Funk, M., Whitelam, G. C. and Schäfer, E. (1999) Fuctional interaction of cryptochrome 1 and phytochrome D. *Plant J.*, **20**, 289.
30. Wagner, D., Fairchild, C. D., Kuhn, R. M. and Quail, P. H. (1996) Chromophore-bearing NH_2-terminal domains of phytochromes A and B determine their photosensory specificity and differential light lability. *Proc. Natl Acad. Sci. USA.* **93**, 4011.
31. Kendrick, R. E. and Kronenberg, G. H. M., eds. (1994) *Photomorphogenesis in Plants.* Kluwer Academic Publishers, Dordrecht, The Netherlands
32. Kraml, M. (1994) Light direction and polarization. In Kendrick, R. E. and Kronenberg, G. M. H. (eds), *Photomorphogenesis in Plants*, 2nd edn. Kluwer Academic Publishers, Dordrecht, The Netherlands, p. 417.
33. Yeh, K.-C., Wu, S.-H., Murphy, J. T. and Lagarias, J. C. (1997) A cyanobacterial phytochrome two-component light sensory system. *Science*, **277**, 1505.
34. Harter, K., Frohnmeyer, H., Kircher, S., Kunkel, T., Mühlbauer, S. and Schäfer, E. (1994) Light induces rapid changes of the phosphorylation pattern in the cytosol of evacuolated parsley protoplasts. *Proc. Natl Acad. Sci. USA*, **91**, 5038.
35. Yeh, K.-C. and Lagarias, J.C. (1998) Eukaryotic phytochromes: light-regulated serine/threonine protein kinases with histidine kinase ancestry. *Proc. Natl Acad. Sci. USA*, **95**, 13976.
36. Neuhaus, G., Bowler, C., Kern, R. and Chua, N. H. (1993) Calcium/calmodulin-dependent and -independent phytochrome signal transduction pathways. *Cell*, **73**, 937.
37. Bowler, C., Neuhaus, G., Yamagata, H. and Chua, N.-H. (1994) Cyclic GMP and calcium mediate phytochrome phototransduction. *Cell*, **77**, 73.
38. Kunkel, T., Neuhaus, G., Batschauer, A., Chua, N.-H. and Schäfer, E. (1996) Functional analysis of yeast-derived phytochrome A and B phycocyanobilin adducts. *Plant J.*, **10**, 625.
39. Koorneef, M., Rolff, E. and Spruit, C.J.P. (1980) Genetic control of light-inhibited hypocotyl elongatin in *Arabidopsis thaliana* (L.) Heynh. *Z. Pflanzenphysiol.*, **1008**, 147.
40. Deng, X. W. and Quail, P. H. (1999) Signaling in light-controlled development. *Semin. Cell Dev. Biol.*, **10**, 121.
41. Fankhauser, C. and Chory, J. (1997) Light control of plant development. *Annu. Rev. Cell Dev. Biol.*, **13**, 203.
42. Neff, M., Fankhauser, C. and Chory, J. (2000) Light: an indicator of time and place. *Genes Dev.*, **14**, 257.
43. Wagner, D., Hoecker, U. and Quail, P. H. (1997) RED1 is necessary for phytochrome B-mediated red light-specific signal transduction in *Arabidopsis*. *Plant Cell*, **9**, 731.
44. Ahmad, M. and Cashmore, A. R. (1996) The pef mutants of *Arabidopsis thaliana* define lesions early in the phytochrome signaling pathway. *Plant J.*, **10**, 1103.
45. Reed, J. W., Nagpal, P., Poole, D. S., Furuya, M. and Chory, J. (1993) Mutations in the gene for the red/far-red light receptor phytochrome B alter cell elongation and physiological responses throughout *Arabidopsis* development. *Plant Cell*, **5**, 147.
46. Kretsch, T., Emmler, K. and Schäfer, E. (1995) Spatial and temporal pattern of light-regulated gene expression during tobacco seedling development: the photosystem II related genes Lhcb (Cab) and PsbP (Oee 2) *Plant J.*, **7**, 715.
47. Whitelam, G. C., Johnson, E., Peng, J., Carol, P., Anderson, M. L., Cowl, J. S. and Harberd,

N. P. (1993) Phytochrome A null mutants of *Arabidopsis* display a wild-type phenotype in white light. *Plant Cell*, **5**, 757.

48. Hoecker, U., Xu, Y. and Quail, P. H. (1998) SPA1: a new genetic locus involved in phytochrome A-specific signal transduction. *Plant Cell*, **10**, 19.

49. Hoecker, U., Teppermann, J. M. and Quail, P. H. (1999) SPA1, a WD-repeat protein specific to phytochrome A signal transduction. *Science*, **284**, 496.

50. Hudson, M., Ringli, C., Boylan, M. T. and Quail, P. H. (1999) The FAR1 locus encodes a novel nuclear protein specific to phytochrome A signaling. *Genes Dev.*, **13** , 2017.

51. Kretsch, T., Poppe, C. and Schäfer, E. (2000) A new type of mutation in the plant photo-receptor phytochrome B causes loss of photoreversibility and an extremely enhanced light sensitivity. *Plant J.*, **22** (3), 177.

52. Bolle, C., Koncz, C. and Chua, N.H. (2000) PAT1, a new member of the GRAS family, is involved in phytochrome A signal transduction. *Genes Dev*, **14**, 1269.

53. Pysh, L. D., Wysocka-Diller, J. W., Camilleri, C., Bouchez, D. and Benfey, P. N. (1999) The GRAS gene family in *Arabidopsis*: sequence characterization and basic expression analysis of the SCARECROW-LIKE genes. *Plant J.*, **18**, 111.

54. Ni, M., Tepperman, J. M. and Quail, P. H. (1998) PIF3, a phytochrome-interacting factor necessary for normal photoinduced signal transduction, is a novel basic helix-loop-helix protein. *Cell*, **95**, 657.

55. Fankhauser, C., Yeh, K. C., Lagarias, J. C., Zhang, H., Elich, T. D. and Chory, J. (1999) PKS1, a substrate phosphorylated by phytochrome that modulates light signaling in *Arabidopsis*. *Science*, **284**:1539.

56. Choi, G., Yi., H., Lee, J., Kwon, Y. K., Soh, M. S., Shin, B., Luka, Z., Hahn, T. R. and Song, P. S. (1999) Phytochrome signaling is mediated through nucleoside diphosphate kinase 2. *Nature*, **401**, 610.

57. Ni, M., Tepperman, J. M. and Quail, P. H. (1999) Binding of phytochrome B to its nuclear signalling partner PIF3 is reversibly induced by light. *Nature*, **400**, 781.

58. Martinez-Garcia, J. F., Huq, E. and Quail, P. H. (2000) Direct targeting of light signals to a promoter element-bound transcription factor. *Science*, **288**, 859.

59. Quail, P. H., Boylan, M. T., Parks, B. M., Short, T. W., Xu, Y. and Wagner, D. (1995) Phyto-chromes: photosensory perception and signal transduction. *Science*, **268**, 675.

60. D'Agostino , I. B. and Kieber, J. J. (1999) Posphorelay signal transduction: the emerging family of plant response regulators. *Trends Biochem. Sci.*, **24**, 452.

61. Sweere, U., Eichenberg, K., Lohrmann, J., Bäurle, I., Kudla, J., Schäfer, E. and Harter, K. (2000) Functional interaction of the plant photoreceptor phytochrome B with the response regulator ARR4. Submitted.

62. Nagy, F. and Schäfer, E. (2000) Nuclear and cytosolic events of light-induced, phytochrome-regulated signaling in higher plants. *EMBO J.*, **19**, 157.

63. Yamaguchi, R., Nakamura, M., Mochizuki, N., Kay, S. A. and Nagatani, A. (1999) Light-dependent translocation of a phytochrome B-GFP fusion protein to the nucleus in transgenic *Arabidopsis*. *J. Cell Biol.*, **145**, 437.

64. Kircher, S., Kozma-Bognar, L., Kim, L., Adam, E., Harter, K., Schäfer, E. and Nagy, F. (1999) Light quality-dependent nuclear import of the plant photoreceptors phytochrome A and B. *Plant Cell*, **11**, 1445.

65. Kim, L., Kircher, S., Toth, R., Adam, E., Schäfer, E. and Nagy, F. (2000) Light induced nuclear import of phytochrome-A: GFP fusion proteins is differentially regulated in transgenic tobacco and *Arabidopsis*. *Plant J.*, **22** (2), 125.

66. Gil, P., Kircher, S., Adam, E., Bury, E., Kozma-Bognar, L., Schäfer E. and Nagy F. (2000)

Photocontrol of subcellular partitioning of phytochrome-B:GFP fusion protein in tobacco seedlings. *Plant J.*, **22** (2), 135.

67. Speth, V., Otto, V. and Schäfer, E. (1987) Intracellular localisation of phytochrome and ubiquitin in red-light-irradiated oat coleoptiles by electron microscopy. *Planta*, **171**, 332.

68. Kircher et al., unpublished

69. Gil et al., unpublished

70. Kircher, S., Ledger, S., Hayashi, H., Weisshaar, B., Schäfer, E. and Frohnmeyer, H. (1998) CPRF4a, a novel plant bZIP protein of the CPFR family: comparative analysis of light dependent espression, post-transcriptional regulation, nuclear import and hetero-dimerisation. *Mol. Gen. Genet.*, **257**, 595.

71. Wellmer, F., Kircher, S., Rügner, A., Frohnmeyer, H., Schäfer, E. and Harter, K. (1999) Phosphorylation of the parsley bZIP transcription factor CPRF2 is regulated by light. *J. Biol. Chem.*, **274** (41), 29476.

72. Wei, N. and Deng, X. W. (1999) Making sense of the COP9 signalosome a regulatory complex conserved from *Arabidopsis* the human. *Trends Gen.*, **15**, 98.

73. Chamovitz, D. A., Wei, N., Osterlund, M. T., von Arnim, A. G., Staub, J. M. and Deng, X. W. (1996) The COP9 complex, a novel multisubunit nuclear regulator involved in light control of a plant developmental switch. *Cell*, **86**, 115.

74. Oyama, T., Shimura, Y. and Okada, K. (1997) The *Arabidopsis* HY5 gene encodes a bZIP protein that regulates stimulus-induced development of root and hypocotyl. *Genes Dev.*, **11**, 2983.

75. Ang, L.-H., Chattopadhyay, S., Wei, N., Oyama, T., Okada, K., Batschauer, A. and Deng, X.-W. (1998) Molecular interaction between COP1 and HY5 defines a regulatory switch for light control of *Arabidopsis* development. *Mol. Cell*, **1**, 213.

76. Osterlund, M. T., Hardtke, C. S., Wei, N., Deng, X.-W. (2000) Targeted destabilization of HY5 during light-regulated development of *Arabidopsis*. *Nature*, **405**, 462.

77. Hardtke, C. S., Gohda, K., Osterlund, M. T., Oyama, T., Okada, K. and Deng, X. W. (2000) HY5 stability and activity in *Arabidopsis* is regulated by phosphorylation in its COP1 binding domain. *EMBO J.*, **19**, 4997.

3 | Wound- and mechanical signalling

FLORIAN SCHALLER AND ELMAR W. WEILER

1. Introduction

Organisms are constantly exposed to fluctuations in various environmental factors such as light, temperature, and humidity, and they are exposed to biotic and abiotic (e.g. mechanical and chemical) stresses. All organisms thus evolved possibilities to sense environmental changes and to react to them. Unlike animals, most plants are unable to change their location to escape adverse environmental conditions. Instead, plant cells possess highly regulated stimulus-response systems by which they detect and respond to unfavourable or potentially damaging situations in a rapid and appropriate manner. Although most plants are not able to move entirely from one location to a more favourable one, the movement of plant organs such as leaves or flowers helps to orient plants in an optimal way relative to environmental factors, such as light. Various forms of intracellular movement, such as the orientation movement of chloroplasts, cytoplasmic streaming, and the traumatotactic or premitotic migration of the nucleus, have been described. These movements are thought to be essential for accomplishment of efficient photosynthesis, cell division, and appropriate delivery of substances that are required for growth and differentiation of plant cells.

The reaction of a cell to a stimulus is assumed to be composed of, at least, three elementary processes (Fig. 1):

- the perception of the stimulus, which includes physical and chemical interactions between the stimulus and a receptor;
- the transduction of changes in the receptor to a specific effector system; and
- responses, initiated by the effector system, which result in a specific reaction of the cell to the stimulus.

Plant development is a complex biological process, especially because it is unusually plastic and can be dramatically altered depending on environmental conditions (1). Plasticity of development is thought to be particularly important for sessile organisms because it allows adaptation to inescapable environmental stresses. Plant

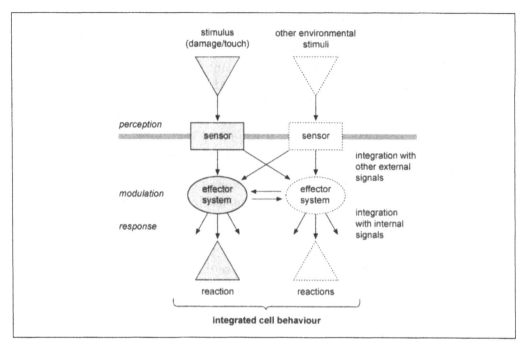

Fig. 1 A model of the reactions of a plant cell to a stimulus. The behaviour of a cell – and the whole plant – is conditioned on the concerted processing of many external and internal signals.

growth and differentiation may be considered the equivalent of animal locomotion. Growing plants are permanently embryogenic, and many plant cells in the mature plant body remain unspecialized, accounting for the remarkable regenerative capabilities of these organisms. Plants can grow rather than move into the more hospitable regions of their local environment, as evident from the strongly asymmetric growth of root systems of plant individuals in adaptation to soil conditions. An understanding of how the environment affects plant growth is essential for the understanding of plant development.

Mechanical stimuli such as rain, wind, changes in osmotic conditions, obstructions in the soil, and gravity are ubiquitous and are known to result in morphological changes in plants (2–5). The fascinating touch responses of specialized plants (6) such as the Venus fly trap (*Dionaea muscipula*) and *Mimosa pudica* are easily recognized because they occur quickly. The responses of most plants to touch (non-injurious mechanical stimulation) is a change in development, generally including a decrease in longitudinal growth and an increase in radial expansion (7) called thigmo-morphogenesis. Other rapid responses in plants following mechanical stimulation include an immediate drop in electrical resistance in the stem (8), the appearance of small voltage transients (9), callose synthesis (10), localized blockage of phloem transport (11), and, within an hour, an increase in ethylene production (12). Plants grown under this kind of stress tend to develop a shorter and more stunted phenotype and are thus less easily damaged by wind. Jaffe (8) showed that within 3 minutes of mechanical stimulation of beans (*Phaseolus vulgaris*), there is a brief increase in growth rate, followed by a complete inhibition of elongation for 15–30 minutes. Normal growth resumes after 3–4 days. Mechanical stimulation also inhibits growth of portions of the plant, especially young tissues that are not stimulated directly. This observation led to the speculation that a signal is communicated to non-stressed areas of the plant (13).

The existence of mechanisms which transduce signals over long distance in plants could be detected in defence reactions as well (14). Proteinase inhibitor genes, together with other defence-related genes, are induced in response to wounding, both at the site of damage and systemically, i.e. more distal in unwounded leaves (15). In tomato plants, systemic induction of defence genes in response to herbivory and/or mechanical wounding is regulated by the 18-amino-acid peptide systemin. Activation of defence genes by systemin involves a signal transduction pathway whose molecular components are only partially understood, but include jasmonic acid (16–20), ethylene and abscisic acid (21–23). Whether Ca^{2+} (24, 25) and a 48 kDa protein kinase (26, 27) are components of this defence pathway or whether they are elements of other signalling pathways that are mediated by wounding and systemin is yet an open question (28). Recent findings indeed suggested the involvement of octadecanoid independent signal transduction pathways in wound-induced gene expression (29, 30).

This chapter reviews recent progress in the understanding of plant sensory functions and signal transduction pathways, which lead to specific reactions of plants to their environment, with focus on wound- and mechanical signalling.

2. Plant signal transduction in response to wound- and mechanical stimuli

The developmental plasticity of higher plants (31) necessitates an accurate sensing and integration of the incoming environmental signals such as light, gravity, wind, minerals, water, gases, and soil status (32). This external information must be conveyed to the cell and integrated with internal signals reaching the cell from the surrounding tissue in order to generate optimal growth and differentiation (Fig. 1). Furthermore, plants have to protect themselves against herbivores and pathogens. These enemies, like hostile environmental conditions, have to be recognized as fast as possible to mount appropriate defence and/or protective reactions. Today we know that Ca^{2+}, protein kinases, inositol lipids, phospholipases, as well as fatty acid-, peptide-, and oligosaccharide signalling molecules, are involved in plant signal transduction.

2.1 Fatty acid signalling

The first fatty acid derivative with presumed biological activity that was identified in plants was traumatin, proposed by English *et al.* (33) to be *trans*-2-dodecenedioic acid

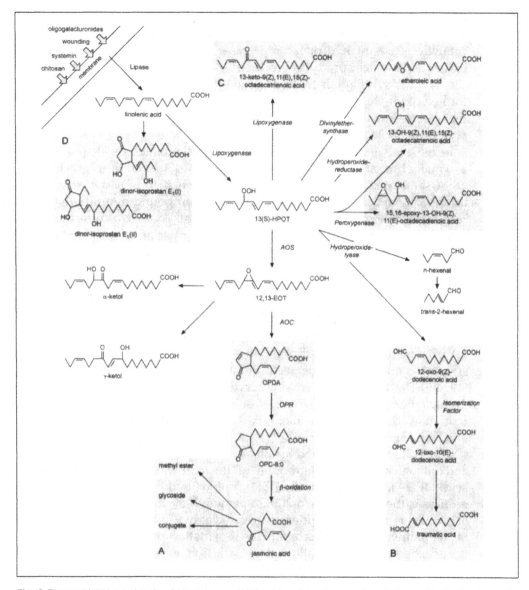

Fig. 2 Fatty acid-derived signals of plants are subdivided into four classes: A, octadecanoids; B, the traumatin family; C, other products of various enzymes; D, non-enzymatically formed phytoprostanes.

(traumatic acid), but later shown to represent an oxidation artefact of 12-oxo-*trans*-10-dodecenoic acid (34) derived from linolenic acid (LA) (Fig. 2). At the same time, this fatty acid derivative was among the first regulating molecules ever isolated directly from plant tissue. The so-far-known fatty acid-derived signals in plants can be divided into several categories (Fig. 2): (a) octadecanoids (octadecanoid-derived cyclopentenones and cyclopentanones, i.e. 12-oxo-phytodienoic acid (OPDA), 10,11-

dihydro-12-oxo-phytodienoic acid (OPC-8:0), jasmonic acid (JA), as well as their methyl esters; (b) the traumatin family and related alkenals resulting from the catalytic action of hydroperoxide lyases on fatty acid hydroperoxides; (c) other classes of fatty acid products produced by the activity of hydroperoxide-reductase, divinylether-synthase or hydroperoxide-peroxygenases; and (d) non-enzymatically formed dinor-isoprostanes, now called phytoprostanes (35). Among these, JA and its metabolites (methyl esters, glycosides, or conjugates) are the best-examined fatty acid signal molecules.

In 1962, the methyl ester of jasmonic acid was isolated by Demole and co-workers as a major fragrance in the ethereal oil of jasmine (36), and the free acid was identified at the same time in culture filtrates of the fungus *Botryodiplodia theobromae*. Jasmonic acid is known to be involved as a signalling compound in multiple aspects of plant responses to their biotic and abiotic environment. Today we know that jasmonic acid, levels of which range from 0.01 μg up to 3.0 μg per gram fresh weight *in planta* (37, 38), is only one of several bioactive compounds within the class of octadecanoids which regulate a broad spectrum of plant responses (39).

The pathway of jasmonic acid (JA) biosynthesis is shown in Fig. 2. Biosynthesis is believed to start with the oxygenation of free α-linolenic acid (LA), which is converted to (9Z,11E,15Z,13S)-13-hydroperoxy-9,11,15-octadecatrienoic acid [13(S)-HPOT] in a reaction catalysed by a 13-lipoxygenase (LOX). The presumed release of LA from membrane lipids through the action of a lipase may be triggered by local or systemic signals like oligogalacturonides, chitosan, systemin (18, 40) or wounding (19, 40). The 13(S)-hydroperoxide serves as a substrate for several enzymes such as divinylether synthase, peroxygenase, hydroperoxide lyase, and hydroperoxide reductase (41, 42) (Fig. 2) or allene oxide synthase (AOS) (43, 44). AOS converts 13(S)-HPOT to an unstable epoxide [12,13(S)-epoxy-9(Z),11,15(Z)-octadecatrienoic acid, 12,13-EOT] which is cyclized by allene oxide cyclase (AOC) (45–49) to the first cyclic and biologically active compound of the pathway, 12-oxo-10,15(Z)-phytodienoic acid (OPDA). Reduction of the 10,11-double bond by a NADPH-dependent OPDA-reductase then yields 3-oxo-2(2'(Z)-pentenyl)-cyclopentane-1-octanoic acid (OPC-8:0) which is believed to undergo three cycles of β-oxidation to yield the end product of the pathway, JA. Evidence for this β-oxidation process has emerged from radio-tracer experiments, which allowed the detection of 3-oxo-2-(2'-pentenyl) cyclopentane hexanoic acid (OPC-6:0) and 3-oxo-2-(2'-pentenyl)cyclopentane butanoic acid (OPC-4:0) as intermediates in the conversion of OPDA to JA (50). While this biosynthetic pathway results in the production of JA with (3R,7S)-configuration [i.e. (+)-7-iso-JA] (51), JA extracted from plant tissue is predominantly in the thermodynamically more favourable (3R,7R)-*trans*-configuration [i.e. (−)-JA] (52). The biosynthesis of JA involves different compartments: The conversion of LA to OPDA is localized in the chloroplasts (44, 53, 54), while the reduction of racemic OPDA to OPC-8:0 probably takes place in the cytoplasm (55, 56), but the *cis*(+)-OPDA reducing acivity may also be localized in the microbodies. It has been shown recently that the *OPR3* gene of *Arabidopsis thaliana* encodes the isoenzyme involved in JA biosynthesis (57) and a mutant (*delayed dehiscence* 1) defective in *OPR3* has been found to be male sterile (58,

59), a phenotype also associated with JA insensitivity or the inability to synthesize LA (60). Finally, the postulated steps of β-oxidation, i.e. conversion of OPC-8:0 to JA, are believed to occur in peroxisomes (61, 62). However, recent reports suggest that, contrary to earlier assumptions (61), mitochondrial β-oxidation might occur in higher plants (62). Thus, the cellular localization of the biosynthetic route from OPDA to JA still needs to be worked out definitely. Like JA, OPDA may accumulate to substantial amounts in plant tissue (63). By contrast, OPC-8:0 occurs only in trace amounts.

2.2 Peptide signalling

(Poly)peptide hormones are a common feature of signalling systems in animals and yeast, and hundreds of polypeptide hormones and growth factors are known in higher animals (64, 65). Polypeptide hormones are nearly always derived from larger precursors and proteolytically processed in secretory vesicles by members of a family of site-specific subtilisin-related proteinases (64). The present arsenal of endogenous plant signal peptides includes four groups involved in wound signal transduction, in cell proliferation, and in the regulation of salt/water homeostasis, i.e. systemin (66), phytosulphokines (67, 68), enod40 (69, 70), and natriuretic peptides (71).

In addition to these endogenous plant peptides, there exist microbial elicitor-peptides, which play a central role in the recognition of microbial pathogens and the induction of active defence responses by plants. For example, an oligopeptide fragment (Pep-13) from a 42 kDa *Phytophthora sojae* cell wall glycoprotein stimulates the transcriptional activation of defence-related genes, and phytoalexin production in *Petroselium crispum* (72). Peptides corresponding to the most conserved domains of eubacterial flagellin induce callose deposition, induction of genes encoding for pathogenesis-related proteins, and a strong inhibition of growth in *Arabidopsis thaliana* seedlings (73).

Furthermore, the hrpZ gene product of the bean halo-blight pathogen, *Pseudomonas syringae* pv. *phaseolicola*, possesses pore-forming activity which allows nutrient release and/or delivery of virulence factors during bacterial colonization of host plants (74). Elicitors thus may not require receptor-based mechanisms in order to be active (75). That ion-channel-forming peptides (alamethicin) are potent elicitors of the biosynthesis of volatile compounds, which act principally via the octadecanoid pathway, could be demonstrated (76).

Plants appear to possess multiple receptors for both endogenous and exogenous (i.e. microbial) peptide signals. A large number of receptor-like kinases (RLKs) have been identified possessing extracellular domains which are likely to be involved in protein/protein interaction (77). Thus, RLKs were hypothesized to be receptors for (poly)peptide ligands.

The identification of (poly)peptide signalling molecules in plants began in tomatoes where a signal originates at the wound site that is transported throughout the plant where it activates the synthesis of defensive proteins that interfere with the digestive system of the attacking herbivores. The mobile wound signal could be identified as the 18-AS peptide 'systemin' (78) and the inducible defensive proteins were identified as serine-, cysteine- and aspartyl-proteinase inhibitors (79–81), and polyphenol oxidase (82). These proteins interact with the proteins and proteinases of herbivore guts and adversely affect proteolysis of the ingested food, reducing the availability of essential amino acids and retarding the growth and development of the herbivores (79, 82). More recently, proteinase inhibitors in sedges and grasses have been found to be a major factor in regulating fluctuating populations of lemmings (83). The mobility of systemin, together with its powerful inducing activity, identified it as a systemic signal and the first oligopeptide signal of higher plants.

The finding that exposure of tomato plants to methyl jasmonate and jasmonic acid also strongly activated the synthesis of proteinase inhibitors (18, 84) led to the proposal of a model in which wounding and systemin activate a lipase in receptor cell membranes, resulting in the release of linolenic acid, the production of jasmonic acid, and, finally, the activation of proteinase inhibitor genes (18). In this model, oligosaccharides are localized signals, whereas systemin is a systemic signal that activates the defence signalling pathway in the entire plant.

2.3 Oligosaccharide signalling

Oligosaccharins are complex carbohydrates that can function in plants as molecular signals that regulate growth, development, and survival of pathogen attacks (85). Studies of plant–microorganism interactions yielded the first evidence that oligosaccharins could serve as biological signals. More recent research has suggested the involvement of oligosaccharins in normal plant growth and development (86–88). Oligogalacturonides derived from the plant cell wall were identified as inducers of defensive proteinase inhibitor genes in excised tomato leaves (89). Subsequently, chitin and chitosan oligomers derived from fungal cell walls were also found to be active inducers (90). Acylated (and sometimes sulphated) chitooligosaccharides of symbiotic Rhizobia, the nod-factors, induce the formation of nodule primordia in host plants and show a remarkable host selectivity (91), demonstrating the plant's ability to recognize oligosaccharide signals with a high degree of specificity.

3. Wounding/herbivore attack

Plants, over hundreds of millions of years, have evolved defensive strategies to protect themselves against herbivores and pathogens. It seems that all the so-far-discovered and mentioned signalling molecules/pathways are involved in this protection scenario. Furthermore, it could be demonstrated that, similar to the fungal infection hyphae in the non-host resistance of parsley to *Phytophthora sojae*, a local mechanical stimulus alone induced morphological and biochemical changes involved in the plant's pathogen resistance response [i.e. generation of intracellular

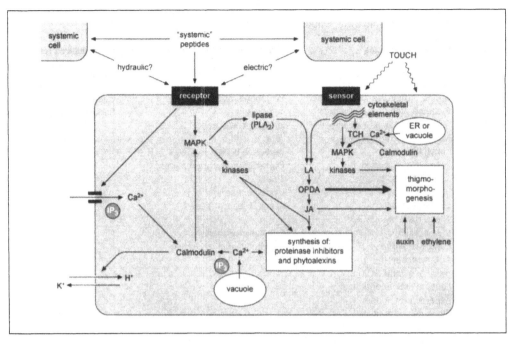

Fig. 3 A synopsis of local and systemic reactions involved in wound and touch sensing of plant cells.

reactive oxygen intermediates and translocation of cytoplasm and the nucleus to the site of stimulation (92)]. Furthermore, the generation of an oxidative burst could be identified in cultured soybean cells after resuspending them in solutions of reduced osmolarity and after subjecting them to direct physical pressure (93). This indicates overlaps in a plant's response to wounding, mechanical challenge, and pathogen attack (Fig. 3; Table 1). Indeed, it has been shown for a range of species that:

- JA does induce the accumulation of defence-related antimicrobial compounds (94, 95); and
- elicitors of pathogen defences may also induce the production of JA by the challenged cell (95).

3.1 Fatty acid signalling/octadecanoids

Farmer and Ryan (96) showed that JA is a signal transducer in herbivore defence and Gundlach *et al.* (95) described an involvement of JA in the induction of low M_r secondary metabolites, which mostly may be defence-related. The crucial function of octadecanoids in herbivore defence is underlined by several facts:

- the LA-deficient *fad3-2 fad 7-2 fad8* mutant of *Arabidopsis* (60) lacks the ability to accumulate JA after wounding and is more susceptible to attacks of chewing insects than the wild-type;

Table 1 Jasmonic acid-, wound- and touch-induced genes

Wound-induced genes	Jasmonic acid-induced genes	Touch-induced genes
Defence genes	Serine-, cysteine- and aspartic	Calmodulin (TCH1)[b]
Serine proteinase inhibitor I[a]	proteinase inhibitor	Calmodulin-related (TCH2)
Serine proteinase inhibitor II[a]	Polyphenol oxidase	Calmodulin-related (TCH3)
Cysteine proteinase inhibitor[a]	Thionins	Endotransglycosylase (TCH4)
Aspartic proteinase inhibitor[a]	HMG-CoA-reductase	
Metallo-carboxypeptidase	JIPs	
inhibitor	Berberine bridge enzyme	
Polyphenol oxidase[a]	Cyt P450 monooxygenases	
Signal pathway genes	Lipase	
Prosystemin	Lipoxygenase[a]	
Lipoxygenase[a]	Allene oxide synthase[a]	
Allene oxide synthase[a]		
Calmodulin[b]	Chalcone synthase	
Polygalacturonase	Dihydroflavonol reductase	
ACC synthase		
Proteinase genes		
Leucine aminopeptidase		
Aspartic proteinase		
Cysteine proteinase		
Carboxypeptidase		

[a] Genes induced by wounding and jasmonic acid.
[b] Genes induced by wounding and touch.

- the *defenceless* mutant (*def1*) of tomato does not raise the JA level after wounding, produces less proteinase inhibitor and is more susceptible to herbivore attack than the wild-type (97);
- the JA-insensitive mutant (*coi1*) of *Arabidopsis* is more susceptible to herbivore attack than wild-type plants (98);
- there exist many defence-related proteins in higher plants, which are JA-induced: serine proteinase inhibitors I and II, cysteine proteinase inhibitor, aspartic proteinase inhibitor, polyphenol oxidase, thionins, ribosome-inactivating protein (JIP 60), berberine bridge enzyme, cytochrome P450 monooxygenases in the sanguinarine biosynthetic pathway, lipases, chalcone synthase, dihydroflavonol reductase, HMG-CoA reductase (see Table 1).

Because of the fact that – in contrast to local wounding – neither JA nor OPDA, when exogenously applied to a leaf, led to a systemic induction of the AOS promoter (99), it seems clear that:

- these octadecanoids are not involved in systemic defence signalling, rather they behave like local response regulators; and
- the systemic activation of the AOS promoter after wounding of a leaf or a root does involve systemic wound factor(s) released by these organs.

Several chemical and/or physical messengers have been discussed which might fulfil such a role:

- the octadecapeptide systemin in tomato (100);
- hydraulic (101); and
- electrical (102) signals (Fig. 3).

These might act alone or in concert, a somewhat controversial issue that is at present not resolved.

3.2 Protein kinases and signal transduction

Reversible protein phosphorylation is a widespread regulatory mechanism in eukaryotic signalling pathways and also plays a crucial role in wound-activated gene expression. It has been demonstrated that the jasmonate-dependent wound-signalling pathway in tobacco involves the action of a rapidly wounding-induced protein kinase (WIPK) (103, 104). In WIPK-antisense plants, both the production of wound-induced jasmonic acid and the accumulation of wound-inducible gene transcripts were inhibited. By contrast, the levels of salicylic acid and transcripts for pathogen-inducible, acidic pathogenesis-related proteins were increased upon wounding in the antisense plants (103). Moreover, reversible protein phosphorylation regulates both JA-dependent and JA-independent, wound-induced signalling in *Arabidopsis*. The JA-dependent pathway is negatively regulated by a protein kinase and positively by a protein phosphatase (98), whereas protein phosphatase inhibitors activate wound-responsive genes in tomato cell culture whilst protein kinase inhibitors block wound-responsive gene induction (105). Okadaic acid-sensitive protein phosphatases are also required for the wound-induced activation of wound-responsive genes like *pin* in tomato and potato (106) and the induction of a phosphatase in tobacco (107). Moreover, the cross-talk between wound-induced MAP kinase and the octadecanoid pathway in alfalfa may be mediated by the action of linolenic acid on a regulatory protein phosphatase 2C (108). Abscisic acid, methyl jasmonate, and electrical activity are not able to activate this kinase in the absence of wounding, suggesting its independence from these signals (109). On the other hand, the activity of plant MAP kinases is known to be modulated by touch stimuli (110), which also induce a MAP kinase kinase (25), by cold and drought (111, 112), by pathogens and several phytohormones (113–117). In leaves of young tomato plants, a 48 kDa myelin basic protein (MBP) kinase, which belongs to the family of MAP kinases, is activated within 2 minutes, both locally and systemically after wounding. MBP-kinase activity also increases in response to polygalacturonic acid and chitosan but not in response to jasmonic acid or 12-oxo-phytodienoic acid (26). That cytoplasmic phospholipases A_2 (PLA_2) are MAP-kinase targets (118) may be of major importance in connection with octadecanoid wound signalling: PLA_2 activity cleaves phospholipids and releases fatty acids that can in turn serve as precursors of signal molecules such as JA (Fig. 3).

3.3 Ca^{2+}-/calmodulin-based signal transduction

The Ca^{2+}-concentration rises in the cytoplasm of *Arabidopsis thaliana*, either as a result of uptake from the extracellular space or of release from internal stores, such as the vacuoles, or the endoplasmic reticulum (ER), in an inositol-1,4,5-trisphosphate-induced process (119). The second messenger Ca^{2+} acts through activation of Ca^{2+}-binding proteins. A growing number of target proteins have now been identified and characterized in plants (120). The best studied are, of course, calmodulin, calmodulin-related proteins and Ca^{2+}-dependent protein kinases and phosphatases. Calmodulin regulates the activities of many different enzymes, such as kinases, phosphatases, phosphodiesterases, adenylate cyclase and Ca^{2+}-ATPases (121). However, calmodulin also seems to have Ca^{2+}-independent regulatory functions (122).

It is well known that wound-signal molecules promote rapid membrane-associated events such as depolarization of the membrane potential and proton influx (24, 123). Increases of intracellular Ca^{2+}-levels (20) and changes in protein phosphorylation (25) upon wounding are well analysed in tomato cells. The systemic wound factor systemin of tomato triggers an increase of cytoplasmic calcium in tomato mesophyll cells, which might be an early step in the systemin-signalling pathway (Fig. 3). Unlike systemin, other inducers of proteinase inhibitors and of wound-induced protein synthesis, such as JA, could not be found to trigger a corresponding increase of cytoplasmic Ca^{2+}. Ca^{2+} is a major regulator of responses to several different stimuli that act via many plant cell-signalling pathways (for review, see 124). Using wound- and/or JA-inducible *Arabidopsis* cDNAs that exhibit various patterns of induction (30) and are differently regulated by reversible phosphorylation events (98). León *et al.* (125) concluded that the JA-dependent and JA-independent wound-signal transduction pathways are differentially regulated by Ca^{2+}:

- Mobilization of intracellular Ca^{2+} pools blocked the induction of JA-responsive (JR) genes by both wounding and JA.
- By contrast, the induction of wound-responsive (WR) genes by wounding could not be blocked by mobilization of intracellular Ca^{2+}.

In addition, calmodulin antagonists blocked the expression of JR genes and up-regulated WR gene expression. The authors conclude therefore that Ca^{2+} and calmodulin act downstream of JA in the JA-dependent pathway, and downstream of reversible phosphorylation events of the JA-dependent and JA-independent wound-signal transduction pathways.

3.4 Peptide signalling

Many plants respond to herbivore attacks by activating defence genes in leaves whose products inhibit digestive proteases of herbivores and reduce the food quality, thereby retarding growth of, or even killing, herbivores (79). In tomato plants, wounding results in the synthesis of over 20 defence-related proteins (80, 126–129). These defensive proteins can be classified in three groups that include:

- antinutritional proteins (proteinase inhibitors and polyphenol oxidases);
- signalling pathway proteins; and
- proteinases (see Table 1).

While the exact function of the proteinases remains obscure, the induction of prosystemin (130), LOX (131, 132) and AOS (133) allows the plant to mount a maximal defence response against chewing predators. A primary wound signal for this signalling cascade is the 18-amino-acid polypeptide hormone systemin (66) that is released at wound sites by chewing herbivores or wounding. Systemin is processed from a larger prohormone protein, called prosystemin, by proteolytic cleavage. A wound-inducible systemin cell surface receptor, regulating an intra-cellular cascade of signalling processes which finally leads to the activation of a PLA_2 to release linolenic acid from membranes, has recently been identified (Fig. 3). The binding of systemin to its receptor led to depolarization of the plasma membrane (24), opening of ion-channels (24, 25, 27), increasing levels of intracellular Ca^{2+} (27), inactivation of the plasma membrane H^+-ATPase (119), the activation of a MAP kinase (26, 106), the synthesis of calmodulin, and the activation of a PLA_2 (20, 134). Release of linolenic acid leads via the octadecanoid pathway to the oxylipins OPDA and JA that regulate the defence genes described above.

4. Mechanotransduction

Plants are rooted in soil and immobile and therefore unable to seek shelter or escape from environmental stress. Alterations in physiology or development often enable plants to withstand even harsh environmental conditions. For example, in response to wind or touch, plants generally grow shorter and more stunted and acquire altered mechanical properties. These mechanostimulus-induced developmental changes are termed thigmomorphogenesis and make a plant less susceptible to damage by subsequent mechanical stresses (135, 136).

The perception of mechanical forces, in this case contact stimuli, is critical for survival for the winding and climbing plants that grow on support structures rather than investing in the generation of massive supportive tissues in order to stand on their own. The ecological advantage of this habit is faster growth which allows these plants to expose their photosynthetic organs to the sunlight very effectively and to grow out of shade very fast. Many other plant organs are able to move relative to the plant's body: Flowers open and close, leaves unfold in the daylight and fold up by night or turn to or away from direct sunlight. In *Mimosa pudica*, for example, leaves collapse suddenly when touched and the modified leaves of the Venus flytrap snap shut on doomed insect prey. Among the several thousand climbing or winding species of the Old and the New World, the tendril climbers are perhaps the most effective.

Tendrils may respond within seconds to a thigmic stimulus. This response involves reversible and irreversible components. During the first and reversible phase (contact coiling), the tip of the tendril curves because of loss of turgor on one side and increase of turgor on the opposite side (137, 138). If the contact stimulus persists, the tip of the tendril coils around the support three to four times while the more basal part of the tendril forms a screw (free coiling). This results in a tight but flexible attachment of the shoot to the support. The reactions involved in the perception of the mechanical stimulus and in signal transmission are still largely unknown. Mechanosensitive epidermal cells, many of them characterized by the differentiation of tactile bleps (139), have been identified since 1924 and are thought to be involved in the early and rapid changes of the osmotic potential leading to contact coiling. In the transmission of the perceived signal, IAA (140–142), ethylene (143, 144), JA (145), and OPDA (146, 147), respectively, were proposed to be of relevance (Fig. 3). Calmodulin and other Ca^{2+}-binding proteins and Ca^{2+} itself participate in the transmission of touch and wind forces (148, 149). In *Arabidopsis*, wind or touch stimulation results in the enhanced expression of several touch (TCH) genes. TCH gene mRNAs accumulate very rapidly, within 10 minutes of touching a plant (22). *TCH1* encodes calmodulin, *TCH2*, and *TCH3* (148, 150, 151) encode calmodulin-releated proteins, and *TCH4* (152) encodes a xyloglucan endotransglycosidase capable of modifying cell wall xyloglucans. Thus, the TCH genes may collaborate in cell wall biosynthesis (153). Cell-wall-associated microtubules themselves are discussed to be:

- sensing elements for mechanical forces; and
- biophysical response elements to mechanical stresses (154).

Alterations of the properties of the cell wall may contribute to the process of thigmomorphogenesis (155–157).

4.1 Fatty acid signalling / octadecanoids

Tendril coiling in *Bryonia dioica* can be elicited by airborne JA-methyl ester without any mechanical contact, and the chemically induced reaction is morphologically and biochemically indistinguishable from the process elicited mechanically (158). Furthermore, the induction of a lipoxygenase after touch stimulation could be observed (159). Evidence has been obtained that the endogenous mechanotransducer is not JA, but rather its biosynthetic precursor, OPDA. OPDA levels rise transiently and highly correlated with the nastic growth response in tendrils of *Bryonia dioica* after mechanical stimulation (63). JA levels remain low and constant during this phase of the response. In bryony tendrils, mechanical force is perceived by tactile bleps, half-dome-shaped protrusions of the epidermis, serving as vectorial mechano-transducers and amplifiers, rendering the organ more sensitive (160). Tactile bleps are dispensable for touch perception *per se*, however, because:

- tendrils lacking specialized surface protrusions of their epidermal cells occur in other species; and
- touch sensitivity appears to be a general feature of plant cells (161).

The thigmomorphogenetic response of bean internodes, a model system for the response of a plant to mechanical stress, also proceeds with increases in the level of OPDA, but not JA (63). It is likely that the response of plants to mechanical stress (e.g. wind, soil resistance, strain, etc.) involves the concerted action of auxins, ethylene, and the octadecanoid OPDA.

4.2 Protein kinases and signal transduction

Mechanical manipulation of *Arabidopsis* leaves can induce transcription of MAPK and MAPKKK genes (110). Furthermore, touching alfalfa leaves for 2 seconds is sufficient to induce a transient activation of a MAPK. *In vitro* phosphorylation assays demonstrated activation within 1–5 minutes of stimulation (109, 162); moreover, the expression levels of mitogen-activated protein kinases and other protein kinases increase 30 minutes following wounding or mechanical stimulation (44, 109, 111). This suggests that these protein kinases, some of which are Ca^{2+}-dependent enzymes (163) or ribosomal protein S6 kinase (111), are involved in the mechanotransduction pathway leading to gene expression and morphological changes.

4.3 Ca^{2+}-/calmodulin-based signal transduction

An early, if not the primary, intracellular event following mechanical stimulation of plant cells appears to be a transient change in the level of cytoplasmic free Ca^{2+} (164), and touch-induced genes have been shown to encode calcium-binding proteins (148, 161). Dramatic up-regulation of the expression of the CaM and CaM-related TCH genes following mechanical stimulation implicates the involvement of Ca^{2+}-binding proteins in touch responses of plants (148). Rapid increases in concentrations of cytoplasmic calcium occur in plants subjected to touch or wind stimulation (149, 161). Moreover, externally applied Ca^{2+} (165) and Ca^{2+}-channel antagonists (166) have been shown to affect the expression of touch-induced genes. In the case of *Bryonia dioica*, this mechanosensitive pool of calcium is mobilized from intracellular stores (164), most probably from the endoplasmic reticulum (167, 168). Two calcium transporters in the ER membranes from *B. dioica* tendrils have been characterized at the molecular level: a primary active, calmodulin, and KCl-stimulated, Ca^{2+}-ATPase that loads the ER lumen, thereby lowering the level of cytoplasmic free Ca^{2+} (167) and a voltage-dependent rectifying calcium-release channel that delivers the ion to the cytoplasm (168). On the basis of the known properties of pump and channel, it has been proposed that they constitute an intracellular oscillator allowing for frequency encoding of calcium signals (168). The *B. dioica* tendril may thus dispose of an extremely versatile calcium-signalling system with high dynamic, temporal, and spatial resolution. Likewise, calcium is involved in the thigmomorphogenetic response of *B. dioica* internodes (149, 169). The events downstream of the calcium signal are still unknown. Calcium-dependent protein kinases (CDPKs) are likely to be involved, at least in the tendrils of *Bryonia dioica* (163).

4.4 Polypeptide signalling

There is no evidence so far for an involvement of peptide signals in plant mechano-transduction.

5. Conclusion

Recent years have seen rapid progress in our understanding of many aspects of wound- and mechanical signalling. Although considerably differing in detail, both processes have many things in common, in particular the involvement of lipid-derived octadecanoid signals and the – almost universal – messenger calcium and the involvement of protein phosphorylation. The large number of protein kinases in plants certainly differs a lot in substrate specificity, although for most enzymes, cellular substrates have not yet been identified. It is more difficult to understand how calcium and octadecanoids could be involved in a specific manner in different signalling pathways. As for calcium, it is now clear that it is a short-range messenger, not very mobile in the plant and even in a single cell. Spatial and temporal patterns of fluctuations in calcium levels might thus be part of the specificity of cellular calcium signalling. Evidence for this has been obtained recently:

- all elements of a calcium oscillator have been identified in *Bryonia dioica* tendrils (168); and
- local membrane domains, most likely of the endoplasmic reticulum (160, 168) with large calcium fluxes have been visualized in the tactile bleps of *B. dioica*;
- using FRET technology, calcium oscillations in guard cells of *Arabidopsis thaliana* have meanwhile been recorded directly (169).

The inability to transform *B. dioica* has precluded direct measurements of such oscillations in the tendril system. However, it has been proposed that the ability of the tendril to sense the surface texture of a support structure, i.e. a spatial and/or temporal pattern of mechanical stimulation, is related to calcium oscillators as an internal clock (168).

The issue of octadecanoid selectivity has been resolved partially by Blechert *et al.*, who have shown that the structural requirements for activity in different octadecanoid-dependent processes is quite different (170, 171). According to these data, OPDA is the signal transducer in mechanotransduction (171, 172), whilst JA is active only indirectly, by increasing the endogenous level of OPDA (172). These assumptions are backed by measurements of octadecanoid levels in mechanically stimulated *B. dioica* tendrils and bean internodes (63, 171). It has been argued that octadecanoid signatures rather than individual compounds might be responsible for specificity in the fatty acid signalling of higher plants (173), a proposal that seems more and more likely given the multitude of physiological responses to be considered.

Acknowledgements

The authors are grateful to the Deutsche Forschungsgemeinschaft, Bonn, for financial support and to Fonds der Chemischen Industrie, Frankfurt, for literature provision.

References

1. Walbot, V. (1985) On the life strategies of plants and animals. *Trends Genet.*, **1**, 165.
2. Darwin, C. (1881) In *The Power of Movement in Plants.* Appleton, New York, p. 129.
3. Morgan, J. M. (1984) Osmoregulation and water stress in higher plants. *Annu. Rev. Plant Physiol.*, **35**, 299.
4. Biddington, N. L. (1986) The effect of mechanically-induced stress in plants – a review. *Plant Growth Regul.*, **4**, 103.
5. Evans, M. L., Moore, R. and Hasenstein, K.-H. (1986) How roots respond to gravity. *Sci. Amer.*, **255**, 112.
6. Sibaoka, T. (1969) Physiology of rapid movements in higher plants. *Annu. Rev. Plant Physiol.*, **20**, 165.
7. Jaffe, M. J. (1973) Thigmomorphogenesis: the response of plant growth and development to mechanical stimulation. *Planta*, **114**, 143.
8. Jaffe, M. J. (1976) Thigmomorphogenesis: electrical resistance and mechanical correlates of the early events of growth retardation due to mechanical stimulation in beans. *Z. Pflanzenphysiol.*, **78**, 24.
9. Pickard, B. G. (1971) Action potentials resulting from mechanical stimulation of pea epicotyls. *Planta*, **97**, 106.
10. Jaffe, M. J., Huberman, M., Johnson, J. and Telewski, F. W. (1985) Thigmomorphogenesis: the induction of callose formation and ethylene evolution by mechanical pertubation in bean stems. *Physiol. Plant.*, **64**, 271.
11. Jaeger, C. H., Goeschl, J. D., Magnuson, C. E., Fares, Y. and Strain, B. R. (1988) Short-term responses of phloem transport to mechanical pertubation. *Physiol. Plant.*, **72**, 588.
12. Biro, R. L. and Jaffe, M. J. (1984) Thigmomorphogenesis: ethylene evolution and its role in the changes observed in mechanically pertubed bean plants. *Physiol. Plant.*, **62**, 289.
13. Jaffe, M. J. (1976) Thigmomorphogenesis: a detailed characterization of the response of beans to mechanical stimulation. *Z. Pflanzenphysiol.*, **77**, 437.
14. Green, T. R. and Ryan, C. A. (1972) Wound-induced proteinase inhibitor in plants leaves: a possible defense mechanism against insects. *Science*, **175**, 776.
15. Ryan, C. A. (1990) Proteinase inhibitors in plants: genes for improving defenses against insects and pathogens. *Annu. Rev. Phytopathol.*, **28**, 425.
16. Ryan C. A. and Pearce, G. (1998) Systemin: a polypeptide signal for plant defensive genes. *Annu. Rev. Cell Dev. Biol.*, **14**, 117.
17. Laudert, D. and Weiler, E. W. (1998) Allene oxide synthase: a major control point in *Arabidopsis thaliana* octadecanoid signalling. *Plant J.*, **15**, 675.

18. Farmer, E. E. and Ryan, C. A. (1992) Octadecanoid precursors of jasmonic acid activate the synthesis of wound-inducible proteinase inhibitors. *Plant Cell*, **4**, 129.
19. Conconi, A., Miquel, M., Browse, J. A. and Ryan, C. A. (1996) Intracellular levels of free linolenic and linoleic acids increase in tomato leaves in response to wounding. *Plant Physiology*, **111**, 797.
20. Lee, S., Suh, S., Kim, S., Crain, R. C., Kwak, J. M., Nam, H.-G. and Lee, Y. (1997) Systemic elevation of phosphatidic acid and lysophospholipid levels in wounded plants. *Plant J.*, **12**, 547.
21. O'Donnell, P. J., Calvert, C., Atzorn, R., Wasternack, C., Leyser, H. M. O. and Bowles, D. J. (1996) Ethylene as a signal mediating the wound response of tomato plants. *Science*, **274**, 1914.
22. Penninckx, I. A. M. A., Thomma, B. P. H. J., Buchala, A., Metraux, J.-P. and Broekaert, W. F. (1998) Concomitant activtion of jasmonate and ethylene response pathways is required for induction of a plant defensin gene in *Arabidopsis*. *Plant Cell*, **10**, 2103.
23. Birkemeier, G. F. and Ryan, C. F. (1998) Wound signaling in tomato plants: evidence that ABA is not a primary signal for defense gene activation. *Plant Physiol.*, **117**, 687.
24. Moyen, C. and Johannes, E. (1996) Systemin transiently depolarizes the tomato mesophyll cell membrane and antagonizes fusicoccin-induced extracellular acidification of mesophyll tissue. *Plant, Cell and Environment*, **19**, 464.
25. Moyen, C., Hammond-Kosack, K. E., Jones, J., Knight, M. R. and Johannes, E. (1998) Systemin triggers an increase of cytoplasmic calcium in tomato mesophyll: Ca^{2+} mobilization from intra- and extracellular compartments. *Plant, Cell Environ*, **21**, 1101.
26. Stratmann, J. W. and Ryan, C. A. (1997) Myelin basic protein kinase activity in tomato leaves is induced systemically by wounding and increases in response to systemin and oligosaccharide elicitors. *Proc. Natl Acad. Sci. USA*, **94**, 11085.
27. Meindl, T., Boller, T. and Felix, G. (1998) The plant wound hormone systemin binds with the N-terminal parts to its receptor but needs the C-terminal part to activate it. *Plant Cell*, **10**, 1561.
28. Howe, G. A. and Ryan, C. A. (1999) Suppressors of systemin signaling identify genes in the tomato wound response pathway. *Genetics*, **153**, 1411.
29. McConn, M., Creelmann, R. A., Bell, E., Mullet, J. E. and Browse, J. (1997) Jasmonate is essential for insect defense in *Arabidopsis*. *Proc. Natl Acad. Sci. USA*, **94**, 5473.
30. Titarenko, E., Rojo, E., León, J. and Sánchez-Serrano, J. J. (1997) Jasmonic acid-dependent and -independent signalling pathways control wound induced gene activation in *Arabidopsis thaliana*. *Plant Physiol.*, **115**, 817.
31. Trewavas, A. J. (1986) In Jenning, D. H. and Trewavas, A. J. (eds), *Plasticity in Plants*. Company of Biologists, Cambridge, p. 31.
32. Gilroy, S. and Trewavas, A. J. (1990) In Larsson, C. and Møller, J. M. (eds), *The Plant Plasma Membrane*. Springer Verlag., Berlin, p. 204.
33. English, J., Bonner, J. and Haagen-Smit, A. J. (1939) The wound hormones of plants. II. The isolation of a crystalline active substance. *Proc. Natl Acad. Sci. USA*, **25**, 323.
34. Zimmerman, D. C. and Coudron, C. A. (1979) Identification of traumatin, a wound hormone, as 12-oxo-*trans*-10-dodecenoic acid. *Plant Physiol.*, **63**, 536.
35. Imbusch, I. and Mueller, M. J. (2000) Formation of isoprostane F(2)-like compounds (phytoprostanes F(1)) from alpha-linolenic acid in plants. *Free Radic. Biol. Med.*, **28**, 720.
36. Demole, E., Lederer, E. and Mercier, D. (1962) Isolement et détermination de la structure du jasmonate de methyle, constituant odorant characteristique de l'essence de jasmin. *Helvetica Chimica Acta*, **XLV**, 675.

37. Mueller, M. J., Brodschelm, W., Spannagl, E. and Zenk, M. H. (1993) Signaling in the elicitation process is mediated through the octadecanoid pathway leading to jasmonic acid. *Proc. Natl Acad. Sc. USA*, **90**, 7490.

38. Mueller, M. J. and Brodschelm, W. (1994) Quantification of jasmonic acid by capillary gas chromatography-negative chemical ionization-mass spectrometry. *Anal. Biochem.*, **218**, 425.

39. Sembdner, G. and Parthier, B. (1993) The biochemistry and the physiological and molecular actions of jasmonates. *Annu. Rev. Plant Physiol. Plant Mol. Biol.*, **44**, 569.

40. Narváez-Vásquez, J., Florin-Christensen, J. and Ryan, C. A. (1999) Positional specificity of a phospholipase A activity induced by wounding, systemin, and oligosaccharide elicitors in tomato leaves. *The Plant Cell*, **11**, 2249.

41. Blée, E. and Joyard, J. (1996) Envelope membranes from spinach chloroplasts are a site of metabolism of fatty acid hydroperoxides. *Plant Physiol.*, **110**, 445.

42. Maréchal, E., Block, M. A., Dorne, A. J., Douce, R. and Joyard, J. (1997) Lipid synthesis and metabolism in the plastid envelope. *Physiol. Plant.*, **100**, 65.

43. Vick, B. A. and Zimmerman, D. C. (1981) Lipoxygenase, hydroperoxide isomerase, and hydroperoxide cyclase in young cotton seedlings. *Plant Physiol.*, **67**, 92.

44. Vick, B. A. and Zimmerman, D. C. (1987) Pathways of fatty acid hydroperoxide metabolism in spinach leaf chloroplasts. *Plant Physiol.*, **85**, 1073.

45. Hamberg, M. (1988) Biosynthesis of 12-oxo-10,15(Z)-phytodienoic acid: identification of an allene oxide cyclase. *Biochem. Biophys. Res. Commun.*, **156**, 543.

46. Hamberg, M. and Fahlstadius, P. (1990) Allene oxide cyclase: a new enzyme in plant lipid metabolism. *Arch. Biochem. Biophys.*, **276**, 518.

47. Ziegler, J., Stenzel, I., Hause, B., Maucher, H., Hamberg, M., Grimm, R., Ganal, M. and Wasternack, C. (2000) Molecular cloning of allene oxide cyclase. *J. Biol. Chem.*, **275**, 19132.

48. Ziegler, J., Hamberg, M., Miersch, O. and Parthier, B. (1997) Purification and characterization of allene oxide cyclase from dry corn seeds. *Plant Physiol.*, **114**, 565.

49. Ziegler, J., Wasternack, C. and Hamberg, M. (1999) On the specificity of allene oxide cyclase. *Lipids*, **34**, 1005.

50. Vick, B. A. and Zimmerman, D. C. (1983) The biosynthesis of jasmonic acid: a physiological role for plant lipoxygenase. *Biochem. Biophys. Res. Commun.*, **111**, 470.

51. Vick, B. A. and Zimmerman, D. C. (1984) Biosynthesis of jasmonic acid by several plant species. *Plant Phys.*, **75**, 458.

52. Quinkert, G., Adam, F. and Dürner, G. (1982) Asymmetrische Synthese von Methyljasmonat. *Angewandte Chemie*, **94**, 866.

53. Song, W. C., Funk, C. D. and Brash, A. R. (1993) Molecular cloning of an allene oxide synthase: a cytochrome P450 specialized for the metabolism of fatty acid hydroperoxides. *Proc. Natl Acad. Sci.USA*, **90**, 8519.

54. Laudert, D., Hennig, P., Stelmach, B. A., Müller, A., Andert, L. and Weiler, E. W. (1997) Analysis of 12-oxo-phytodienoic acid enantiomers in biological samples by capillary gas chromatography-mass spectrometry using cyclodextrin stationary phases. *Anal. Biochem.*, **246**, 211.

55. Schaller, F. and Weiler, E. W. (1997) Enzymes of octadecanoid biosynthesis in plants. 12-Oxo-phytodienoate 10,11-reductase. *Eur. J. Biochem.*, **245**, 294.

56. Schaller, F. and Weiler, E. W. (1997) Molecular cloning and characterization of 12-oxophytodienoate reductase, an enzyme of the octadecanoid signaling pathway from *Arabidopsis thaliana*. *J. Biol. Chem.*, **272**, 28066.

57. Schaller, F., Biesgen, C., Müssig, C., Altmann, T. and Weiler, E. W. (2000) 12-Oxophyto-dienoate reductase 3 (OPR3) is the enzyme involved in jasmonic acid biosynthesis. *Planta*, **210**, 979.
58. Sander, P. M., Lee, P. Y., Biesgen, C., Boone, J. D., Beals, T. P., Weiler, E. W. and Goldberg, R. B. (2000) The Arabidopsis DELAYED DEHISCENCE1 gene encodes an enzyme in the jasmonic acid synthesis pathway. *Plant Cell*, **12**, 1041.
59. Stinzi, A. and Browse, J. (2000) The *Arabidopsis* male-sterile mutant, opr3, lacks the 12-oxophytodienoic acid reductase required for jasmonate synthesis. *Proc. Natl Acad. Sci. USA*, **97**, 10625.
60. McConn, M., Creelman, R. A., Bell, E., Mullet, J. E. and Browse, J. (1997) Jasmonate is essential for insect defense in *Arabidopsis*. *Proc. Natl Acad. Sci. USA*, **94**, 5473.
61. Gerhardt, B. (1983) Localization of β-oxidation enzymes in peroxisomes isolated from nonfatty plant tissues. *Planta*, 159, 238.
62. Masterson, C. and Wood, C. (2000) Mitochondrial β-oxidation of fatty acids in higher plants. *Phys. Plant.*, **109**, 217.
63. Stelmach, B. A., Müller, A., Hennig, P., Laudert, D., Andert, L. and Weiler, E. W. (1998) Quantitation of the octadecanoid 12-oxo-phytodienoic acid, a signalling compound in plant mechanotransduction. *Phytochemistry*, **47**, 539.
64. Steiner, D. F., Smeekens, S. P., Ohagi, D. and Chan, S. J. J. (1992) The new enzymology of precursor processing endoproteases. *J. Biol. Chem.*, **267**, 23435.
65. Docherty, K. and Steiner, D. F. (1982) Post-translational proteolysis in polypeptide hormone biosynthesis. *Ann. Rev. Physiol.*, **44**, 625.
66. Pearce, G., Johnson, S. and Ryan, C. A. (1991) A polypeptide from tomato leaves induces wound-inducible proteinase inhibitor protein. *Science*, **253**, 895.
67. Stuart, R. and Street, H. E. (1969) Studies on the growth in culture of plant cells. IV. The initiation of division in suspensions of stationary phase cells of *Acer pseudoplatanus* L. *J. Exp. Bot.*, **20**, 556.
68. Matsubayashi, Y. and Sakagami, Y. (1996) Phytosulfokine, sulfated peptides that induce the profileration of single mesophyll cells of *Asparagus officinalis* L. *Proc. Natl Acad. Sci. USA*, **93**, 7623.
69. Franssen, H. J. (1998) Plants embrace a stepchild: the discovery of peptide growth regulators. *Curr. Opin. Plant Biol.*, **1**, 384.
70. Schaller, A. (1999) Oligopeptide signalling and the action of systemin. *Plant Mol. Biol.*, **40**, 763.
71. Gehring, C. A. (1999) Natriuretic peptides – a new class of plant hormones? *Ann. Botany*, **83**, 329.
72. Nürnberger, T., Nennstiel, D., Jabs, T., Sacks, W. R., Hahlbrock, K. and Scheel, D. (1994) High affinity binding of a fungal oligopeptide elicitor to parsley plasma membranes triggers multiple defense responses. *Cell*, **78**, 449.
73. Gómez-Gómez, L., Felix, G. and Boller, T. (1999) A single locus determines sensitivity to bacterial flagellin in *Arabidopsis thaliana*. *Plant J.*, **18**, 277.
74. Lee, J., Klüsener, B., Tsiamis, G., Stevens, C., Neyt, C., Tampakak, A., Panopoulos, N., Nöller, J., Weiler, E. W., Cornelis, G. R., Mansfield, J. W. and Nürnberger, T. (2001) HrpZ$_{Psph}$ from the plant pathogen *Pseudomonas syringae* pv. *phaseolicola* binds to lipid bilayers and forms an ion-conducting pore *in vitro*. *Proc. Natl Acad. Sci. USA*, **98**, 289.
75. Klüsener, B. and Weiler, E. W. (1999) Pore-forming properties of elicitors of plant defense reactions and cellulolytic enzymes. *FEBS Lett*, **459**, 263.
76. Engelberth, J., Koch, T., Schüler, G., Bachmann, N., Rechtenbach, J. and Boland W. (2001) Ion channel-forming alamethicin is a potent elicitor of volaile biosynthesis and tendril

coiling. Cross talk between jasmonate and salicylate signaling in lima beans. *Plant Physiol.*, **125**, 369.

77. Xiaorong, Z. (1998) Leucin-rich repeat receptor-like kinases in plants. *Plant Mol. Biol. Rep.*, **16**, 301.

78. Ryan, C. A. (1992) The search for the proteinase inhibitor-inducing factor, PIIF. *Plant Mol. Biol.*, **19**, 123.

79. Ryan, C. A. (2000) The systemin signaling pathway: differential activation of plant defensive genes. *Biochim. Biophys. Acta.*, **1477**, 112.

80. Hildmann, T., Ebneth, M., Peña-Cortès, H., Sánchez-Serrano, J. J., Willmitzer, L. and Prat, S. (1992) General roles for abscisic and jasmonic acids in gene activation as a result of mechanical wounding. *Plant Cell*, **4**, 1157.

81. Bolter, C. J. (1993) Methyl jasmonate induces papain inhibitor(s) in tomato leaves. *Plant Physiol.*, **103**, 1347.

82. Constable, C. P., Bergey, D. R. and Ryan, C. A. (1995) Systemin activates synthesis of wound-inducible tomato leaf polyphenol oxidase via the octanoid defense signaling pathway. *Proc. Natl Acad. Sci. USA*, **92**, 407.

83. Seldal, T. andersen, D.-J. and Hogstedt, G. (1994) Grazing-induced proteinase inhibitors: a possible cause for lemming population cycles. *Oikos*, **70**, 3.

84. Farmer, E. E. and Ryan, C. A. (1990) Interplant communication: airborne methyl jasmonate induces synthesis of proteinase inhibitors in plant leaves. *Proc. Natl Acad. Sci. USA*, **87**, 7713.

85. Albersheim, P. and Anderson, A. J. (1971) Proteins from plant cell walls inhibit polygalacturonases secreted by plant pathogens. *Proc. Natl Acad. Sci. USA*, **68**, 1815.

86. Fry, S. C., Aldington, S., Hetherington, P. R. and Aitken, J. (1993) Oligosaccharides as signals and substrates in the plant cell wall. *Plant Physiol.*, **103**, 1.

87. Darvill, A., Augur, C., Bergmann, C., Carlson, R. W., Cheong, J.-J., Eberhard, S., Hahn, M. G., Lo, V.-M., Marfa, V., Meyer, B., Mohnen, D., O'Neil, M. A., Spiro, M. D., van Halbeek, H., York, W. W.S. and Albersheim, P. (1992) Oligosaccharins – oligosaccharides that regulate growth, development and defense responses in plants. *Glycobiology*, **2**, 181.

88. Ryan, C. A. and Farmer, E. E. (1991) Oligosaccharide signals in plants: a current assessment. *Annu. Rev. Plant Physiol. Plant Mol. Biol.*, **42**, 651.

89. Bishop, P., Makus, D. J., Pearce, G. and Ryan, C. A. (1981) Proteinase inhibitor induction factor activity in tomato leaves resides in oligosaccharides enzymatically released from cell walls. *Proc. Natl Acad. Sci. USA*, **78**, 3536.

90. Walker-Simmons, M. and Ryan, C. A. (1984) Protease inhibitor synthesis in tomato leaves. *Plant Physiol.*, **76**, 787.

91. Mylona, P., Pawlowski, K. and Bisseling, T. (1995) Symbiotic nitrogen fixation. *Plant Cell*, **7**, 869.

92. Gus-Mayer, S., Naton, B., Hahlbrock, K. and Schmelzer, E. (1998) Local mechanical stimulation induces components of the pathogen defense response in parsley. *Proc. Natl Acad. Sci. USA*, **95**, 8398.

93. Yahraus, T., Chandra, S., Legendre, L. and Low, P. S. (1995) Evidence for a mechanically induced oxidative burst. *Plant Physiol.*, **109**, 1259.

94. Baldwin, I. T. (1996) Methyl jasmonate-induced nicotine production in *Nicotiana attenuata*: inducing defenses in the field without wounding. *Entomol. Exp. Appl.*, **80**, 213.

95. Gundlach,, H., Mueller, M. J., Kutchan, T. M. and Zenk, M. H. (1992) Jasmonic acid is a signal transducer in elicitor-induced plant cell cultures. *Proc. Natl Acad. Sci. USA*, **89**, 2389.

96. Farmer, E. E. and Ryan, C. A. (1990) Interplant communication: airborne methyl jasmonate induces synthesis of proteinase inhibitors in plant leaves. *Proc. Natl Acad. Sci USA*, **87**, 7713.

97. Howe, G. A., Lightner, J., Browse, J. and Ryan, C. A. (1996) An octadecanoid pathway mutant (JL5) of tomato is compromised in signaling for defense against insect attack. *Plant Cell*, **8**, 2067.

98. Rojo, E., Titarenko, E., León, J., Berger, S., Vancanneyt, G. and Sánchez-Serrano, J. J. (1998) Reversible protein phosphorylation regulates jasmonic acid-dependent and acid-independent wound signal transduction pathways in *Arabidopsis thaliana*. *Plant J.*, **13**, 153.

99. Kubigsteltig, I., Laudert, D. and Weiler, E. W. (1999) Structure and regulation of the *Arabidopsis thaliana* allene oxide synthase gene. *Planta*, **208**, 463.

100. Schaller, A. and Ryan, C. A. (1996) Systemin – a polypeptide defense signal in plants. *BioEssays*, **18**, 27.

101. Malone, M. and Alarcon, J. J. (1995) Only xylem borne factors can account for systemic wound signalling in the tomato plant. *Planta*, **196**, 740.

102. Wildon, D. C., Thain, J. F., Minchin, P. E.K., Gubb, I. R., Reily, A. J., Skipper, Y. D., Doherty, H. M., O'Donnel, P. J. and Bowles, D. J. (1992) Electrical signalling and systemic proteinase inhibitor induction in the wounded plant. *Nature*, **360**, 62.

103. Seo, S., Okamoto M., Seto, H., Ishizuka, K., Sano, H. and Ohashi, Y. (1995) Tobacco MAP kinase: a possible mediator in wound signal transduction pathways. *Science*, **270**, 1988.

104. Seo, S., Sano, H. and Ohashi, Y. (1999) Jasmonate-based wound signal transduction requires activation of WIPK, a tobacco mitogen-activated protein kinase. *Plant Cell*, **11**, 289.

105. Schaller, A. and Oecking, C. (1999) Modulation of the plasma membrane H^+-ATPase activity differentially activates wound and pathogen defense responses in tomato plants. *Plant Cell*, **11**, 263.

106. Dammann, C., Rojo, E. and Sánchez-Serrano, J. J. (1997) Abscisic acid and JA activate wound-inducible genes in potato through separate, organ-specific signal transduction pathways. *Plant J.*, **11**, 773.

107. Kenton, P., Mur, L. A.J. and Draper, J. (1999) A requirement for calcium and protein phosphatase in the jasmonate-induced increase in tobacco leaf acid phosphatase specific activity. *J. Exp. Bot.*, **199**, 1331.

108. Baudouin, E., Meskienne, I. and Hirt, H. (1999) Unsaturated fatty acids inhibit MP2C, a protein phosphatase 2C involved in the wound-induced MAP kinase pathway regulation. *Plant J.*, **20**, 343.

109. Börge, L., Ligterink, W., Meskienne, I., Barker, P. J., Heberle-Bors, E., Huskisson, N. S. and Hirt, H. (1997) Wounding induces the rapid and transient activation of a specific MAP kinase pathway. *Plant Cell*, **9**, 75.

110. Bögre, L., Ligterink, W., Heberle-Bors, E. and Hirt, H. (1996) Mechanosensors in plants. *Nature*, **383**, 489.

111. Mizoguchi, T., Irie, K., Hayashida, N., Yamaguchi-Shinozaki, K., Matsumoto, K. and Shinozaki, K. (1996) A gene encoding a mitogen-activated protein kinase kinase kinase is induced simultaneously with genes for a mitogen-activated protein kinase and S6 ribosomal protein kinase by touch, cold, and water stress in *Arabidopsis thaliana*. *Proc. Natl Acad. Sci. USA*, **93**, 765.

112. Jonak, C., Kiegerl, S., Ligterink, W., Barker, P. J., Huskisson, N. S. and Hirt, H. (1996) Stress signaling in plants: a MAP kinase pathway is activated by cold and drought. *Proc. Natl Acad. Sci. USA*, **93**, 11274.

113. Knetsch, M. L.W., Wang, M., Snaar-Jagalska, B. E. and Heimovaara-Dijkstra, S. (1996) Abscisic acid induces mitogen-activated protein kinase activation in barley aleurone protoplasts. *Plant Cell*, **8**, 1061.

114. Huttly, A. and Phillips, A. L. (1995) Gibberellin-regulated expression in oat aleurone cells of two kinases that show homology to MAP kinase and a ribosomal protein kinase. *Plant Mol. Biol.*, **27**, 1043.

115. Mizoguchi, T., Gotoh, Y., Nishida, E., Yamaguchi-Shinozaki, K., Hayashida, N., Iwasaki, T., Kamada, H. and Shinozaki, K. (1994) Characterization of two cDNAs that encode MAP kinase homologues in *Arabidopsis thaliana* and analysis of the possible role of auxin in activating such kinase activities in cultured cells. *Plant J.*, **5**, 111.

116. Kieber, J. J., Rothenberg, M., Roman, G., Feldmann, K. A. and Ecker, J. R. (1993) CTR1, a negative regulator of the ethylene response pathway in *Arabidopsis* encodes a member of the Raf family of protein kinases. *Cell*, **72**, 427.

117. Petersen, M., Brodersen, P., Naested, H., Andreasson, E., Lindhart, U., Johansen, B., Nielsen, H. B., Lacy, M., Austin, M. J., Parker, J. E., Sharma, S. B., Klessig, D. F., Martienssen, R., Mattsson, O., Jensen, A. B. and Mundy, J. (2000) *Arabidopsis* MAP kinase 4 negatively regulates systemic acquired resistance. *Cell*, **103**, 1111.

118. Lin, L. L., Wartmann, M., Lin, A. Y., Knopf, J., Seth, A. and Davis, R. J. (1993) cPLA2 is phosphorylated and activated by MAP kinase. *Cell*, **72**, 269.

119. Gilroy, S., Read, N. D. and Trewavas, A. J. (1990) Elevation of cytoplasma calcium by caged calcium or caged inositol triphosphate initiate stomatal closure. *Nature*, **346**, 769.

120. Roberts, D. M. and Harmon, A. C. (1992) Calcium-modulated proteins: targets of intra-cellular calcium signals in higher plants. *Annu. Rev. Plant Physiol. Plant Mol. Biol.*, **43**, 375.

121. Klee, C. B. and Vanaman, T. C. (1982) Calmodulin. *Adv. Protein Chem.*, **35**, 213.

122. Geiser, J. R., van Tinnen, D., Brockerhoff, S. E., Neffm, M. M. and Davis, T. N. (1991) Can calmodulin function without binding calcium? *Cell*, **65**, 949.

123. Felix, G. and Boller, T. (1995) Systemin induces rapid ion fluxes and ethylene biosynthesis in *Lycopersicon peruvianum* cells. *Plant J.*, **7**, 381.

124. Bush, D. S. (1995) Calcium regulation in plant cells and its role in signalling. *Annu. Rev. Plant Physiol. Plant Mol. Biol.*, **46**, 95.

125. León, J., Titarenko, E. and Sánchez-Serrano, J. J. (1998) Jasmonic acid-dependent and -independent wound signal transduction pathways are differentially regulated by Ca^{2+}/calmodulin in *Arabidopsis thaliana*. *Mol. Gen. Genet.*, **258**, 412.

126. Bergey, D. R., Howe, G. A. and Ryan, C. A. (1996) Polypeptide signaling for plant defensive genes exhibits analogies to defense signaling in animals. *Proc. Natl Acad. Sci. USA*, **93**, 12053.

127. Bergey, D. R., Orozco-Cardenas, D., De Moura, D. S. and Ryan, C. A. (1999) A wound- and systemin-inducible polygalacturonase in tomato leaves. *Proc. Natl Acad. Sci. USA*, **96**, 1756.

128. Pautôt, V., Holzer, F. M., Reisch, B. and Walling L. L. (1993) Leucine aminopeptidase: an inducible component of the defense response in *Lycopersicon esculentum* (tomato) *Proc. Natl Acad. Sci. USA*, **90**, 9906.

129. Bolter, C. J. (1993) Methyl jasmonate induces papain inhibitor(s) in tomato leaves. *Plant Physiol.*, **103**, 1347.

130. McGurl, B., Pearce, G., Orozco-Cardenas, M. and Ryan, C. A. (1992) Structure, expression, and antisense inhibition of the systemin precursor gene. *Science*, **255**, 1570.

131. Royo, J., Vacanney, G., Pérez, A. G., Sanz, C., Störmann, K., Rosahl, S. and Sánchez-Serrano, J. J. (1996) Characterization of three potato lipoxygenases with distinct

enzymatic activities and different organ-specific and wound-regulated expression patterns. *J. Biol. Chem.*, **271**, 21012.

132. Heitz, T., Bergey, D. and Ryan, C. A. (1997) A gene encoding a chloroplast-targeted lipoxygenase in tomato leaves is transiently induced by wounding, systemin, and methyl jasmonate. *Plant Physiol.*, **114**, 1805.

133. Howe, G. A., Lee, G. I., Itoh, A., Li, L. and DeRocher, E. (2000) Cytochrome P450-dependent metabolism of oxylipins in tomato. Cloning and expression of allene oxide synthase and fatty acid hydroperoxide layse. *Plant Physiol.*, **123**, 711.

134. Narváez-Vásquez, J., Florin-Christensen, J. and Ryan, C. A. (1999) Positional specificity of a phospholipase A activity induced by wounding, systemin, and oligosaccharide elicitors in tomato leaves. *Plant Cell*, **11**, 2249.

135. Jaffe, M. J. and Forbes, S. (1993) Thigmomorphogenesis: the effect of mechanical pertubation on plants. *Plant Growth Regul.*, **12**, 313.

136. Mitchell, C. A. and Meyers, P. N. (1995) Mechanical stress regulation of plant growth and development. *Hort. Rev.*, **17**, 1.

137. Jaffe, M. J. and Galston, A. W. (1966) Physiological studies on pea tendrils. 1. Growth and coiling following mechanical stimulation. *Plant Physiol.*, **41**, 1014.

138. Jaffe, M. J. and Galson, A. W. (1968) The physiology of tendrils. *Annu. Rev. Plant Physiol.*, **19**, 417.

139. Haberlandt, G. (1924) In *Physiologische Pflanzenanatomie*, Wilhelm Engelmann Verl., Leipzig.

140. Galun, E. (1959) The cucumber tendril: a new test organ for gibberellins. *Experientia*, **15**, 184.

141. Reinhold, L. (1967) Induction of coiling in tendrils by auxin and carbon dioxide. *Science*, **158**, 791.

142. Junker, S. (1977) Thigmotropic coiling of tendrils of *Passiflora quadrangularis* L. is not caused by lateral redistribution of auxin. *Physiol. Plant.*, **41**, 51.

143. Jaffe, M. J. (1970) Physiological studies of pea tendrils. VII. Evaluation of a technique for the asymmetrical application of ethylene. *Plant Physiol.*, **41**, 631.

144. Bangerth, F. (1974) Interaktionen von Auxin und Äthylen bei der thigmotropen Bewegung der Ranken von *Cucumis sativus*. *Planta*, **117**, 329.

145. Falkenstein, E., Groth, B., Mithöfer, A. and Weiler, E. W. (1991) Methyl jasmonate and α-linolenic acid are potent inducers of tendril coiling. *Planta*, **185**, 316.

146. Weiler, E. W., Albrecht, T., Groth, B., Xia, Z.-Q., Luxem, M., Liss, H., Andert, L. and Sprengler, P. (1993) Evidence for the involvement of jasmonates and their octadecanoid precursors in the tendril coiling response of *Bryonia dioica*. *Phytochemistry*, **32**, 591.

147. Weiler, E. W., Kutchan, T. M., Gorba, T., Brodschelm, W., Niesel, U. and Bublitz, F. (1994) The *Pseudomonas* phytotoxin coronatine mimics octadecanoid signalling molecules of higher plants. *FEBS Lett.*, **345**, 9.

148. Braam, J. and Davies, R. W. (1990) Rain-induced, wind-induced, and touch-induced expression of calmodulin and calmodulin-related genes in *Arabidopsis*. *Cell*, **50**, 357.

149. Knight, M., Campbell, A., Smith, S. M. and Trewavas, A. J. (1991) Transgenic plant aequorin reports the effects of touch, cold-shock and elicitors on cytoplasmic calcium. *Nature*, **352**, 524.

150. Sistrunk, M. L., Antosiewicz, D. M., Puruggganan, M. M. and Braam, J. (1994) *Arabidopsis* TCH3 encodes a novel Ca^{2+} binding protein and shows environmentally induced and tissue-specific regulation. *Plant Cell*, **6**, 1553.

151. Khan, A., Johnson, K. A., Braam, J. and James, M. (1997) Comparative modeling of the three-dimensional structure of the calmodulin-related TCH2 protein from *Arabidopsis*. *Proteins*, **27**, 144.

152. Xu, W., Purugganan, M. M., Polisensky, D. H., Antosiewicz, D. M., Fry. S. C. and Braam, J. (1995) *Arabidopsis TCH4*, regulated by hormones and the environment, encodes a xyloglucan endotransglycosylase. *Plant Cell*, **7**, 1555.

153. Redgwell, R. J. and Fry, S. C. (1993) Xyloglucan endotransglycosylase activity increases during kiwifruit (*Actinidia deliciosa*) ripening: Implications for fruit softening. *Plant Physiol.*, **103**, 1399.

154. Wymer, C. L., Wymer, S. A., Cosgrove, D. J. and Cyr, R. J. (1996) Plant cell growth responds to external forces and the response requires intact microtubules. *Plant Physiol.*, **110**, 425.

155. Antosiewicz, D. M., Polisensky, D. H. and Braam, J. (1995) Cellular localization of the Ca^{2+} binding TCH3 protein of *Arabidopsis*. *Plant J.*, **8**, 623.

156. Braam, J., Sistrunk, M. L., Polisensky, D. H., Xu, W., Purugganan, M. M., Antosiewicz, D. M., Campbell, P. and Johnson, K. A. (1996) Life in a changing world: *TCH* gene regulation of expression and responses to environmental signals. *Physiol. Plant.*, **98**, 909.

157. Xu, W., Campbell, P., Vargheese, A. K. and Braam, J. (1996) The *Arabidopsis* XET-related gene family: environmental and hormonal regulation of expression. *Plant J.*, **9**, 879.

158. Kaiser, I., Engelberth, J., Groth, B. and Weiler, E. W. (1994) Touch- and methyl jasmonate-induced lignification in tendrils of *Bryonia dioica* Jacq. *Bot. Acta*, **107**, 24.

159. Mauch, F., Kmecl, A., Schaffrath, U., Volrath, S., Görlach, J., Ward, E., Ryals, J. and Dudler, R. (1997) Mechanosensitive expression of a lipoxygenase gene in wheat. *Plant Physiol.*, **114**, 1561.

160. Engelberth, J., Wanner, G., Groth, B. and Weiler, E. W. (1995) Functional anatomy of the mechanoreceptor cells in the tendrils of *Bryonia dioica* Jacq. *Planta*, **196**, 316.

161. Galaud, J.-P., Lareyre, J.-J. and Boyer, N. (1993) Isolation, sequencing and analysis of the expression of *Bryonia* calmodulin after mechanical pertubation. *Plant Mol. Biol.*, **23**, 839.

162. Suzuki, K. and Shinshi, H. (1995) Transient activation and tyrosine phosphorylation of a protein kinase in tobacco cells treated with a fungal elicitor. *Plant Cell*, **7**, 639.

163. Piotrowski, M., Liß, H. and Weiler, E. W. (1996) Touch-induced protein phosphorylation in mechanosensitive tendrils of *Bryonia dioica* Jacq. *J. Plant Physiol.*, **147**, 539.

164. Haley, A. A., Russell, A. J., Wood, N., Allian, A. C., Knight, M., Campbell, K. and Trewavas, A. T. (1995) Effects of mechanically signaling on plant cell cytosolic calcium. *Proc. Natl Acad. Sci. USA*, **92**, 4124.

165. Braam, J. (1992) Regulated expression of the calmodulin-related TCH genes in cultured *Arabidopsis* cells: induction by calcium and heat shock. *Proc Natl Acad. Sci. USA*, **89**, 3213.

166. Polisensky, D. H. and Braam, J. (1992) Cold-shock regulation of *Arabidopsis TCH* genes and the effects of modulating intracellular calcium levels. *Plant Physiol.*, **111**, 1271.

167. Liss, H. and Weiler, E. W. (1994) Ion-translocating ATPases in tendrils of *Bryonia dioica* Jacq. *Planta*, **194**, 169.

168. Klüsener, B., Boheim, G., Liss, H., Engelberth, J. and Weiler, E. W. (1995) Gadolinium-sensitive, voltage dependent calcium release channels in the endoplasmic reticulum of a higher plant mechanoreceptor organ. *EMBO J.*, **14**, 2708.

169. Pei, Z.-M., Murata, Y., Benning, G., Thomine, S., Klüsener, B., Allen, G. J., Grill, E. and Schroeder, J. I. (2000) Calcium channels activated by hydrogen peroxide mediate abscisic acid signalling in guard cells. *Nature*, **406**, 731.

170. Blechert, S., Bockelmann, C., Brümmer, O., Füsslein, M., Gundlach, H., Haider, G., Hölder, S., Kutchan, T. M., Weiler, E. W. and Zenk, M. H. (1997) Structural separation of biological activities of jasmonates and related compounds. *J. Chem. Soc. Perkin I*, **23**, 3549.

171. Blechert, S., Bockelmann, C., Füsslein, M., von Schrader, T., Stelmach, B. A., Niesel, U. and Weiler, E. W. (1999) Structure-activity relationships reveal two sub-groups of active octadecanoids in elicitation of the tendril coiling response of *Bryonia dioica*. *Planta*, **207**, 470.

172. Stelmach, B. A., Müller, A. and Weiler, E. W. (1999) 12-Oxo-phytodienoic acid and indole-3-acetic acid in jasmonic acid-treated tendrils of *Bryonia dioica*. *Phytochemistry*, **51**, 187.

173. Weber, A., Vick, B. A. and Farmer, E. E. (1997) Dinor-oxo-phytodienoic acid: a new hexadecanoid signal in the jasmonate family. *Proc. Natl Acad. Sci. USA*, **94**, 10473.

4 | The role of active oxygen species in plant signal transduction

EVA VRANOVÁ, FRANK VAN BREUSEGEM, JAMES DAT,
ENRIC BELLES-BOIX AND DIRK INZÉ

1. Active oxygen species in plants

1.1 Biochemical properties

Plants, as other aerobic organisms, require oxygen for the efficient production of energy. During the reduction of O_2 to H_2O, active oxygen species (AOS), namely superoxide radical ($O_2^{\bullet-}$), hydrogen peroxide (H_2O_2) and hydroxyl radical ($OH^{\bullet}$) can be formed (Fig. 1). The reaction chain requires input of energy initially, whereas subsequent steps are exothermic and can occur spontaneously, either catalysed or not (1). Acceptance of excess energy by O_2 can additionally lead to the formation of singlet oxygen ($^1O_2$), a highly reactive molecule as compared to O_2. $^1O_2$ can last for nearly 4 μs in water and 100 μs in a polar environment (2). $^1O_2$ can either transfer its excitation energy to other biological molecules or combine with them, thus forming endoperoxides or hydroperoxides (3). $O_2^{\bullet-}$ is a moderately reactive, short-lived AOS with a half-life of approximately 2–4 μs (1). $O_2^{\bullet-}$ cannot cross biological membranes and is readily dismutated to H_2O_2. Alternatively, $O_2^{\bullet-}$ can reduce quinones and transition metal complexes of Fe^{3+} and Cu^{2+}, thus affecting the activity of metal-containing enzymes. Hydroperoxyl radicals ($HO_2^{\bullet}$) that are formed from $O_2^{\bullet-}$ by protonation in aqueous solutions can cross biological membranes and subtract hydrogen atoms from polyunsaturated fatty acids and lipid hydroperoxides, thus initiating lipid auto-oxidation (3). H_2O_2 is moderately reactive and a relatively long-lived molecule (half-life of 1 ms) that can diffuse some distances from its site of production (1). H_2O_2 may inactivate enzymes by oxidizing their thiol groups. For example, enzymes of the Calvin cycle, copper/zinc superoxide dismutase, and iron superoxide dismutase are inactivated by H_2O_2 (4, 5). The most reactive of all AOS is the hydroxyl radical ($OH^{\bullet}$) that is formed from H_2O_2 by the so-called Haber–Weiss or Fenton reactions by using metal catalysts (3). $OH^{\bullet}$ can potentially react with all biological molecules, and because cells have no enzymatic mechanism to eliminate this highly reactive AOS, its excess production leads ultimately to cell death.

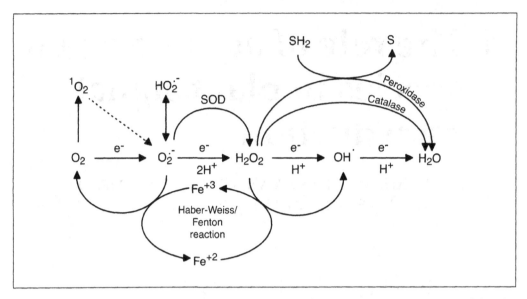

Fig. 1 Interconversion of active oxygen species (AOS) derived from O_2. Ground state molecular oxygen (O_2) can be activated by excess energy, reversing the spin of one of the unpaired electrons to form singlet oxygen ($^1O_2$). Alternatively, one-electron reduction leads to the formation of superoxide radical ($O_2^{\bullet-}$). $O_2^{\bullet-}$ exists in equilibrium with its conjugate acid, hydroperoxyl radical ($HO_2^{\bullet-}$). Subsequent reduction steps then form hydrogen peroxide (H_2O_2), hydroxyl radical ($OH^\bullet$), and water (H_2O). Metal ions that are mainly present in cells in the oxidized form (Fe^{3+}) are reduced in the presence of $O_2^{\bullet-}$ and consequently may catalyse the conversion of H_2O_2 to $OH^\bullet$ by the Fenton or Haber–Weiss reactions. Superoxide dismutases (SODs), catalases, and peroxidases enzymatically reduce AOS. Peroxidases require a reducing substrate (SH_2) for the reaction.

1.2 Sources of AOS in plant cells

Most cellular compartments have the potential to become a source of AOS (Fig. 2). It is generally agreed, however, that chloroplasts are the most powerful source of AOS in plants. Light energy is absorbed by the antennae of photosystem II (PSII) and photosystem I (PSI), and transferred via the electron transport chain to the final acceptor, $NADP^+$. The reducing power of NADPH is then consumed in the Calvin–Benson cycle upon fixation of CO_2 (6). Conditions that limit CO_2 fixation, such as drought, salt stress, ozone, and high or low temperatures, reduce the regeneration of $NADP^+$ by the Calvin–Benson cycle, consequently over-reducing the photosynthetic electron transport chain. High light exacerbates these conditions and $O_2^{\bullet-}$ is formed by auto-oxidation of components of the electron transport chain by using O_2 as a reductant (1). Excess light excitation energy that is not coupled to the electron transport chain might, in addition, generate triplet chlorophyll in the antennae of PSII or PSI and this excitation energy can be passed to O_2, forming $^1O_2$ (7). To prevent over-reduction of the electron transport chain and subsequent production of AOS in chloroplasts under conditions limiting CO_2 fixation, plants have evolved a photorespiratory pathway (8). During photorespiration, phospho-glycolate that is produced by oxygenase activity of ribulose-1,5-bisphosphate

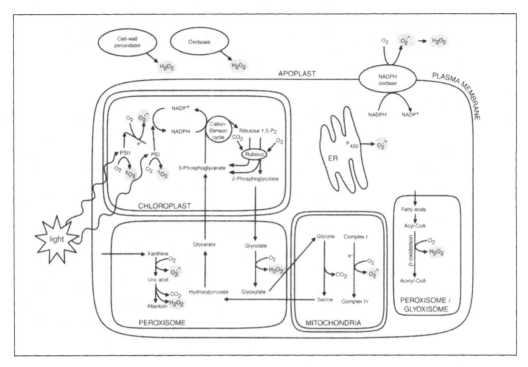

Fig. 2 Sources of AOS in plants. For details, see text.

carboxylase (Rubisco) is metabolized in multiple cellular compartments, forming phosphoglycerate and CO_2, which are believed to drive the Calvin–Benson cycle and thus to regenerate $NADP^+$ (8, 9). As part of the photorespiratory pathway, H_2O_2 is formed in the peroxisomes during the conversion of glycolate to glyoxylate (9). H_2O_2 can also be produced in peroxisomes during catabolism of lipids as a byproduct of β-oxidation of fatty acids (10). Catabolism of purines probably takes place in the peroxisomes as well (11). The first reaction of this catabolic chain, the oxidation of xanthine to uric acid by xanthine oxidase, generates $O_2^{\bullet-}$, while uric acid is oxidized to allantoin, yielding H_2O_2 and CO_2. When growth and other energy-requiring processes in plants are reduced or cease as a consequence of stress, the electron transport chain in the mitochondria may become over-reduced, favouring the generation of $O_2^{\bullet-}$ (12). Other important source of AOS in plants that have received little attention are the detoxification reactions catalysed by cytochromes, in particular cytochrome P450 in the cytoplasm and the endoplasmic reticulum (ER). During these reactions, electron leakage to oxygen and the decomposition of the intermediate oxygenate of cytochrome P450 can form $O_2^{\bullet-}$ (13). AOS are also generated in plants at the plasma membrane level or extracellularly in the apoplast. Plasma membrane NADPH-dependent oxidase (NADPH oxidase) has recently received a lot of attention as a source of AOS for the oxidative burst, which is typical of incompatible plant–pathogen interactions. In phagocytes, plasma membrane-localized NADPH oxidase was identified as a major contributor to their bactericidal capacity (14). NADPH oxidase is composed of the heterodimeric membrane-located flavo-cytochrome (gp91^{phox} and p22^{phox}) that forms an electron transport chain responsible for the reduction of molecular oxygen to $O_2^{\bullet-}$:

$$O_2 + NADPH \rightarrow O_2^{\bullet-} + NADP^+ + H^+.$$

Cytosolic subunits of the NADPH oxidase (p47^{phox}, p67^{phox} and p40^{phox}) associate with the plasma membrane upon activation (15). $O_2^{\bullet-}$ accumulates at the external surface of the plasma membrane and is then further dismutated to H_2O_2 by rapid enzyme-catalysed dismutation (16). In plants, proteins immunologically related to p22^{phox} (17), p47^{phox} and p67^{phox} (18), and genes homologous to gp91^{phox} have been identified (19). Moreover, chemical inhibitors of NADPH oxidase, such as diphen-ylene iodonium (DPI), have been shown to block or severely reduce AOS production upon a number of biotic and abiotic stresses (20–25). In addition to NADPH oxidase, pH-dependent cell wall peroxidases, germin-like oxalate oxidases and amine oxidases have been proposed as sources of H_2O_2 in the apoplast (20). pH-dependent cell wall peroxidases are activated by alkaline pH and, in the presence of a reductant, H_2O_2 is formed. Alkalinization of the apoplast upon elicitor recognition precedes the oxidative burst and the production of H_2O_2 by pH-dependent cell wall peroxidases has been proposed as an alternative way of AOS production during biotic stress (15, 20). Oxalate oxidase catalyses the conversion of oxalate to CO_2 and H_2O_2 and the activity of this enzyme may also be significant in certain plant–pathogen interactions (20). Amine oxidases catalyse oxidation of a wide variety of biogenic amines to their corresponding aldehydes with release of NH_3 and H_2O_2 (20). H_2O_2 formed by oxidation of amines may be directly utilized by wall-bound peroxidases in ligni-fication and cell wall strengthening, both during normal growth and in response to external stimuli, such as wounding and pathogenesis (20, 24).

1.3 Mechanisms that modulate AOS levels in plants

1.3.1 Antioxidant systems

Many, if not all, biotic and abiotic stresses stimulate AOS production (for review, see reference 1). Due to the highly cytotoxic and reactive nature of AOS, their accumu-lation must be under tight control. Plants possess very efficient enzymatic and non-enzymatic antioxidant defence systems that allow detoxification of AOS and protection of plant cells from oxidative damage (Fig. 3; for review, see references 3, 4, 11, 26–29).

Superoxide dismutases (SOD) catalyse the dismutation of $O_2^{\bullet-}$ to H_2O_2:

$$2O_2^{\bullet-} + 2H^+ \rightarrow H_2O_2 + O_2.$$

SODs are localized in organelles with the highest rate of $O_2^{\bullet-}$ production, such as chloroplasts and mitochondria, as well as in the cytosol. Peroxisomal, nuclear and extracellular isoforms have also been reported (4, 30).

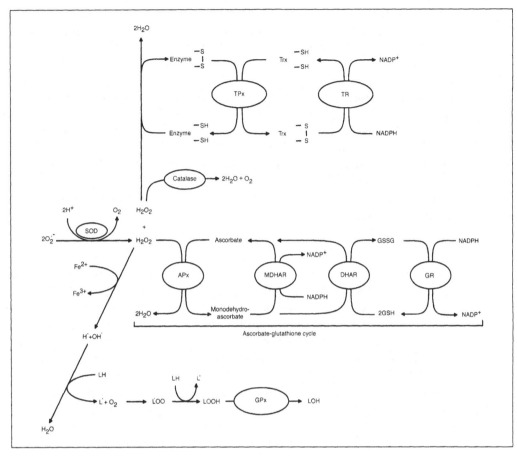

Fig. 3 Plant antioxidant systems. Superoxide dismutases (SODs) catalyse the dismutation of $O_2{}^{\bullet-}$ to H_2O_2. H_2O_2 is reduced mainly by catalases or by ascorbate peroxidases (APx). Reduction by APx requires a substrate (ascorbate) that is reduced via a cycle of coupled oxidation and reduction reactions catalysed by mono-dehydroascorbate reductase (MDHAR) or dehydroascorbate reductase (DHAR) and glutathione reductase (GR), known as the ascorbate-glutathione cycle. Oxidized enzymes can be reduced by thioredoxin peroxidase (TPx) using thioredoxin [Trx-$(SH)_2$] as a reducing substrate. Oxidized thioredoxin (Trx-S-S) is then regenerated by thioredoxin reductase (TR) at the expense of NADPH or other reducing substrates, such as ferredoxin. Lipid hydroperoxides (LOOH) that are formed by oxidation of lipids (LH) are reduced by glutathione peroxidase (GPx).

H_2O_2 is reduced either by catalases:

$$2H_2O_2 \rightarrow 2H_2O + O_2,$$

or by peroxidases that require reducing substrates:

$$H_2O_2 + 2XH_2 \rightarrow 2H_2O + 2X.$$

The reducing substrate for ascorbate peroxidases (APx) is ascorbate that is regenerated by either monodehydroascorbate reductase at the expense of NADPH or by glutathione (GSH)-dependent dehydroascorbate reductase using GSH as a reducing substrate. Oxidized glutathione (GSSG) is then reduced by glutathione reductase (GR) at the expense of NADPH. This cycle of oxidation/ reduction reactions is called the ascorbate-glutathione cycle and operates in the chloroplasts, the cytosol, and the peroxisomes. Glutathione peroxidases (GPx) may reduce lipid hydroperoxides that are formed by lipid peroxidation during the oxidative stress using GSH as a reducing substrate. Knowledge on plant GPx is still limited. However, citrus GPx, the best-studied GPx in plants, has higher affinity for lipid hydroperoxides than for H_2O_2 and is localized in the soluble cellular fraction (28). Catalases are predominantly localized in peroxisomes, where they remove H_2O_2 that is produced by photorespiration, by catabolism of fatty acids, and possibly also by purines. Oxidized proteins can be reduced in the cell by thioredoxin peroxidases using thioredoxin [Trx-(SH)$_2$] as a reducing substrate:

$$X\text{-}S\text{-}S + Trx\text{-}(SH)_2 \rightarrow X\text{-}(SH)_2 + Trx\text{-}S\text{-}S.$$

Oxidized thioredoxin (Trx-S-S) is then regenerated by thioredoxin reductase at the expense of NADPH or other substrates, such as ferredoxin (3, 26). In addition to enzymatic antioxidants, a number of antioxidant molecules also effectively scavenge AOS within the cell. These include glutathione, ascorbate, thioredoxin, carotenoids, tocopherols, phenolics, and other compounds (29). The distinct subcellular localization and biochemical properties of antioxidant enzymes (4), their differential inducibility at the enzyme/gene expression level (31), and the plethora of non-enzymatic scavengers (29) render the antioxidant systems a very versatile and highly modulable unit that can control AOS accumulation temporally and spatially. Such a controlled modulation of AOS levels is significant in light of the recent evidence for a signalling capacity of AOS, where the AOS concentration can determine the type of response (32, 33). However, despite the extensive knowledge on the plant antioxidant systems and their role in detoxifying AOS, little is known on their involvement in modulation of AOS levels for redox signalling (34–36).

1.3.2 NADPH oxidase complex

Controlled synthesis of AOS is an alternative mechanism for the modulation of cellular AOS levels. Among the different AOS generating enzymes in plants (see Section 1.2), NADPH oxidase is considered the main source of the early and sustained accumulation of AOS, the so-called oxidative burst, typical of a hypersensitive response (HR), and necessary for defence gene induction and establishment of immunity during incompatible plant–pathogen interactions (32, 37). AOS accumulation induced by hypo-osmotic stress (21), wounding (22), ozone (23–25), and UV-B (24) is sensitive to NADPH oxidase inhibitors, suggesting that this enzyme plays a ubiquitous role in AOS synthesis and/or AOS signal amplification during both biotic and abiotic stress responses. Further supporting this role, results from our group indicate that the NADPH oxidase can be activated by a rise in endogenous H_2O_2 and is sensitive to the redox balance of the cell (Dat *et al.*, in preparation). Thus, regulation of the activity of NADPH oxidase may be of importance for controlling temporal and spatial induction of the oxidative burst.

Components of the signal transduction pathway that lead to the plant oxidative burst are beginning to emerge. Ca^{2+} influx, K^+/Cl^- efflux, alkalinization of the apoplast, and protein phosphorylation precede the initiation of the oxidative burst (21, 38–40). In support, Ca^{2+} chelators or Ca^{2+} channel blockers, omission of extracellular Ca^{2+}, ion channel blockers (K^+/Cl^-) and protein kinase inhibitors suppress the oxidative burst (21, 32, 39), while the ionophore amphotericin B that stimulates K^+/Cl^- efflux and Ca^{2+}/H^+ influx, as well as inhibitors of the protein phosphatases 1 and 2A initiate an oxidative burst even in the absence of the stress signal (38, 39). Alkalinization of the apoplast probably only plays a modulatory role on the oxidative burst, as the alkalinization of the extracellular medium enhances cryptogein-induced oxidative burst in tobacco cells, but by itself has no effect on AOS production (41). Interestingly, calcium fluxes not only precede the oxidative burst, but H_2O_2 itself can transiently increase cytosolic Ca^{2+} level in a concentration-dependent manner (42). Similarly, DPI can decrease K^+/Cl^- effluxes and extracellular alkalinization induced upon hypo-osmotic stress (21). These observations are indicative for a positive feedback regulation of the signalling pathway that activates NADPH oxidase by its own product, which is further supported by observation that H_2O_2 itself induces an oxidative burst (37, 43; Dat et al., in preparation). Accordingly, expression of the *Arabidopsis* homologue of the NADPH oxidase subunit gp91[phox] is induced by H_2O_2 (44). How the components of this signal transduction pathway participate in the induction of the oxidative burst is at present not fully understood. However, several possible ways have been proposed. The plant homologue of the gp91[phox] protein possesses EF hands that bind Ca^{2+}, which indicates a direct modulation of the NADPH oxidase activity by Ca^{2+} (19). Ca^{2+} is required for the phosphorylation and the translocation of the p47[phox] and p67[phox] subunits to the plasma membrane, a process that probably employs Ca^{2+} and/or calmodulin-dependent protein kinases (18). The small GTP-binding protein p21[rac] is another cognate protein of the mammalian NADPH oxidase complex that, upon activation, translocates from cytosol to the plasma membrane. Thus, GTP binding and its subsequent hydrolysis may modulate the activity of the enzyme and, as such, AOS generation (14). A small GTPase immunologically related to the mammalian Rac2 protein has been identified in tomato and tobacco (18, 45), and has been shown to translocate to the plasma membrane upon treatment with elicitor (18). A modulatory role of Rac proteins on AOS synthesis in plants has been demonstrated in rice overproducing constitutively active and dominant negative forms of OsRac1, a homologue of human Rac1 and Rac2 proteins. Whereas constitutively active OsRac1 induces DPI-sensitive AOS production and promotes cell death in transformed rice cells, dominant negative OsRac1 has an opposite effect (46). Additionally, the oxidative burst in plants can be inhibited by inhibitors of phospholipase A_2 (40) and stimulated by the protein kinase C agonist, phorbol 12-myristate 13-acetate (47). The data suggest multiple control points for the NADPH oxidase-mediated AOS production in plants, where each of them may become either positively or negatively regulated.

2. AOS as signal molecules

2.1 Signalling role of AOS in defence responses

With the identification of catalase as a salicylic acid (SA)-binding protein, together with a set of experiments suggesting that H_2O_2 is downstream from SA in pathogenesis-related (*PR-1*) gene induction (48), attention was focused on the putative signalling role of H_2O_2 in plant defence responses. In an elegant series of experiments, Levine *et al.* (32) demonstrated that H_2O_2 is indeed a diffusible signal in the induction of plant defence genes, such as *GST* and *GPx*. A catalase trap, placed between cells inoculated with an avirulent pathogen and uninfected cells, blocked the diffusible signal that originated from the infected cells and was necessary for defence gene induction. Other *in planta* experiments demonstrated that H_2O_2 is a local, as well as a systemic signal in pathogen defence. Transgenic plants with elevated levels of H_2O_2 due to the constitutive overproduction of glucose oxidase or repression of peroxisomal catalase were more resistant to pathogens, accumulated SA, and expressed *PR* genes and proteins (49, 50). Moreover, accumulation of H_2O_2 in leaves of catalase-deficient tobacco plants was sufficient to induce expression of defence proteins (GPx, PR-1) not only locally, but also systemically (50). Although H_2O_2 is a diffusible molecule, its half-life is only 1 ms, which essentially excludes it from being the mobile signal for induction of defence responses in distal tissues. This problem may be overcome by a relay of H_2O_2-generating microbursts, including the NADPH oxidase as a mechanism for the reiteration of these microbursts (37, 43). Such a model was proposed based on the observation of microscopic HR lesions that appear throughout *Arabidopsis* plants upon infection with avirulent bacterial pathogens. These micro-HRs correlated with acquirement of resistance, expression of defence genes (*GST, PR-2*), and could be blocked by DPI, an inhibitor of NADPH oxidase. Moreover, co-application of glucose/glucose oxidase (generating H_2O_2) to the plants was sufficient to induce these responses (37). Such a system of H_2O_2 spreading requires signal amplification, and SA has been proposed to play this agonistic role (37, 51). Evidence is also accumulating for a signalling role of AOS in defence responses to abiotic stresses. Pretreatment of maize seedlings with H_2O_2 or menadione, a superoxide-generating compound, induces chilling tolerance (52). Plants regenerated from potato nodal explants treated with H_2O_2 are significantly more thermotolerant than control plants (53). Partial exposure of *Arabidopsis* plants to excess light results in systemic acclimation of unexposed leaves to photooxidative stress (54). Acclimation correlates with the expression of the *APx2* gene and is proposed to be mediated by H_2O_2, because the induction of *APx2* is sensitive to catalase. *Arabidopsis* leaves pretreated with H_2O_2 also become tolerant to excess light (54). H_2O_2 accumulates systemically also in wounded tomato plants, and this H_2O_2 accumulation can be blocked by DPI, suggesting an NADPH-dependent mechanism of H_2O_2 spreading. However, whether or not this H_2O_2 accumulation is associated with the acquirement of an enhanced resistance was not analysed (22). Although H_2O_2 is generally considered a signalling molecule in defence responses, $O_2^{\bullet-}$ (or a derived product) likely plays this role as well. Phytoalexin synthesis in soybean cells in response to pathogens or elicitors is blocked by DPI and SOD, but not by catalase (39). Similarly, $O_2^{\bullet-}$, but not H_2O_2, is necessary and sufficient to induce lesion formation and PR-1 mRNA accumulation in the 'lesion-simulating disease resistance response' mutant (*lsd1*) of *Arabidopsis* (55). Furthermore, one of the members of a tomato multigene family encoding extensin is transcriptionally induced upon treatment with the $O_2^{\bullet-}$-generating compounds digitonin or xanthine oxidase, but not with H_2O_2 (56). Conversely, induction by paraquat treatment of a rice gene encoding cytosolic APx is mediated via H_2O_2 (promoted by inhibition of catalase or APx) rather than $O_2^{\bullet-}$ (reduced by SOD inhibitors) (57). Bacteria and yeast induce distinct defence proteins in response to either $O_2^{\bullet-}$ or H_2O_2, although a considerable overlap exists between the two responses (58, 59). Thus, it is likely that in defence responses $O_2^{\bullet-}$ acts as a signalling molecule executing its function independently of H_2O_2.

2.2 Signalling role of AOS in cell death

Cell death is an essential process during the plant's life cycle. Two modes of cell death have been described in plants: programmed cell death (PCD) and necrosis. PCD is genetically controlled and has characteristic features of the apoptotic cell death in animal cells, such as cell shrinkage, cytoplasmic and nuclear condensation, chromatin condensation and DNA fragmentation. Necrosis results from severe and persistent trauma and is considered not to be genetically orchestrated (60, 61). PCD and necrosis have been suggested to be just two distinct ends of the same process that can be initiated by the same signal, AOS (62).

Plant cell death is best studied during the HR typical of incompatible plant–pathogen interactions (16). During the HR, the oxidative burst coincides with the induction of cell death at the site of the pathogen attack. This localized cell death limits the spread of the invading pathogen. The source of the oxidative burst is considered to be an NADPH oxidase complex. However, modulation of the activity of antioxidant enzymes probably contributes to AOS generation during the HR as well. In tobacco cells that undergo HR upon infiltration with fungal elicitors, the Cat1 and Cat2 mRNA and protein levels decrease and catalase activity is suppressed, which is paralleled by a strong H_2O_2 accumulation (35). Similarly, virus-induced HR-like cell death is accompanied by the suppression of cytosolic *APx* expression (36). This suppression probably contributes to the accumulation of H_2O_2 and activation of the cell death program.

The first evidence that AOS act as signals that initiate a transduction pathway towards plant cell death rather than kill the cells by reaching toxic levels came from experiments in soybean cell cultures, in which a short pulse of H_2O_2 was sufficient to activate a hypersensitive cell death mechanism (32). Accordingly, exogenous H_2O_2 (> 5 mM) initiated an active cell death pathway (requiring DNA and protein synthesis) in *Arabidopsis* suspension cultures (33). Concentration of H_2O_2 that activated the cell death pathway was in both cases higher than the one inducing expression of defence genes (32, 33). H_2O_2 (10 mM) needed to be present for 60 minutes to initiate an irreversible cell death process in *Arabidopsis* cells (33). It is believed that within these 60 minutes, H_2O_2 initiates a cell death programme via an interplay with other signalling molecules, such as ethylene and SA (23, 33, 63).

The ability of H_2O_2 to induce cell death was also demonstrated *in planta*. In transgenic plants with lower H_2O_2-scavenging capacities or others overproducing H_2O_2-generating enzymes, cell death appeared spontaneously or could be easily induced by a stress (50, 64). Transient exposure of catalase-deficient tobacco plants to conditions promoting H_2O_2 synthesis (high light) was sufficient to activate a PCD programme similar to that observed during incompatible plant–pathogen interactions (Dat *et al.*, in preparation).

$O_2^{\bullet-}$, but not H_2O_2, has been shown to initiate a runaway cell death phenotype in the *Arabidopsis lsd1* mutant, providing genetic evidence for the role of $O_2^{\bullet-}$ in plant cell death (55). *lsd1* grown under long days spontaneously forms necrotic lesions on leaves and is unable to stop the spreading of cell death. In front of the spreading zone of cell death, $O_2^{\bullet-}$ is drastically accumulating. Hence, $O_2^{\bullet-}$ seems to be the critical signal in the cell death process that is monitored via a 'rheostat' LSD1. Despite rather controversial data on $O_2^{\bullet-}$ versus H_2O_2 in cell death activation, AOS are indisputably the signal that activates genetically controlled cell death programme(s) in plants.

2.3 Signalling role of AOS in growth and morphogenesis

Under stress conditions, one of the strategies that plants have adopted is to slow down growth. The ability to reduce cell division under unfavourable conditions may not only allow conservation of energy for defence purposes, but may also limit the risk of heritable damage (65). AOS, as ubiquitous messengers of stress responses, likely play a signalling role in these adaptive processes. Low concentrations of menadione, a redox cycling compound, impairs the G1-to-S transition, slows DNA replication, and delays entry into mitosis (66). Accordingly, exogenous application of micromolar concentrations of GSH raises the number of meristematic cells undergoing mitosis, whereas depletion of GSH has the opposite effect (67). While cell cycle progression is under negative control of AOS, H_2O_2 stimulates somatic embryogenesis (68) and is essential for root gravitropism (69). However, the role of AOS in plant growth and development is still poorly understood and requires further research.

2.4 AOS and redox signalling

2.4.1 Redox-sensitive proteins

Plants can sense, transduce, and translate the AOS signal into appropriate cellular responses. This process requires the presence of redox-sensitive proteins that can undergo reversible oxidation/reduction and may switch 'on' and 'off' depending on the cellular redox state. AOS can oxidize the redox-sensitive proteins directly (3, 70) or indirectly, via the ubiquitous redox-sensitive molecules, such as glutathione or thioredoxins, which control the cellular redox state (3, 71). Redox-sensitive metabolic enzymes may directly modulate cellular metabolism, whereas redox-sensitive signalling proteins execute their function via downstream signalling components, such as kinases, phosphatases, and transcription factors. Two molecular mechanisms of redox-sensitive regulation of protein function prevail in living organisms. One mechanism employs oxidation of thiol (–SH) groups of proteins. The thiol group (–SH) can gain oxygen atoms to yield sulphenic (–SOH), sulphinic (–SO_2H) or sulphonic (–SO_3H) moieties or may form intramolecular or intermolecular disulphide bonds. Changes in the chemistry of –SH groups modify electronic and steric conformation of the cysteine residue, thereby affecting protein conformations and/or protein–protein interactions. All these changes alter the functionality of proteins that have cysteine residues at strategic positions (71). The other mechanism for redox control utilizes oxidation of iron–sulphur clusters (Fe–S) within Fe–S-containing proteins. Oxidation of Fe–S clusters by $O_2^{\bullet-}$ inactivates the cluster and affects enzyme activity. Additionally, free Fe^{2+} released from an oxidized Fe–S cluster can promote the formation of $OH^{\bullet}$ in cells (72). The activity of many chloroplastic enzymes, such as the key enzymes of the Calvin cycle, glycolytic glucose-6-phosphate dehydrogenase and the coupling factor that provides ATP for biosynthetic reactions, are under redox control. Reduction of their regulatory disulphide bridges leads to activation of the enzymes, except for glucose-6-phosphate dehydrogenase that is inactivated. The reducing power used to modulate the activity of these enzymes originates from the chloroplast electron transport chain, namely from ferredoxin that is oxidized by ferredoxin-thioredoxin reductases. This oxidation is coupled to reduction of thioredoxins that ultimately reduce regulatory cysteines in metabolic enzymes (26). Many prokaryotic and eukaryotic metabolic enzymes, such as aconitase, succinate dehydrogenase, fumarase, sulphite reductase, nitrite reductase, and enzymes of purine metabolism, contain Fe–S clusters, suggesting that they may be also subject to redox regulation (72).

The best-studied example of a redox-sensing receptor in plants is the cytochrome *bf* complex of the photosynthetic electron transport chain located in chloroplasts (Fig. 4). Because PSII and PSI act in sequence during linear electron flow, the amount of light energy delivered to the two reaction centres must be controlled. When light harvesting is not balanced by light energy utilization and dissipation, toxic radicals are formed, leading to oxidative damage. One of the control mechanisms regulates dissociation of the light-harvesting complex from PSII, a process controlled by phosphorylation. The kinase responsible for that phosphorylation is activated by reduction of the plastoquinone pool, a signal that is transduced to kinase activation via a structural change of the Fe–S protein associated with the cytochrome *bf* complex (73). Additionally, by a yet unresolved mechanism, the redox state of plastoquinone controls the rate of transcription of the chloroplast genes encoding PSII and PSI reaction centre apoproteins (Fig. 4; 74), as well as their mRNA stability and translation rate (75, 76). The redox state of plastoquinone also controls nuclear gene expression (77).

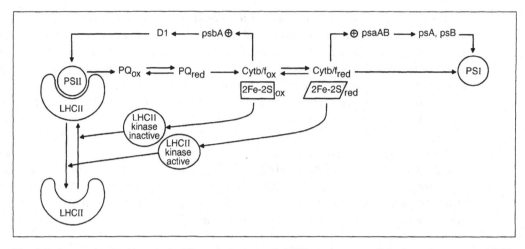

Fig. 4 Redox sensing in chloroplasts. When photosystem II (PSII) receives more light than photosystem I (PSI), plastoquinone (PQ) becomes predominantly reduced, resulting in the activation of the light-harvesting complex II (LHCII) kinase via structural changes around the Rieske iron–sulphur (2Fe–2S) protein of the cytochrome *bf* (cyt *bf*) complex. The kinase phosphorylates LHCII and PSII, resulting in the migration of LHCII away from the PSII and decreasing the light absorption by PSII. Oxidation of PQ reverses the structural changes of cyt *bf* and leads to the kinase inactivation. Phosphatase-mediated dephosphorylation of the mobile LHCII leads to its reassociation with PSII and an increase in the amount of light absorbed by PSII (73). Alternatively, the redox state of PQ controls the adjustment of the stoichiometry of PSI and II by transcriptional regulation of chloroplast genes that encode apoproteins of the PSI (psA and psB proteins) and PSII (D1 protein) reaction centres (74).

2.4.2 AOS and redox-regulated gene expression

In bacteria and yeast, AOS induces expression of at least 80 and 115 proteins, respectively (58, 78). It is believed that most of these responses are regulated at the transcriptional level (58, 70, 79). Similarly, at least 80 genes are induced by AOS in plants (Vranová *et al.*, in preparation). Bacterial genes are organized in regulons that are controlled by specific transcription factors. $O_2^{\bullet-}$-induced genes are controlled by the SoxR protein that has Fe–S clusters and, upon oxidation, induces the expression of a downstream transcription factor called SoxS. H_2O_2-induced genes are controlled by the oxidation of thiol groups present in the transcription factor OxyR (70). Alternatively, two-component systems may activate the expression of bacterial genes upon redox changes. A redox sensor, a membrane-associated phosphoprotein, becomes phosphorylated on histidine when either oxidized or reduced by components of the electron transport chain. Its substrate, the redox response regulator, is a sequence-specific DNA-binding protein that becomes phosphorylated at an aspartate residue, regulating transcription (80). In yeast, genes induced by redox signals consist of a complex network of different regulons, so-called stimulons (81). Activity of one of the best-studied redox-sensitive transcription factors, yAP1, is controlled by redox signals at the level of nuclear localization and DNA binding (82). Upon imposition of oxidative stress, yAP1 relocalizes from the cytoplasm to the nucleus and its DNA-binding capacity increases. In mammalian systems, many studies point to the significance of two classes of transcription factors that are sensitive to redox

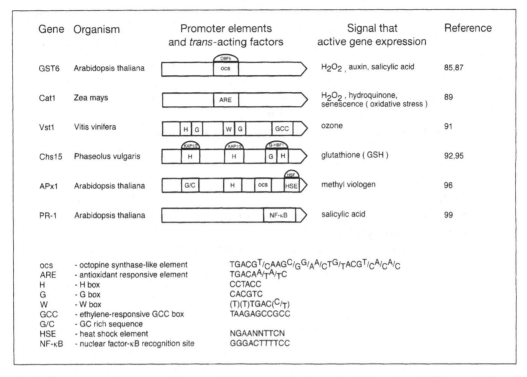

Fig. 5 *Cis* promoter elements and *trans*-acting factors putatively implicated in redox-controlled expression of plant genes. For a detailed description, see text.

signals: nuclear factor κB (NF-κB) and activator protein-1 (AP1). Mammalian redox-sensitive transcription factors are regulated similarly to that in yeast. First, either the pro-oxidant state in the cytoplasm (determined by the GSSG/GSH ratio) or AOS activate transcription factors and induce their translocation to the nucleus, where the reducing environment is required for proper DNA binding. Thioredoxins and the redox factor Ref-1 provide the reducing power for DNA binding (71). Thus, two major steps in the transcriptional activation of eukaryotic transcription factors seem to be influenced by redox balance. Although AOS and the cellular redox state are known to control expression of plant genes (32, 49, 50, 77, 83), neither signalling pathway(s) nor transcription factors and promoter elements specific for the redox regulation have been identified in plants to date. There are, however, several candidates for promoter elements as well as for DNA-binding factors that may act as redox response elements (Fig. 5).

Glutathione *S*-transferases (GSTs) catalyse the conjugation of GSH to a variety of hydrophobic electrophilic compounds, which in activated form may attack cellular macromolecules. Compounds with bound GSH are then subject to cellular detoxification pathways (84). A range of factors, such as growth factors, pathogens, herbicides, hormones, and cellular stress agents, induces expression of *GST* genes (85). The signal by which electrophilic compounds regulate *GST* gene expression is believed to be a pro-oxidant state in the cells, probably resulting from a reduced GSH content (84). The promoter element responsible for the induction of the Ya subunit in mouse *GST* by electrophilic compounds consists of two adjacent AP1-like sites (86). The consensus sequence of this site is TGACA(A/T)(A/T)GC and is called an antioxidant-responsive element or electrophile-responsive element. Two adjacent AP1-like sites are also present in the *Arabidopsis GST6* gene and constitute the *ocs*-like promoter element (85). This promoter element is, at least in part, required for *GST6* induction by auxin, H_2O_2, and SA (87). Because the *ocs* element can also be activated by inactive analogues of auxin or SA, it was suggested that stress brought about by these compounds rather than by the 'true hormonal effect' is responsible for the gene activation (88). A single antioxidant-responsive element has recently been identified in the promoter of a maize catalase gene (*Cat1*) and was found to bind nuclear factors from senescing scutella that accumulate *Cat1* transcripts, possibly as a result of oxidative stress (89).

WRKY proteins are a large family of recently discovered transcriptional regulators that are specific to plants (90). WRKY transcription factors are induced by several stresses and during senescence. They possess a redox-sensitive zinc-finger DNA-binding domain in which two cysteines together with two histidines interact electrostatically with a zinc atom to form a 'zinc finger', which makes them excellent candidates for redox regulation (71). WRKY proteins bind W boxes [consensus sequence (T)(T)TGAC(C/T)] present in promoters of many defence genes. Interestingly, the W box is the only common motif in promoters of *Arabidopsis* genes that are coordinately regulated with the *PR-1* gene, a marker of systemic acquired resistance (SAR) (90). The W box is also present in a minimal promoter of the stilbene synthase gene (*Vst1*) that is required for ozone inducibility (91). Additional promoter elements, such as the G box, H box, and ethylene-responsive GCC box, are present in the ozone-responsive part of the promoter.

The G box (CACGTG) is a ubiquitous *cis* element present in many plant genes and is thought to mediate responses to diverse environmental stimuli, including light, elicitors, and redox changes (92, 93). Together with the H box (CCTACC), the G box functions in the activation of phenylpropanoid biosynthetic genes. Transcription of at least two of these genes that encode phenylalanine ammonia-lyase (*PAL*) and chalcone synthase (*CHS*) is under redox control (induced by GSH) (94). Both a G box and an adjacent H box in the bean *chs15* promoter were found to bind a bZIP protein G/HBF-1. The binding was enhanced by phosphorylation of the G/HBF-1 that was triggered by GSH (92). Similarly, elicitation of a bean cell suspension with GSH increased specific nuclear activities of KAP-1 and KAP-2 protein factors that recognize an H box motif (95).

Heat-shock factors and heat-shock elements can also participate in redox-regulated gene expression. Activation of a heat-shock factor is characterized by the conversion of this factor from a monomer to trimer state, a process induced by heat shock and a large variety of conditions that generate abnormally folded proteins. Disulphide-linked aggregates of cellular proteins are formed as a consequence of disturbed intracellular redox homeostasis and are one of the signals required for heat-shock factor trimerization (71). A mutation of the heat-shock element in the promoter of the *Arabidopsis Apx1* gene delays the inducibility by oxidative stress (methyl viologen) (96).

There are also some indications for a mammalian redox-regulated transcription factor NF-κB in plant cells. In mammals, NF-κB resides normally in the cytoplasm where it is associated with a transcriptional repressor IκB. AOS stimulates IκB phosphorylation and its dissociation from NF-κB, leading to the nuclear localization of NF-κB. The *Arabidopsis* mutants, *npr1* and *nim1*, which are compromised in inducing SAR, are mutated in the gene that encodes a protein highly similar to the transcriptional repressor IκB (97, 98). Moreover, a sequence similar to the NF-κB recognition site was identified in the *Arabidopsis PR-1* promoter (99).

Additionally, a novel homeodomain protein of the HD-Zip class isolated from tomato and the zinc-finger protein LSD1 from *Arabidopsis* are negative regulators of oxidative cell death, and have been proposed to act as transcriptional regulators downstream of the AOS signal (100, 101).

2.4.3 Kinases and phosphatases in redox signal transduction

A cascade of three protein kinases [mitogen-activated protein kinase kinase kinase (MAPKKK), protein kinase kinase (MAPKK) and protein kinase (MAPK)] is a conserved functional module in a variety of signal transduction pathways in diverse organisms (for review, see reference 102). Activation of a mammalian redox-sensitive transcription factor AP1 requires *de novo* transcription of its subunits c-jun and c-fos. The signal that activates *c-jun* expression via phosphorylation of the cognate transcription factor is transduced via the MAPK pathway (103). Recently, a MAPK module that senses the H_2O_2 signal and translates it to the expression of defence genes (*GST6, HSP18.2*) was identified in *Arabidopsis* (104). This module consists of an upstream kinase ANP1 (MAPKKK) and the downstream kinases AtMPK3 and AtMPK6 (MAPKs). H_2O_2, but not auxin, cold, or abscisic acid (ABA), activate this kinase cascade (104). At the amino acid level, AtMPK3 and AtMPK6 are highly similar to the wound-induced protein and SA-induced protein kinases (WIPK and SIPK) of tobacco MAPKs, respectively (102). However, only SIPK, and not WIPK, is activated by ozone, H_2O_2 and xanthine/xanthine oxidase (generating $O_2^{\bullet-}$) in tobacco (105). Activation of SIPK by the avirulence gene product Avr9 in tobacco cells carrying the corresponding resistance gene (*Cf9*) is, however, independent of AOS (106). DPI that blocks the oxidative burst in elicited cells has no effect on the activity of SIPK, suggesting that a signal other than AOS activates SIPK upon elicitor treatment.

Type 2C protein phosphatase (PP2C) has been implicated in a negative feedback loop, which controls the wound-induced MAPK pathway in alfalfa, whereby expression of the corresponding gene (*MP2C*) is induced by this pathway (107). Recently, a gene encoding PP2C (*NtPP2C1*) that is transcriptionally responsive to different oxidative stress stimuli has been isolated in tobacco (108). However, *NtPP2C1* expression is suppressed by oxidative stress, suggesting that NtPP2C1 is not part of a negative feedback loop in the AOS signalling pathway. High homology of NtPP2C1 to PP2Cs implicated in the negative feedback control of the ABA signal transduction pathway suggests a possible interference of the AOS with the ABA signalling pathway (for details, see Section 3.4). In addition to MAPKs, a receptor-like protein kinase gene that is transcriptionally activated by AOS has also been identified in *Arabidopsis* (109).

3. AOS: part of a signalling network

To have such a profound effect on plant growth and metabolism, AOS must utilize and/or interfere with other signalling pathways or molecules. There is evidence that plant hormones are positioned downstream of the AOS signal. H_2O_2 induces accumulation of stress hormones, such as SA (50, 110) and ethylene (50). Tobacco plants exposed to ozone accumulate ABA (111) and induction of the *PDF1.2* gene by paraquat is impaired in *Arabidopsis* mutants insensitive to jasmonates (*coi1*) and ethylene (*ein2*) (112). Plant hormones are not only downstream of the AOS signal, but AOS themselves are secondary messengers in many hormone-signalling pathways (22, 48, 113). Therefore, it is conceivable that feedback/feedforward interactions between hormones and AOS occur.

3.1 Salicylic acid

The relationship between AOS and SA is well documented. SA is a plant hormone mainly associated with the establishment of SAR (for review, see reference 114). Upon pathogen infection, levels of SA increase in both challenged and non-challenged leaves, and plants become more resistant to secondary infection. SAR is compromised in plants that do not accumulate SA because of the over-expression of the bacterial SA-dehydrogenase gene (*nahG*) (115). Originally, H_2O_2 was proposed as the downstream signal of SA in SAR (48). SA and its active analogues inactivate catalase, whereas treatment of tobacco plants with SA increases H_2O_2 levels, and H_2O_2 or H_2O_2-producing compounds induce the expression of genes (*PR-1*) associated with SAR (48). However, later studies revealed that H_2O_2 and H_2O_2-inducing chemicals are unable to induce *PR-1* gene and protein expression in nahG plants (50, 116, 117). The debate on the location of H_2O_2 versus SA in the plant defence signalling pathway was put to rest with the proposition that both molecules work in unison as a self-amplifying system (51). In support of such a model, H_2O_2 induces SA accumulation (50, 110) and SA enhances H_2O_2 accumulation (48, 118). This self-amplification loop involving H_2O_2 and SA may generate microbursts that amplify the H_2O_2 signal required for oxidative cell death and establishment of acquired resistance against pathogens (37, 51). Evidence is accumulating that both H_2O_2 and SA are also secondary messengers in acclimation to abiotic stresses. SA and H_2O_2 levels are enhanced during acclimation to heat (119, 120). Treatment of plants with either of the molecules acclimate plants to heat and chilling stress (52, 53, 121) and defence responses to ozone are compromised in nahG plants (23). Thus, similar mechanisms that employ both SA and H_2O_2 may operate to establish the plant immunity against abiotic environmental cues.

3.2 Ethylene

Ethylene is another well-established signalling molecule that has long been recognized in plant stress responses. Tobacco plants with reduced peroxisomal catalase activity produce ethylene as early as 2 h upon exposure to high light (50). Accordingly, exogenous application of H_2O_2 activates in pine needles ethylene production in a concentration-dependent manner (122). Moreover, ozone, which is believed to form AOS in the apoplast, induces accumulation of the ethylene in tobacco plants as early as within 1 h, and this early induction positively correlates with ozone sensitivity (123). The ozone-tolerant tobacco cultivar Bel B lacks this early ethylene burst (123). The ethylene burst seems to originate from *de novo* ethylene synthesis, as the content of 1-aminocyclopropane-1-carboxylic acid (ACC), a precursor of ethylene biosynthesis, increases concomitantly with ethylene production (122). The latter is also blocked by inhibitors of the ethylene biosynthetic enzymes, ACC synthase and ACC oxidase (122, 124). Recently, ozone-induced $O_2^{\bullet-}$ accumulation and cell death have been demonstrated to be substantially reduced in the ethylene-insensitive *Arabidopsis* mutant *ein2* (63), whereas the ethylene-overproducing mutant *eto1* is hypersensitive to ozone (125). These results suggest that ethylene has a potentiating role in oxidative cell death by controlling $O_2^{\bullet-}$ accumulation.

3.3 Jasmonic acid

Jasmonic acid (JA) is a stress hormone primarily implicated in wound stress responses. When compared with ethylene, JA seems to have an opposite effect on oxidative cell death. The JA-insensitive *Arabidopsis* mutant *jar1* and the JA-biosynthetic mutant *fad3-2 fad7-2 fad8* show an increased magnitude of ozone-induced oxidative burst, SA accumulation, and HR-like cell death. Accordingly, when the ozone-sensitive *Arabidopsis* ecotype Cvi-0 is pretreated with methyl jasmonate, ozone-induced H_2O_2 accumulation, SA content, and PR-1 mRNA level are reduced, and ozone-induced cell death is completely abolished (63, 126). This result is in agreement with the observation that JA potentiates GSH synthesis (127), thereby controlling AOS accumulation and, subsequently, cell death. However, this hypothesis is in contrast to the observation that treatment of tomato leaves with methyl jasmonate promotes H_2O_2 accumulation (22). Therefore, the relationship between these two molecules remains unresolved.

3.4 Abscisic acid

ABA is implicated in a number of abiotic stress responses associated with water loss, such as drought, cold, salinity, and heat shock (128). One of the downstream ABA responses is a stomatal closure that prevents excessive transpiration. As a consequence, availability of CO_2 for the fixation in the Calvin–Benson cycle is limited, which may, in turn, enhance AOS production in the chloroplasts (for details, see Section 1.2. and reference 1). Despite this well-known causal link between ABA and

AOS, little attention has been drawn on the relationship between these two signalling molecules and/or their signalling pathways. Earlier studies have shown that ABA can modulate the activity of antioxidant enzymes and/or affect expression of genes encoding antioxidant enzymes (129–131). Furthermore, ozone promotes accumulation of ABA in tobacco (111). Both H_2O_2 and ABA independently acclimate maize seedlings to chilling; however, the relative position of both messengers in that process has not been determined (132). The antioxidant gene *Cat1*, encoding a seed-specific catalase in maize, is also induced by both H_2O_2 and ABA. Because H_2O_2 levels rapidly increase upon ABA treatment, H_2O_2 was proposed to be downstream of ABA in the *Cat1* induction (133). A model for the signalling pathway leading to downstream ABA responses starts to emerge and is, in part, based on studies of regulation of stomatal closure by ABA (for review, see reference 128). Briefly, stomatal closure is mediated via a reduction in Cl^-, K^+, and organic solute content in the two guard cells adjacent to the stomatal pore. ABA initiates this process via an increase in cytoplasmic Ca^{2+} that can be released from internal stores via cyclic ADP-ribose and from external stores via influx across the plasma membrane (128, 134). The ABA signalling pathway is negatively regulated by PP2C-like phosphatases (ABI1, ABI2) that are transcriptionally activated by ABA, creating a negative feedback loop (135). It was demonstrated that ABA induces the generation of H_2O_2 in stomatal guard cells and that H_2O_2 activates Ca^{2+} influx as well as stomatal closure. Accordingly, ABA-induced stomatal closure was inhibited by catalase and DPI (113). These results place H_2O_2 downstream of ABA in the regulation of stomatal closure. Alternatively, H_2O_2 may modulate ABA responses by compromising the negative effect of PP2Cs on the ABA pathway, because a tobacco homologue of PP2Cs, implicated in ABA signalling, is transcriptionally down-regulated by a number of oxidative stress stimuli (108). Thus, the scenario where AOS enhance and/or prolong downstream ABA responses or activate the pathway by decreasing the amount of the negative regulator PP2C may be envisaged. A simplified model of the ABA signal transduction pathway, with depicted places of AOS intervention, is presented in Fig. 6.

3.5 Growth-stimulating hormones

In addition to stress responses, AOS also modulate plant growth and development (for details, see Section 2.3) and a relationship between AOS and growth hormones or their signalling pathways can be anticipated. Recently, it has been demonstrated that the H_2O_2-induced MAPK cascade in *Arabidopsis* represses auxin-inducible gene expression (104; Fig. 7). This antagonistic effect on auxin-responsive gene expression is in line with the opposite effect of H_2O_2 and auxin on cell cycle progression (66, 137). Only a few studies point to possible links between AOS and cytokinins or gibberellins. Both hormones modulate the expression of antioxidant enzymes (131) and endogenous cytokinins were shown to have an AOS-scavenging capacity (138) and to increase tolerance of plants to oxidative stress (139). Down-regulation by gibberellins of the H_2O_2-scavenging peroxiredoxin (*Per1*) gene in barley aleurone has been suggested as a signal for release from dormancy (140).

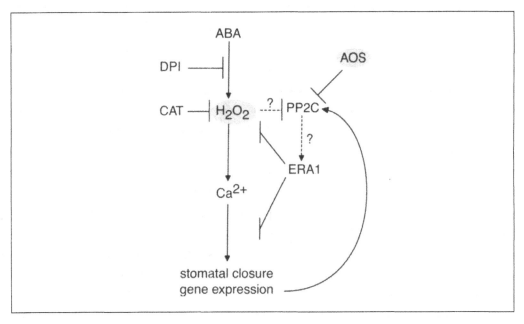

Fig. 6 Model of AOS intervention with the ABA signalling cascade. ABA stimulates an increase in free cytosolic Ca^{2+} and subsequently promotes stomatal closure. Both processes are stimulated by externally added H_2O_2. ABA-stimulated stomatal closure is blocked by an inhibitor of the NADPH oxidase (DPI) and by catalase (CAT) (113). Protein phosphatases 2C (PP2C) affect the ABA signalling pathway negatively, either upstream or downstream of Ca^{2+} fluxes (128, 134). PP2Cs may act via farnesyl transferase (ERA1) that negatively regulates ABA signalling and is genetically positioned downstream or in parallel to the PP2Cs (136). ABA elevates PP2C mRNA, thus creating a negative feedback loop (135). AOS negatively affect expression of PP2C (108), thereby possibly enhancing and/or prolonging the downstream responses of ABA. Alternatively, activation of the ABA pathway by decreasing the amount of the negative regulator PP2C by H_2O_2 may be envisaged.

3.6 Nitric oxide

Nitric oxide (NO$^\bullet$) is a signal molecule used by mammals to regulate various biological processes of the immune, nervous, and vascular systems (141). It is becoming apparent that NO$^\bullet$ is a ubiquitous signal also in plants. NO$^\bullet$ promotes leaf expansion, seed germination, and de-etiolation. However, at the same time, it may inhibit hypocotyl and internode elongation, induce defence genes and phytoalexin production, and potentiate the induction of hypersensitive cell death (142–147). Two sources of NO$^\bullet$ production are proposed in plants:

- from nitrate by nitrate reductase enzymatically or non-enzymatically; or
- from L-arginine by NO synthase as in mammals (148).

NO$^\bullet$ can be either beneficial or harmful for the cells and this dual role probably depends on the cellular NO$^\bullet$ concentration and its interaction with AOS (142, 144, 147). Because of the presence of an unpaired electron, NO$^\bullet$ readily interacts with $O_2^{\bullet-}$ to form peroxynitrite (ONOO$^-$) that can oxidize DNA, lipids, protein thiols, and iron clusters, resulting in loss of enzyme activities and cellular damage (70, 147).

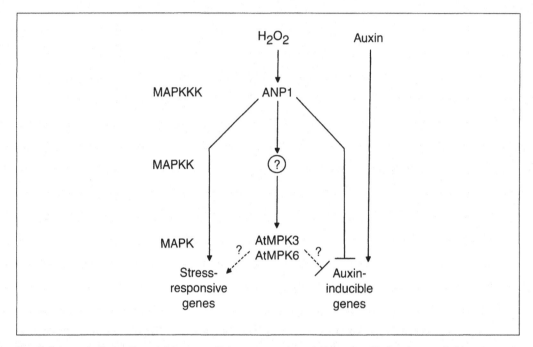

Fig. 7 Schematic illustration of the cross-talk between auxin and H_2O_2 signalling pathways. H_2O_2 activates a specific *Arabidopsis* mitogen-activated protein kinase kinase kinase (MAPKKK) ANP1, which initiates a phosphorylation cascade leading to the activation of the downstream mitogen-activated protein kinases (MAPK) AtMPK3 and AtMPK6 and to the induction of specific stress-responsive genes (*GST6* and *HSP18.2*). The same cascade represses auxin-induced expression of auxin-responsive genes (*GH3*) (104).

However, interaction of NO• with lipid alcoxyl or lipid peroxyl radicals is useful, because it breaks the self-perpetuating chain reaction that normally leads to membrane damage (147). In some cases, inhibition of enzyme activity by NO• may also prove beneficial. NO•-mediated inhibition of aconitase activity may reduce the electron flow through the mitochondrial electron transport chain, thereby decreasing the generation of AOS (149). Moreover, NO• converts the cytosolic aconitase into an iron-regulatory protein that controls iron homeostasis (149). As iron is the catalyst of the Fenton reaction that produces harmful OH•, a limited availability of iron prevents oxidative damage. Thus, NO• may affect the oxidative metabolism at multiple levels either by exacerbating or counteracting effects of AOS.

4. Conclusions

Besides exacerbating cellular damage, AOS are also capable of acting as ubiquitous signal molecules in plants. AOS are a central component in stress responses and the level of AOS determines the type of response. Whereas at low concentrations AOS induce defence genes and adaptive responses, at high concentrations cell death is initiated. To allow for this dual role, cellular levels of AOS must be tightly controlled.

The numerous AOS sources and a complex system of oxidant scavengers provide the flexibility necessary for these functions. However, how these systems are regulated to achieve the temporal and spatial control of AOS production is still poorly understood. Sublethal amounts of AOS acclimate plants to biotic and abiotic stress conditions and reduce plant growth, probably as part of an adaptational mechanism. Although a substantial genome response and the activity of many enzymes are known to be affected by AOS, molecular and biochemical mechanisms of adaptation are still not understood and the signalling pathways involved remain elusive. AOS communicate with other signal molecules and pathways forming part of the signalling network that controls responses downstream of AOS. Recently, information on the role of AOS as signal molecules in growth and morphogenesis has emerged, suggesting that AOS are not only stress signal molecules but may also be an intrinsic signal in plant growth and development. Genetic analysis in addition to physiological studies will be required to position AOS signals in the transduction pathway(s) and to understand how the signals are perceived and transduced to specific downstream responses.

Acknowledgements

We wish to acknowledge Martine De Cock, Stijn Debruyne, and Rebecca Verbanck for the excellent help in preparing the manuscript.

References

1. Dat, J., Van Breusegem, F., Vandenabeele, S., Vranová, E., Van Montagu, M. and Inzé, D. (2000) Active oxygen species and catalase during plant stress responses. *Cell. Mol. Life Sci.*, **57**, 779.
2. Foyer, C. H. and Harbinson, J. (1994) Oxygen metabolism and the regulation of photosynthetic electron transport. In Foyer, C. H. and Mullineaux, P. M. (eds.), *Causes of Photooxidative Stress and Amelioration of Defense Systems in Plants*. CRC Press, Boca Raton, FL, p. 1.
3. Halliwell, B. and Gutteridge, J. M. C. (1989) *Free Radicals in Biology and Medicine*. Oxford, Clarendon Press.
4. Bowler, C., Van Camp, W., Van Montagu, M. and Inzé, D. (1994) Superoxide dismutase in plants. *CRC Crit. Rev. Plant Sci.*, **13**, 199.
5. Charles, S. A. and Halliwell, B. (1980) Effect of hydrogen peroxide on spinach (*Spinacia oleracea*) chloroplast fructose bisphosphatase. *Biochem. J.* **189**, 373.
6. Malkin, R. and Niyogi, K. (2000) Photosynthesis. In Buchanan, B. B., Gruissem, W. and Jones, R. L. (eds.), *Biochemistry and Molecular Biology of Plants*. American Society of Plant Physiologists, Rockville, p. 568.
7. Krause, G. H. (1994) The role of oxygen in photoinhibition of photosynthesis. In Foyer, C. H. and Mullineaux, P. M. (eds.), *Causes of Photooxidative Stress and Amelioration of Defense Systems in Plants*. CRC Press, Boca Raton, FL, p. 43.
8. Kozaki, A. and Takeba, G. (1996) Photorespiration protects C3 plants from photo-oxidation. *Nature*, **384**, 557.

9. Siedow, J. N. and Day, D. A. (2000) Respiration and photorespiration. In Buchanan, B. B., Gruissem, W. and Jones, R. L. (eds.), *Biochemistry and Molecular Biology of Plants*. American Society of Plant Physiologists, Rockville, p. 676.

10. Somerville, C., Browse, J., Jaworski, J. G. and Ohlrogge, J. B. (2000) Lipids. In Buchanan, B. B., Gruissem, W. and Jones, R. L. (eds.), *Biochemistry and Molecular Biology of Plants*. American Society of Plant Physiologists, Rockville, p. 456.

11. del Río, L. A., Sandalio, L. M., Corpas, F. J., López-Huertas, E., Palma, J. M. and Pastori, G. M. (1998) Activated oxygen-mediated metabolic functions of leaf peroxisomes. *Physiol. Plant.*, **104**, 673.

12. Purvis, A. C. and Shewfelt, R. L. (1993) Does the alternative pathway ameliorate chilling injury in sensitive plant tissues? *Physiol. Plant.*, **88**, 712.

13. Urban, P., Mignotte, C., Kazmaier, M., Delorme, F. and Pompon, D. (1997) Cloning, yeast expression and characterization of the coupling of two distantly related *Arabidopsis thaliana* NADPH-cytochrome P450 reductases with P450 CYP73A5. *J. Biol. Chem.*, **272**, 19176.

14. Segal, A. W. and Abo, A. (1993) The biochemical basis of the NADPH oxidase of phagocytes. *Trends Biochem. Sci.*, **18**, 43.

15. Wojtaszek, P. (1997) Oxidative burst: an early plant response to pathogen infection. *Biochem. J.*, **322**, 681.

16. Lamb, C. and Dixon, R. A. (1997) The oxidative burst in plant disease resistance. *Ann. Rev. Plant Physiol. Plant Mol. Biol.*, **48**, 251.

17. Tenhaken, R., Levine, A., Brisson, L. F., Dixon, R. A. and Lamb, C. (1995) Function of the oxidative burst in hypersensitive disease resistance. *Proc. Natl Acad. Sci. USA*, **92**, 4158.

18. Xing, T., Higgins, V. J. and Blumwald, E. (1997) Race-specific elicitors of *Cladosporium fulvum* promote translocation of cytosolic components of NADPH oxidase to the plasma membrane of tomato cells. *Plant Cell*, **9**, 249.

19. Keller, T., Damude, H. G., Werner, D., Doerner, P., Dixon, R. A. and Lamb, C. (1998) A plant homolog of the neutrophil NADPH oxidase gp91[phox] subunit gene encodes a plasma membrane protein with Ca^{2+} binding motifs. *Plant Cell*, **10**, 255.

20. Bolwell, G. P. and Wojtaszek, P. (1997) Mechanisms for the generation of reactive oxygen species in plant defence – broad perspective. *Physiol. Mol. Plant Pathol.*, **51**, 347.

21. Cazalé, A.-C., Rouet-Mayer, M.-A., Barbier-Brygoo, H., Mathieu, Y. and Laurière, C. (1998) Oxidative burst and hypoosmotic stress in tobacco cell suspensions. *Plant Physiol.*, **116**, 659.

22. Orozco-Cardenas, M. and Ryan, C. A. (1999) Hydrogen peroxide is generated systemically in plant leaves by wounding and systemin via the octadecanoid pathway. *Proc. Natl Acad. Sci. USA*, **96**, 6553.

23. Rao, M. V. and Davis, K. R. (1999) Ozone-induced cell death occurs via two distinct mechanisms in *Arabidopsis*: the role of salicylic acid. *Plant J.*, **17**, 603.

24. Allan, A. C. and Fluhr, R. (1997) Two distinct sources of elicited reactive oxygen species in tobacco epidermal cells. *Plant Cell*, **9**, 1559.

25. Pellinen, R., Palva, T. and Kangasjärvi, J. (1999) Subcellular localization of ozone-induced hydrogen peroxide production in birch (*Betula pendula*) leaf cells. *Plant J.*, **20**, 349.

26. Ruelland, E. and Miginiac-Maslow, M. (1999) Regulation of chloroplast enzyme activities by thioredoxins: activation or relief from inhibition? *Trends Plant Sci.*, **4**, 136.

27. Willekens, H., Chamnongpol, S., Van Montagu, M., Inzé, D. and Van Camp, W. (1995) Role of H_2O_2 and H_2O_2-scavenging enzymes in environmental stress. *AgBiotech News Inform.*, **7**, 189N.

28. Eshdat, Y., Holland, D., Faltin, Z. and Ben-Hayyim, G. (1997) Plant glutathione peroxidases. *Physiol. Plant.*, **100**, 234.

29. Alscher, R. G. and Hess, J. L. (1993) *Antioxidants in Higher Plants*. CRC Press, Boca Raton, FL.

30. Ogawa, K., Kanematsu, S. and Asada, K. (1996) Intra- and extra-cellular localization of 'cytosolic' CuZn-superoxide dismutase in spinach leaf and hypocotyl. *Plant Cell Physiol.*, **37**, 790.

31. Willekens, H., Van Camp, W., Van Montagu, M., Inzé, D., Sandermann Jr, H. and Langebartels, C. (1994) Ozone, sulfur dioxide, and ultraviolet B have similar effects on mRNA accumulation of antioxidant genes in *Nicotiana plumbaginifolia* (L.). *Plant Physiol.*, **106**, 1007.

32. Levine, A., Tenhaken, R., Dixon, R. and Lamb, C. (1994) H_2O_2 from the oxidative burst orchestrates the plant hypersensitive disease resistance response. *Cell*, **79**, 583.

33. Desikan, R., Reynolds, A., Hancock, J. T. and Neill, S. J. (1998) Harpin and hydrogen peroxide both initiate programmed cell death but have differential effects on defence gene expression in *Arabidopsis* suspension cultures. *Biochem. J.*, **330**, 115.

34. Malan, C., Greyling, M. M. and Gressel, J. (1990) Correlation between CuZn superoxide dismutase and glutathione reductase, and environmental and xenobiotic stress tolerance in maize inbreds. *Plant Sci.*, **69**, 157.

35. Dorey, S., Baillieul, F., Saindrenan, P., Fritig, B. and Kauffmann, S. (1998) Tobacco class I and II catalases are differentially expressed during elicitor-induced hypersensitive cell death and localized acquired resistance. *Mol. Plant-Microbe Interact.*, **11**, 1102.

36. Mittler, R., Feng, X. and Cohen, M. (1998) Post-transcriptional suppression of cytosolic ascorbate peroxidase expression during pathogen-induced programmed cell death in tobacco. *Plant Cell*, **10**, 461.

37. Alvarez, M. E., Pennell, R. I., Meijer, P., Ishikawa, A., Dixon, R. A. and Lamb, C. (1998) Reactive oxygen intermediates mediate a systemic signal network in the establishment of plant immunity. *Cell*, **92**, 773.

38. Pugin, A., Frachisse, J.-M., Tavernier, E., Bligny, R., Gout, E., Douce, R. and Guern, J. (1997) Early events induced by the elicitor cryptogein in tobacco cells: involvement of a plasma membrane NADPH oxidase and activation of glycolysis and the pentose phosphate pathway. *Plant Cell*, **9**, 2077.

39. Jabs, T., Tschöpe, M., Colling, C., Hahlbrock, K. and Scheel, D. (1997) Elicitor-stimulated ion fluxes and O_2^- from the oxidative burst are essential components in triggering defense gene activation and phytoalexin synthesis in parsley. *Proc. Natl Acad. Sci. USA*, **94**, 4800.

40. Piedras, P., Hammond-Kosack, K. E., Harrison, K. and Jones, J. D. G. (1998) Rapid, *Cf-9*- and Avr9-dependent production of active oxygen species in tobacco suspension cultures. *Mol. Plant–Microbe Interact.*, **11**, 1155.

41. Simon-Plas, F., Rustérucci, C., Milat, M.-L., Humbert, C., Montillet, J.-L. and Blein, J.-P. (1997) Active oxygen species production in tobacco cells elicited by cryptogein. *Plant Cell Environ.*, **20**, 1573.

42. Price, A. H., Taylor, A., Ripley, S. J., Griffiths, A., Trewavas, A. J. and Knight, M. R. (1994) Oxidative signals in tobacco increase cytosolic calcium. *Plant Cell*, **6**, 1301.

43. Park, H.-J., Miura, Y., Kawakita, K., Yoshioka, H. and Doke, N. (1998) Physiological mechanisms of a sub-systemic oxidative burst triggered by elicitor-induced local oxidative burst in potato tuber slices. *Plant Cell Physiol.*, **39**, 1218.

44. Desikan, R., Burnett, E. C., Hancock, J. T. and Neill, S. J. (1998) Harpin and hydrogen peroxide induce the expression of a homologue of gp91-*phox* in *Arabidopsis thaliana* suspension cultures. *J. Exp. Bot.*, **49**, 1767.

45. Kieffer, F., Simon-Plas, F., Maume, B. F. and Blein, J.-P. (1997) Tobacco cells contain a protein, immunologically related to the neutrophil small G protein Rac2 and involved in elicitor-induced oxidative burst. *FEBS Lett.*, **403**, 149.
46. Kawasaki, T., Henmi, K., Ono, E., Hatakeyama, S., Iwano, M., Satoh, H. and Shimamoto, K. (1999) The small GPT-binding protein Rac is a regulator of cell death in plants. *Proc. Natl Acad. Sci. USA*, **96**, 10922.
47. Desikan, R., Hancock, J. T., Coffey, M. J. and Neill, N. J. (1996) Generation of active oxygen in elicited cells of *Arabidopsis thaliana* is mediated by a NADPH oxidase-like enzyme. *FEBS Lett.*, **382**, 213.
48. Chen, Z., Silva, H. and Klessig, D. F. (1993) Active oxygen species in the induction of plant systemic acquired resistance by salicylic acid. *Science*, **262**, 1883.
49. Wu, G., Shortt, B. J., Lawrence, E. B., León, J., Fitzsimmons, K. C., Levine, E. B., Raskin, I. and Shah, D. M. (1997) Activation of host defense mechanisms by elevated production of H_2O_2 in transgenic plants. *Plant Physiol.*, **115**, 427.
50. Chamnongpol, S., Willekens, H., Moeder, W., Langebartels, C., Sandermann Jr, H., Van Montagu, M., Inzé, D. and Van Camp, W. (1998) Defense activation and enhanced pathogen tolerance induced by H_2O_2 in transgenic plants. *Proc. Natl Acad. Sci. USA*, **95**, 5818.
51. Van Camp, W., Van Montagu, M. and Inzé, D. (1998) H_2O_2 and NO: redox signals in disease resistance. *Trends Plant Sci.*, **3**, 330.
52. Prasad, T. K., Anderson, M. D., Martin, B. A. and Stewart, C. R. (1994) Evidence for chilling-induced oxidative stress in maize seedlings and a regulatory role for hydrogen peroxide. *Plant Cell*, **6**, 65.
53. Lopez-Delgado, H., Dat, J. F., Foyer, C. H. and Scott, I. M. (1998) Induction of thermo-tolerance in potato microplants by acetylsalicylic acid and H_2O_2. *J. Exp. Bot.*, **49**, 713.
54. Karpinski, S., Reynolds, H., Karpinska, B., Wingsle, G., Creissen, G. and Mullineaux, P. (1999) Systemic signaling and acclimation in response to excess excitation energy in *Arabidopsis*. *Science*, **284**, 654.
55. Jabs, T., Dietrich, R. A. and Dangl, J. L. (1996) Initiation of runaway cell death in an *Arabidopsis* mutant by extracellular superoxide. *Science*, **273**, 1853.
56. Wisniewski, J.-P., Cornille, P., Agnel, J.-P. and Montillet, J.-L. (1999) The extensin multigene family responds differentially to superoxide or hydrogen peroxide in tomato cell cultures. *FEBS Lett.*, **447**, 264.
57. Morita, S., Kaminaka, H., Masumura, T. and Tanaka, K. (1999) Induction of rice cytosolic ascorbate peroxidase mRNA by oxidative stress; the involvement of hydrogen peroxide in oxidative stress signalling. *Plant Cell Physiol.*, **40**, 417.
58. Demple, B. (1991) Regulation of bacterial oxidative stress genes. *Ann. Rev. Genet.*, **25**, 315.
59. Jamieson, D. J. (1992) *Saccharomyces cerevisiae* has distinct adaptive responses to both hydrogen peroxide and menadione. *J. Bacteriol.*, **174**, 6678.
60. Pennell, R. I. and Lamb, C. (1997) Programmed cell death in plants. *Plant Cell*, **9**, 1157.
61. O'Brien, I. E. W., Baguley, B. C., Murray, B. G., Morris, B. A. M. and Ferguson, I. B. (1998) Early stages of the apoptotic pathway in plant cells are reversible. *Plant J.*, **13**, 803.
62. Jabs, T. (1999) Reactive oxygen intermediates as mediators of programmed cell death in plants and animals. *Biochem. Pharmacol.*, **57**, 231.
63. Overmyer, K., Tuominen, H., Kettunen, R., Betz, C., Langebartels, C., Sandermann Jr, H. and Kangasjärvi, J. (2000) Ozone-sensitive Arabidopsis *rcd1* mutant reveals opposite roles for ethylene and jasmonate signaling pathways in regulating superoxide-dependent cell death. *Plant Cell*, **12**, 1849.

64. Kazan, K., Murray, F. R., Goulter, K. C., Llewellyn, D. J. and Manners, J. M. (1998) Induction of cell death in transgenic plants expressing a fungal glucose oxidase. *Mol. Plant–Microbe Interact.*, **11**, 555.

65. May, M. J., Vernoux, T., Leaver, C., Van Montagu, M. and Inzé, D. (1998) Glutathione homeostasis in plants: implications for environmental sensing and plant development. *J. Exp. Bot.*, **49**, 649.

66. Reichheld, J.-P., Vernoux, T., Lardon, F., Van Montagu, M. and Inzé, D. (1999) Specific checkpoints regulate plant cell cycle progression in response to oxidative stress. *Plant J.*, **17**, 647.

67. Sánchez-Fernández, R., Fricker, M., Corben, L. B., White, N. S., Sheard, N., Leaver, C. J., Van Montagu, M., Inzé, D. and May, M. J. (1997) Cell proliferation and hair tip growth in the *Arabidopsis* root are under mechanistically different forms of redox control. *Proc. Natl Acad. Sci. USA*, **94**, 2745.

68. Cui, K., Xing, G., Liu, X., Xing, G. and Wang, Y. (1999) Effect of hydrogen peroxide on somatic embryogenesis of *Lycium barbarum* L. *Plant Sci.*, **146**, 9.

69. Joo, J. H., Bae, Y. S. and Lee, J. S. (2001) Role of auxin-induced reactive oxygen species in root gravitropism. *Plant Physiol.*, **126**, 1055.

70. Storz, G. and Imlay, J. A. (1999) Oxidative stress. *Curr. Opin. Microbiol.*, **2**, 188.

71. Arrigo, A.-P. (1999) Gene expression and the thiol redox state. *Free Rad. Biol. Med.*, **27**, 936.

72. Imsande, J. (1999) Iron-sulfur clusters: formation, perturbation, and physiological functions. *Plant Physiol. Biochem.*, **37**, 87.

73. Vener, A. V., Ohad, I. and Andersson, B. (1998) Protein phosphorylation and redox sensing in chloroplast thylakoids. *Curr. Opin. Plant Biol.*, **1**, 217.

74. Pfannschmidt, T., Nilsson, A. and Allen, J. F. (1999) Photosynthetic control of chloroplast gene expression. *Nature*, **397**, 625.

75. Salvador, M. L. and Klein, U. (1999) The redox state regulates RNA degradation in the chloroplast of *Chlamydomonas reinhardtii*. *Plant Physiol.*, **121**, 1367.

76. Trebitsh, T., Levitan, A., Sofer, A. and Danon, A. (2000) Translation of chloroplast *psbA* mRNA is modulated in the light by counteracting oxidizing and reducing activities. *Mol. Cell. Biol.*, **20**, 1116.

77. Karpinski, S., Escobar, C., Karpinska, B., Creissen, G. and Mullineaux, P. M. (1997) Photosynthetic electron transport regulates the expression of cytosolic ascorbate peroxidase genes in Arabidopsis during excess light stress. *Plant Cell*, **9**, 627.

78. Godon, C., Lagniel, G., Lee, J., Buhler, J.-M., Kieffer, S., Perrot, M., Boucherie, H., Toledano, M. B. and Labarre, J. (1998) The H_2O_2 stimulon in *Saccharomyces cerevisiae*. *J. Biol. Chem.*, **273**, 22480.

79. Stephen, D. W. S., Rivers, S. L. and Jamieson, D. J. (1995) The role of the *YAP1* and *YAP2* genes in the regulation of the adaptive oxidative stress responses of *Saccharomyces cerevisiae*. *Mol. Microbiol.*, **16**, 415.

80. Allen, J. F. (1993) Redox control of transcription: sensors, response regulators, activators and repressors. *FEBS Lett.*, **332**, 203.

81. Jamieson, D. J. (1998) Oxidative stress responses of the yeast *Saccharomyces cerevisiae*. *Yeast*, **14**, 1511.

82. Kuge, S., Jones, N. and Nomoto, A. (1997) Regulation of yAP-1 nuclear localization in response to oxidative stress. *EMBO J.*, **16**, 1710.

83. Hérouart, D., Van Montagu, M. and Inzé, D. (1993) Redox-activated expression of the cytosolic copper/zinc superoxide dismutase gene in *Nicotiana*. *Proc. Natl Acad. Sci. USA*, **90**, 3108.

84. Daniel, V. (1993) Glutathione *S*-transferases: gene structure and regulation of expression. *Crit. Rev. Biochem. Mol. Biol.*, **28**, 173.

85. Chen, W., Chao, G. and Singh, K. B. (1996) The promoter of a H_2O_2-inducible, *Arabidopsis* glutathione *S*-transferase gene contains closely linked OBF- and OBP1-binding sites. *Plant J.*, **10**, 955.

86. Friling, R. S., Bergelson, S. and Daniel, V. (1992) Two adjacent AP-1-like binding sites form the electrophile-responsive element of the murine glutathione *S*-transferase Ya subunit gene. *Proc. Natl Acad. Sci. USA*, **89**, 668.

87. Chen, W. and Singh, K. B. (1999) The auxin, hydrogen peroxide and salicylic acid induced expression of the *Arabidopsis GST6* promoter is mediated in part by an ocs element. *Plant J.*, **19**, 667.

88. Ulmasov, T., Hagen, G. and Guilfoyle, T. (1994) The ocs element in the soybean *GH2/4* promoter is activated by both active and inactive auxin and salicylic acid analogues. *Plant Mol. Biol.*, **26**, 1055.

89. Polidoros, A. N. and Scandalios, J. G. (1999) Role of hydrogen peroxide and different classes of antioxidants in the regulation of catalase and glutathione *S*-transferase gene expression in maize (*Zea mays* L.). *Physiol. Plant.*, **106**, 112.

90. Eulgem, T., Rushton, P. J., Robatzek, S. and Somssich, I. E. (2000) The WRKY superfamily of plant transcription factors. *Trends Plant Sci.*, **5**, 199.

91. Schubert, R., Fischer, R., Hain, R., Schreier, P. H., Bahnweg, G., Ernst, D. and Sandermann Jr, H. (1997) An ozone-responsive region of the grapevine resveratrol synthase promoter differs from the basal pathogen-responsive sequence. *Plant Mol. Biol.*, **34**, 417.

92. Dröge-Laser, W., Kaiser, A., Lindsay, W. P., Halkier, B. A., Loake, G. J., Doerner, P., Dixon, R. A. and Lamb, C. (1997) Rapid stimulation of a soybean protein-serine kinase that phosphorylates a novel bZip DNA-binding protein, G/HBF-1, during the induction of early transcription-dependent defenses. *EMBO J.*, **16**, 726.

93. Menkens, A. E., Schindler, U. and Cashmore, A. R. (1995) The G-box: a ubiquitous regulatory DNA element in plants bound by the GBF family of bZIP proteins. *Trends Biochem. Sci.*, **20**, 506.

94. Wingate, V. P. M., Lawton, M. A. and Lamb, C. J. (1988) Glutathione causes a massive and selective induction of plant defense genes. *Plant Physiol.*, **87**, 206.

95. Yu, L. M., Lamb, C. J. and Dixon, R. A. (1993) Purification and biochemical characterization of proteins which bind to the H-box *cis*-element implicated in transcriptional activation of plant defense genes. *Plant J.*, **3**, 805.

96. Storozhenko, S., De Pauw, P., Van Montagu, M., Inzé, D. and Kushnir, S. (1998) The heat-shock element is a functional component of the Arabidopsis *APX1* gene promoter. *Plant Physiol.*, **118**, 1005.

97. Cao, H., Glazebrook, J., Clarke, J. D., Volko, S. and Dong, X. (1997) The Arabidopsis *NPR1* gene that controls systemic acquired resistance encodes a novel protein containing ankyrin repeats. *Cell*, **88**, 57.

98. Ryals, J., Weymann, K., Lawton, K., Friedrich, L., Ellis, D., Steiner, H.-Y., Johnson, J., Delaney, T. P., Jesse, T., Vos, P. and Uknes, S. (1997) The Arabidopsis *NIM1* protein shows homology to the mammalian transcription factor inhibitor IκB. *Plant Cell*, **9**, 425.

99. Lebel, E., Heifetz, P., Thorne, L., Uknes, S., Ryals, J. and Ward, E. (1998) Functional analysis of regulatory sequences controlling PR-1 gene expression in Arabidopsis. *Plant J.*, **16**, 223.

100. Mayda, E., Tornero, P., Conejero, V. and Vera, P. (1999) A tomato homeobox gene (HD-Zip) is involved in limiting the spread of programmed cell death. *Plant J.*, **20**, 591.

101. Dietrich, R. A., Richberg, M. H., Schmidt, R., Dean, C. and Dangl, J. L. (1997) A novel zinc finger protein is encoded by the Arabidopsis *LSD1* gene and functions as a negative regulator of plant cell death. *Cell*, **88**, 685.

102. Jonak, C., Ligterink, W. and Hirt, H. (1999) MAP kinases in plant signal transduction. *Cell. Mol. Life Sci.*, **55**, 204.

103. Cano, E. and Mahadevan, L. C. (1995) Parallel signal processing among mammalian MAPKs. *Trends Biochem. Sci.*, **20**, 117.

104. Kovtun, Y., Chiu, W.-L., Tena, G. and Sheen, J. (2000) Functional analysis of oxidative stress-activated mitogen-activated protein kinase cascade in plants. *Proc. Natl Acad. Sci. USA*, **97**, 2940.

105. Samuel, M. A., Miles, G. P. and Ellis, B. E. (2000) Ozone treatment rapidly activates MAP kinase signalling in plants. *Plant J.*, **22**, 367.

106. Romeis, T., Piedras, P., Zhang, S., Klessig, D. F., Hirt, H. and Jones, J. D. G. (1999) Rapid Avr9- and Cf-9-dependent activation of MAP kinases in tobacco cell cultures and leaves: convergence of resistance gene, elicitor, wound, and salicylate responses. *Plant Cell*, **11**, 273.

107. Meskiene, I., Bögre, L., Glaser, W., Balog, J., Brandstötter, M., Zwerger, K., Ammerer, G. and Hirt, H. (1998) MP2C, a plant protein phosphatase 2C, functions as a negative regulator of mitogen-activated protein kinase pathways in yeast and plants. *Proc. Natl Acad. Sci. USA*, **95**, 1938.

108. Vranová, E., Langebartels, C., Van Montagu, M., Inzé, D. and Van Camp, W. (2000) Oxidative stress, heat shock and drought differentially affect expression of a tobacco protein phosphatase 2C. *J. Exp. Bot.*, **51**, 1763.

109. Czernic, P., Visser, B., Sun, W., Savouré, A., Deslandes, L., Marco, Y., Van Montagu, M. and Verbruggen, N. (1999) Characterization of an *Arabidopsis thaliana* receptor-like protein kinase gene activated by oxidative stress and pathogen attack. *Plant J.*, **18**, 321.

110. León, J., Lawton, M. A. and Raskin, I. (1995) Hydrogen peroxide stimulates salicylic acid biosynthesis in tobacco. *Plant Physiol.*, **108**, 1673.

111. Ederli, L., Pasqualini, S., Batini, P. and Antonielli, M. (1997) Photoinhibition and oxidative stress: effects of xanthophyll cycle, scavenger enzymes and abscisic acid content in tobacco plants. *J. Plant Physiol.*, **151**, 422.

112. Penninckx, I. A. M. A., Thomma, B. P. H. J., Buchala, A., Métraux, J.-P. and Broekaert, W. F. (1998) Concomitant activation of jasmonate and ethylene response pathways is required for induction of a plant defensin gene in Arabidopsis. *Plant Cell*, **10**, 2103.

113. Pei, Z.-M., Murata, Y., Benning, G., Thomine, S., Klüsener, B., Allen, G. J., Grill, E. and Schroeder, J. I. (2000) Calcium channels activated by hydrogen peroxide mediate abscisic acid signalling in guard cells. *Nature*, **406**, 731.

114. Durner, J., Shah, J. and Klessig, D. F. (1997) Salicylic acid and disease resistance in plants. *Trends Plant Sci.*, **2**, 266.

115. Delaney, T. P., Uknes, S., Vernooij, B., Friedrich, L., Weymann, K., Negrotto, D., Gaffney, T., Gut-Rella, M., Kessmann, H., Ward, E. and Ryals, J. (1994) A central role of salicylic acid in plant disease resistance. *Science*, **266**, 1247.

116. Bi, Y.-M., Kenton, P., Mur, L., Darby, R. and Draper, J. (1995) Hydrogen peroxide does not function downstream of salicylic acid in the induction of PR protein expression. *Plant J.*, **8**, 235.

117. Neuenschwander, U., Vernooij, B., Friedrich, L., Uknes, S., Kessmann, H. and Ryals, J. (1995) Is hydrogen peroxide a second messenger of salicylic acid in systemic acquired resistance? *Plant J.*, **8**, 227.

118. Rao, M. V., Paliyath, G., Ormrod, D. P., Murr, D. P. and Watkins, C. B. (1997) Influence of salicylic acid on H_2O_2 production, oxidative stress, and H_2O_2-metabolizing enzymes. *Plant Physiol.*, **115**, 137.

119. Dat, J. F., Lopez-Delgado, H., Foyer, C. H. and Scott, I. M. (1998) Parallel changes in H_2O_2 and catalase during thermotolerance induced by salicylic acid or heat acclimation in mustard seedlings. *Plant Physiol.*, **116**, 1351.

120. Dat, J. F., Foyer, C. H. and Scott, I. M. (1998) Changes in salicylic acid and antioxidants during induced thermotolerance in mustard seedlings. *Plant Physiol.*, **118**, 1455.

121. Senaratna, T., Touchell, D., Bunn, E. and Dixon, K. (2000) Acetyl salicylic acid (aspirin) and salicylic acid induce multiple stress tolerance in bean and tomato plants. *Plant Growth Regul.*, **30**, 157.

122. Ievinsh, G. and Tillberg, E. (1995) Stress-induced ethylene biosynthesis in pine needles: a search for the putative 1-aminocyclopropane-1-carboxylic-independent pathway. *J. Plant Physiol.*, **145**, 308.

123. Schraudner, M., Langebartels, C. and Sandermann Jr, H. (1996) Plant defence systems and ozone. *Biochem. Soc. Trans.*, **24**, 456.

124. Tuomainen, J., Betz, C., Kangasjärvi, J., Ernst, D., Yin, Z.-H., Langebartels, C. and Sandermann Jr, H. (1997) Ozone induction of ethylene emission in tomato plants: regulation by differential accumulation of transcripts for the biosynthetic enzymes. *Plant J.*, **12**, 1151.

125. Tuominen, H., Overmyer, K. and Kangasjärvi, J. (2000) Regulation of ozone sensitivity and ozone-induced gene expression by ethylene, jasmonic acid and salicylic acid in *Arabidopsis* mutants. Abstract presented at the 11th International Conference on Arabidopsis Research, Madison (WI, USA), June 24-28, 2000 (#119).

126. Rao, M. V., Lee, H.-i., Creelman, R. A., Mullet, J. E. and Davis, K. R. (2000) Jasmonic acid signaling modulates ozone-induced hypersensitive cell death. *Plant Cell*, **12**, 1633.

127. Xiang, C. and Oliver, D. J. (1998) Glutathione metabolic genes coordinately respond to heavy metals and jasmonic acid in Arabidopsis. *Plant Cell*, **10**, 1539.

128. Grill, E. and Ziegler, H. (1998) A plant's dilemma. *Science*, **282**, 252.

129. Gong, M., Li, Y.-J. and Chen, S.-Z. (1998) Abscisic acid-induced thermotolerance in maize seedlings is mediated by calcium and associated with antioxidant systems. *J. Plant Physiol.*, **153**, 488.

130. Bueno, P., Piqueras, A., Kurepa, J., Savouré, A., Verbruggen, N., Van Montagu, M. and Inzé, D. (1998) Expression of antioxidant enzymes in response to abscisic acid and high osmoticum in tobacco BY-2 cell cultures. *Plant Sci.*, **138**, 27.

131. Kurepa, J., Hérouart, D., Van Montagu, M. and Inzé, D. (1997) Differential expression of CuZn- and Fe-superoxide dismutase genes of tobacco during development, oxidative stress and hormonal treatments. *Plant Cell Physiol.*, **38**, 463.

132. Prasad, T. K., Anderson, M. D. and Stewart, C. R. (1994) Acclimation, hydrogen peroxide and abscisic acid protect mitochondria against irreversible chilling injury in maize seedlings. *Plant Physiol.*, **105**, 619.

133. Guan, L. M., Zhao, J. and Scandalios, J. G. (2000) Cis-elements and trans-factors that regulate expression of the maize Cat1 antioxidant gene in response to ABA and osmotic stress: H_2O_2 is the likely intermediary signaling molecule for the response. Plant J., 22, 87.

134. Allen, G. J., Kuchitsu, K., Chu, S. P., Murata, Y. and Schroeder, J. I. (1999) Arabidopsis abi1-1 and abi2-1 phosphatase mutations reduced abscisic acid-induced cytoplasmic calcium rises in guard cells. Plant Cell, 11, 1785.

135. Gosti, F., Beaudoin, N., Serizet, C., Webb, A. A. R., Vartanian, N. and Giraudat, J. (1999) ABI1 protein phosphatase 2C is a negative regulator of abscisic acid signaling. Plant Cell, 11, 1897.

136. Pei, Z.-M., Ghassemian, M., Kwak, C. M., McCourt, P. and Schroeder, J. I. (1998) Role of farnesyltransferase in ABA regulation of guard cell anion channels and plant water loss. Science, 282, 287.

137. Walker, L. and Estelle, M. (1998) Molecular mechanisms of auxin action. Curr. Opin. Plant Biol., 1, 434.

138. Gidrol, X., Lin, W. S., Dégousée, N., Yip, S. F. and Kush, A. (1994) Accumulation of reactive oxygen species and oxidation of cytokinin in germinating soybean seeds. Eur. J. Biochem., 224, 21.

139. Barna, B., Adam, A. L. and Kiraly, Z. (1997) Increased levels of cytokinin induced tolerance to necrotic diseases and various oxidative stress-causing agents in plants. Phyton, 37, 25.

140. Stacey, R. A. P., Munthe, E., Steinum, T., Sharma, B. and Aalen, R. B. (1996) A peroxiredoxin antioxidant is encoded by a dormancy-related gene, Per1, expressed during late development in the aleurone and embryo of barley grains. Plant Mol. Biol., 31, 1205.

141. Hippeli, S., Heiser, I. and Elstner, E. F. (1999) Activated oxygen and free oxygen radicals in pathology: new insights and analogies between animals and plants. Plant Physiol. Biochem., 37, 167.

142. Leshem, Y. Y. (1996) Nitric oxide in biological systems. Plant Growth Regul., 18, 155.

143. Noritake, T., Kawakita, K. and Doke, N. (1996) Nitric oxide induces phytoalexin accumulation in potato tuber tissues. Plant Cell Physiol., 37, 113.

144. Delledonne, M., Xia, Y., Dixon, R. A. and Lamb, C. (1998) Nitric oxide functions as a signal in plant disease resistance. Nature, 394, 585.

145. Durner, J., Wendehenne, D. and Klessig, D. F. (1998) Defense gene induction in tobacco by nitric oxide, cyclic GMP, and cyclic ADP-ribose. Proc. Natl Acad. Sci. USA, 95, 10328.

146. Beligni, M. V. and Lamattina, L. (2000) Nitric oxide stimulates seed germination and de-etiolation, and inhibits hypocotyl elongation, three light-inducible responses in plants. Planta, 210, 215.

147. Beligni, M. V. and Lamattina, L. (1999) Is nitric oxide toxic or protective? Trends Plant Sci., 4, 299.

148. Yamasaki, H., Sakihama, Y. and Takahashi, S. (1999) An alternative pathway for nitric oxide production in plants: new features of an old enzyme. Trends Plant Sci., 4, 128.

149. Navarre, D. A., Wendehenne, D., Durner, J., Noad, R. and Klessig, D. F. (2000) Nitric oxide modulates the activity of tobacco aconitase. Plant Physiol., 122, 573.

5 | Heat-stress-induced signalling

KAPIL BHARTI AND LUTZ NOVER

1. Introduction

1.1 The heat-stress response

In 1962, the pioneering work of the Italian developmental biologist F. Ritossa (1) led to one of the most seminative discoveries in modern cell biology. After a serendipitous increase in the temperature of the incubator with the *Drosophila* cultures, he observed remarkable changes of the puffing, i.e. gene activity patterns of the polytene chromosomes in larval salivary glands. Surprisingly enough, the same reprogramming of transcription was also observed with chemical stressors like salicylate, 2,4-dinitrophenol, and azide. About 10 years later Tissieres and co-workers (2) identified the newly formed heat-stress proteins (Hsps), and McKenzie and Meselson (3) characterized the corresponding mRNAs. Due to the favourable properties, i.e. heat-stress (hs)-inducible mass formation of new mRNAs and proteins, *Drosophila* hs genes were among the first amenable to cloning procedures. Soon, the rapidly developing field included investigations of other eukaryotic organisms and bacteria. It turned out that Ritossa had discovered the central parts of a general stress response system conserved throughout the living world [for references to early literature see reviews by Ashburner and Bonner (4), Nover (5), and Nover *et al.* (6)].

The newly formed heat-stress proteins (Hsps) can be assigned to 11 multiprotein families conserved among bacteria, plants, and animals. Initially, the terminology of Hsps was simply based on their hs inducibility and electrophoretic mobility in an SDS-polyacrylamide gel (M_r in kDa). Hence Hsp70 or Hsp90 referred to proteins with M_rs of 70 or 90 kDa, whose level or rate of synthesis increased under hs conditions. However, with improving knowledge about sequence and functional details of these proteins, the original two criteria became less important and there was a transition to a terminology based on sequence homology. Due to the mosaic of regulatory elements in the promoter regions, e.g. of the *hsp70* multigene family, members can be expressed constitutively, induced by cold stress or heat stress or their expression is under developmental/hormonal control. Moreover, the size (M_r) of the proteins belonging to the Hsp70 family can actually vary between 60 and 110 kDa. This is also true for other Hsp families. Size variability is particularly striking for the multiplicity of proteins found in the plant Hsp20 family (small Hsps), encoded by about 20 different genes. This shift from an operational term to a gene-based nomenclature including aspects of intracellular localization and function of the proteins is now completed for plants by the publication of the *Arabidopsis* genome. A compilation of all related open reading frames (ORFs) encoding putative members of *Arabidopsis* Hsp families can be found in a special issue of Cell Stress and Chaperones (7).

During the last decade more and more details of the biochemical function of Hsps have emerged. As molecular chaperones they help other proteins to maintain or regain their native conformation by stabilizing partially unfolded states. They do not contain specific information for correct folding, but rather prevent unproductive interactions (aggregation) between non-native proteins. Frequently, members of different Hsp families form multisubunit complexes, so-called chaperone machines, and different chaperone complexes may interact to generate networks for protein maturation, assembly and targeting (8–14). As a consequence, separate sets of chaperones/chaperone machines exist in all compartments of eukaryotic cells with ongoing synthesis and/or processing of proteins, i.e. in the cytoplasm, ER, mitochondria, and chloroplasts.

1.2 Complexity of the heat-stress response

In the context of this review on stress-dependent signal transduction pathways in plants, it is essential to stress to levels of complexity.

- Due to their sedentary life, plants are generally more directly affected by unfavourable environmental changes than animals. The usual situation for plant growth and development is best characterized by a daily multistress challenge. As a result of this, a highly integrated and flexible network of stress-response systems has evolved. They are characterized by a number of multivalent or even general stress metabolites and proteins. Hormones, in particular ethylene, abscisic acid, and jasmonic acid, are not only frequently found as stress metabolites, but they are also part of systemic and developmental signalling systems.

- Although the predominant synthesis of Hsps under hs conditions as well as the far advanced analyses of the heat-stress transcription factors are central topics of research in the field, these aspects comprise only a small part of the hs response. Similar to other stress response systems, there is a transient and complex reprogramming of many if not of all cellular activities. As a consequence, cells in the recovery period following a hs treatment differ in many aspects from the cells before the stress. Due to the inherent amplification of a changed cellular phenotype, heat and other stressors are well known to cause severe developmental defects if applied in sensitive stages of a developing organism (5, 15, 16). The whole complexity of stress-induced changes at the gene expression level will soon be apparent using proteomic approaches or microarray techniques (17–20).

When considering heat-stress-induced signal transduction, it is important to recall that, besides the mechanisms triggering Hsp synthesis at the transcriptional level, there are many other Hsf-independent events orchestrated to the whole programme (5). Such are:

- cell signalling due to changes of membrane properties, calcium levels, protein interactions or activity of protein kinases/phosphatases;
- cell cycle arrest;
- ribosome biosynthesis;
- translational reprogramming;
- effects on the photosynthetic apparatus;
- apoptosis.

In view of the complexity of the processes affected and manifold connections between different parts, the picture is far from complete. In the following, we try to summarize hs effects on signalling pathways or components of such pathways in general without direct relation to specific targets. In Section 3, we discuss in detail the well chracterized Hsf system. Finally, examples of other complex cellular functions (translation, cell cycle, ribosome synthesis, photosynthesis, apoptosis) are described, which are markedly influenced by hs. Because of the well documented conservation of essential elements of cell biology between all eukaryotic organisms, we frequently discuss results obtained with yeast and vertebrates or even use them as primary sources of information, helping to evaluate or understand the slowly emerging picture for plants.

2. Signalling systems

2.1 Membranes

Membranes are the first sites for sensing changes in the cell's surrounding, such as changes of temperature, pH, turgor or atmospheric pressure, which all cause modulation of membrane fluidity. As an immediate consequence of temperature transitions, the proportion of phospholipids with hexagonal instead of bilayer arrangement may markedly increase. This and any other changes in the physical state or local phospholipid composition of cellular membranes can directly influence membrane functions such as signal reception and transduction, transport of nutrients, exo- and endocytosis, etc. (21–23). Cells respond to these transient disturbances of the membrane homeostasis by homoviscous adaptation and regeneration of the normal domain structure of membranes (24–28). Similar to the situation with many other cellular functions, the temperature range for membrane homeostasis can be extended by preconditioning (27–34).

Using the cyanobacterium *Synechocystis* as model system for a photosynthetic cell, Vigh, Murata and co-workers nicely demonstrated that the fatty acid composition of membranes deeply influenced the photosynthetic activity during and after the stress period as well as the temperature threshold for maximum expression of Hsps (27, 34,

35). There is evidence that stress adaptation at the membrane level in this organism may also involve direct association of membranes with proteins of the Hsp20 and Hsp60 families (Hsp17 and GroEL). This is reminiscent of the intriguing effects of the eukaryotic Hsp21 and Hsp70 for the protection and function of the photosynthetic apparatus in chloroplasts (36, 37).

For animal cells and yeast, it becomes more and more apparent that sphingolipids, which are complex lipids of plasma membranes, act as heat-stress-induced signalling molecules. Sphingolipids and their derivatives, e.g. ceramide, are involved in signalling pathways regulating stress responses, apoptosis (see Section 4.5), cellular senescence, cell cycle (see Section 4.2), inflammation, tumour development and intracellular Ca^{2+} mobilization (38–40). Initial observations about the potential role of sphingolipids as regulators of cellular stress response stem from temperature-sensitive yeast mutants defective in ceramide synthesis (41, 42). Following this, ceramide and other sphingolipid derivative were found to increase in yeast as the result of a heat-stress-enhanced activity of ceramide synthetase (41, 43, 44).

The direct biological consequences of such an increase in the levels of sphingolipids in the stress response of yeast is not yet clear. However, three effects are worth mentioning:

• Simola *et al.* (45) reported that treatment of cells with dihydrosphingosine caused increased expression of the TPS2 subunit of trehalose synthetase and thus accumulation of trehalose. Trehalose, a well-known thermoprotectant disaccharide of yeast, was shown to assist the Hsp104 chaperone in protecting heat-damaged proteins in the cytosol and ER. But there was no effect on membrane vesicle trafficking, indicating that increase in trehalose formation is not involved in the restoration of membrane functioning.

• By using a *lacZ* reporter construct with the global stress-response element (STRE), the authors also found that STRE promoters can be activated by a dihydrosphingosine-dependent signalling pathway (45).

• Finally, phytosphingosine mediates the heat-stress-induced increase of protein breakdown via the ubiquitin-dependent proteasome pathway (42).

2.2 Calcium

Calcium (Ca^{2+})-dependent signalling pathways operate by transient and localized changes of the Ca^{2+} concentration by influx from extracellular sources or release from intracellular storage sites. The multiplicity of calmodulin (CaM) or related Ca^{2+}-binding proteins in plants emphasize the important role of Ca^{2+}- signalling in plant development and responses to external stimuli, e.g. light, gravity, wounding, salt and others (46–49). Ca^{2+}-binding proteins are regulatory subunits of protein kinases and/or phosphatases controlling not only the time point of activation/deactivation but also the intracellular localization and substrate specificity of these enzymes. The concept of Trewavas (47) of a membrane-associated Ca^{2+}-signalling complex (transducon) with incorporated receptor, ion channel, protein kinase, and/or phosphatase, as well as calmodulin-type regulatory proteins, helps to understand many characteristics of Ca^{2+}-dependent effects.

Evidence for the role of Ca^{2+}-sensing proteins in the stress response is still indirect and fragmentary (5, 49–52). On the one hand, induced heat-stress tolerance in yeast and maize require CaM-dependent processes (50, 53). On the other hand, members of the Hsp families bind CaM. This is well known for the vertebrate and plant Hsp70 (54, 55) and for the chloroplast GroES homologue Cpn10 (56). Finally, CaM-like proteins are themselves hs-inducible proteins (57, 58). In view of the concept that chaperones of the Hsp70 and Hsp90 families control the activity state of Hsfs (see Section 3.3.3), the results about CaM–Hsp70 interaction are particularly striking. Interestingly, the highly conserved motif in the N-terminal ATP-binding domain of Hsp70 required for this interaction (-PRALRRLRTACERAKRTLSST-) is also involved in the nuclear import of the chaperone. It is tempting to speculate that CaM binding to Hsp70 influences the characteristic redistribution of the latter from cytoplasm to the nucleus and/or its interaction with denatured proteins during hs. The CaM/Hsp70 complex could represent a functionally specialized subpopulation of the total cellular Hsp70 pool.

2.3 Protein kinase cascades

Rapid changes in the phosphorylation status of proteins brought about by protein kinases and phosphatases are decisive means for changes of activity, interaction with other components (proteins, nucleic acids), intracellular distribution, and stability of proteins. The modifying enzymes (kinases and phosphatases) are themselves phosphorylated proteins, i.e. subject to the same modification of their properties. They are integrated into cascades or networks of activating and deactivating proteins. In several cases, scaffold proteins with multiple interaction sites help in the proper arrangement of kinases, phosphatases, and substrates (59). More recently, many examples for stress- and hormone-dependent signalling as well as for cell cycle control were also elaborated for plants (60–63). Because protein kinases and phosphatases are integral parts of other paragraphs of this review, we briefly summarize here only heat-stress-induced changes assumed to affect the phosphorylation status of Hsps and Hsfs.

In vertebrates and yeast, two different MAP kinase pathways are known to be activated by heat stress, SAPK/JNK (stress-activated protein kinases/c-jun N-terminal kinases) and p38 kinases. JNK is induced mainly by heat but also by UV irradiation, oxidative stress, inflammatory cytokines, and chemokines. At least in animal cells, the activation of the JNK pathway by heat or oxidative stress coincides with the activation of Hsf1 and the expression of Hsp70 (64). Searching for the heat-sensitive upstream kinase of the SAPK cascade, Dorion *et al.* (65) detected a short-lived protein, which disappears during heat stress. As a result, cells become refractory to reinduction of SAPK by a second stress.

The p38 kinase pathway is activated mainly in response to osmotic stress but also induced by heat. p38 kinase (Hog1), originally identified in budding yeast as a kinase activated by osmotic stress, was subsequently described also in fission yeast as protein kinase Spc1 (Sty1). Spc1 is involved in the responses to heat and oxidative stress, UV irradiation, nutrient limitation, and sex differentiation (66). The activity of Spc1 is controlled by the phosphatase Pyp1. Under normal conditions Pyp1 binds to Spc1, keeping it in an inactive state. Disruption of the Pyp1/Spc1 complex by hs results in rapid activation of Spc1 by upstream protein kinases, e.g. Wik1 or Wis1 (67). Similar to the situation with Spc1 in fission yeast, activity and intracellular distribution between cytoplasm and nucleus of Hog1 in baker's yeast is controlled by interaction with two phosphatases, i.e. Ptp2 in the nucleus and Ptp3 in the cytoplasm (68).

In mammals, the p38 is part of a stress-induced cascade of protein kinases ending with the phosphorylation of substrate proteins, e.g. of Hsp27 (RKK→p38(RK)→ MAPKAP-2→Hsp27). The heat-sensitive primary target of this cascade is unknown. Interestingly, the multimeric structure of Hsp27 is drastically changed by the phosphorylation. Dimers and monomers of phosphorylated Hsp27 cause a reorganization of the actin system (69–72). This intriguing role of Hsp27 for a phosphorylation-controlled rearrangement of the cytoskeleton probably does not exist in plants. The corresponding members of the Hsp20 family, i.e. the cytosolic class CI and class CII small Hsps, are not phosphorylated (73) and interaction of plant small Hsps with actin were never reported. Unfortunately, not much is known about this type of stress-related protein kinase and phosphatase cascades in plants. However, work from our group gave the first evidence for a heat-stress-inactivated 50 kDa protein kinase in tomato suspension culture cells (74) detected by phosphorylation of myelin basic protein (MBP).

Another kinase affected by heat stress is protein kinase C (PKC). PKC1 of budding yeast was shown to be induced by mild heat stress and responsible for induction of thermotolerance in yeast subjected to mild heat stress. Interestingly, yeast defective in PKC1 were not compromised in Hsf-dependent reporter activities, suggesting that thermotolerance dependent on PKC1 activation is not mediated by synthesis of Hsps (75). Another member of the protein kinase C family in mammals (PKCK2) also responds to stress conditions. Among the known substrates of CK2 are proteins associated with the heat-stress response, i.e. Hsp56, Hsp90, Hsf1, Egr-1, DNA repair machinery (topoisomerase II, DNA ligase). Gerber *et al.* (76) reported that after a mild heat stress, CK2 activity was increased about threefold and the kinase is mainly localized in the nucleus.

3. The Hsf network

3.1 Basic structure of Hsfs

Generally, selective transcription of eukaryotic genes requires interaction of sequence-specific transcription factors with the highly complex basal transcription machinery (transcriptosome) assembled from about 100 subunits around the start site of a gene (77–80). To approach the question of signal transduction with respect to the rapid activation of hs genes, it is necessary to start with the description of the regulatory proteins involved. These heat-stress transcription factors (Hsfs) have a modular structure, basically similar to many other proteins regulating gene activity (81–85). Despite a considerable variability in size and sequence, their basic structure is conserved among eukaryotes (see Plate 1).

- Close to the N-terminus, the highly structured **DNA-binding domain** (DBD) is the most conserved part of Hsfs. It is formed of a three-helical bundle (H1, H2, H3) and a four-stranded antiparallel β-sheet (β1, β2, β3, β4). The hydrophobic core of this domain ensures the precise positioning of the central helix–turn–helix motif (H2-T-H3, Plate 1B) required for specific recognition of the heat-stress promoter elements (HSEs; 86, 87). This palindromic binding motif (HSE: 5′-AGAAnnTTCT-3′) was found to be conserved in all eukaryotes (5, 88). The only crystal structure of a Hsf/DNA complex, reported for the DBD of the *Kluyveromyces lactis* Hsf, showed two adjacent monomers of the DBD bound to the palindromic HSE motif. Protein contacts between both subunits are mediated by the 10 amino acid residues of the loop (wing) between β3 and β4 strands (89–91). Because this wing is lacking in plant Hsfs, it will be interesting to elaborate the differences between plant and non-plant Hsfs with respect to the arrangement of Hsf subunits bound to DNA.

- **The oligomerization domain** (HR-A/B region) is connected to the DBD by a flexible linker of variable length (15–80 amino acid residues, see Plate 1). The heptad pattern of hydrophobic amino acid residues in the HR-A/B region suggest a coiled-coil structure characteristic of protein interaction domains of the Leu-zipper type. Unfortunately, precise structural data of this region in the context of a full-length Hsf are lacking. However, for recombinant fragments of the HR-A/B region of the *K. lactis* Hsf, a highly elongated, triple-stranded coiled-coil structure with a break between the A and B parts of the protein domain was reported (92). The B part is essential for the stability of the complete structure.

  In plants, there are three classes of Hsfs clearly separated by peculiarities of their HR-A/B regions (82, 83). Hsfs of the B-type are similar to all non-plant Hsfs (see Plate 1A). In contrast to this, Hsfs of the A- and C-types have extended HR-A/B regions due to an insertion of 21 (class A) and 7 (class C) amino acid residues between the A and B parts of the HR-A/B region (see example for HsfA2 in Plate 1B). The structural significance of this is unclear. But there is evidence that the tomato HsfB1 exists as a dimer, whereas the class A Hsfs exist as trimers and that hetero-oligomerization is only possible between members of the same class.

- The **nuclear localization signals** (NLSs) of Hsfs are so-called bipartite clusters of basic amino acid residues. They are marked by NLS in the block diagrams Plate 1 A, B, and sequence details are given in Plate 1B. In some cases, the nucleo-cytoplasmic distribution can be markedly influenced by nuclear export. Due to a leucine-rich export signal in the HR-C region (NES), the tomato HsfA2 is mostly found in the cytoplasm unless complexed in hetero-oligomers with HsfA1 (93–95).

- The C-terminal domains (CTD) of the Hsfs with the **activator function** are the least conserved in sequence and size. With the exception of the plant class B Hsfs, the CTDs are usually acidic and enriched in proline, serine, threonine, glutamic, acid and aspartic acid residues. Their function as transcription activators evidently resides in short peptide motifs found in or close to the HR-C regions. These activator motifs (AHA motifs, see references 96, 97) are characterized by aromatic (W, F, Y), large hydrophobic (L, I, V) and acidic (E, D) amino acid residues (see example given in Plate 1B). Similar AHA motifs are found in the centre of the activation domains of human, *Drosophila*, and yeast Hsfs as well as in many other transcription factors of mammals, for example VP16, RelA, Sp1, Fos, Jun, steroid receptors, and yeast (e.g. Gal4 and Gcn4; see references 96–98). Most likely, they represent the essential sites of contacts with subunits of the basal transcription complex as shown by pull-down experiments (99, Döring *et al.*, unpublished).

3.2 Multiplicity of Hsfs

Our knowledge on the plant Hsf system mainly stems from the cloning and functional characterization of four tomato Hsfs, i.e. HsfA1, HsfA2, HsfA3 and HsfB1 (87, 93–95, 97, 100–103). However, based on the presence of the conserved DNA binding domain plus the adjacent HR-A/B region (see Plate 1) and the *Arabidopsis* genome sequence, we identified 21 open reading frames encoding putative Hsfs in this model plant (see review 83). For comparison, there are a total of four Hsfs in vertebrates, only one Hsf in *Drosophila* and the nematode *Caenorhabditis elegans* and one Hsf plus three Hsf-related proteins in yeast (82–85, 114). Thus, the complexity of the plant Hsf system far exceeds that of any other organism, whose genomic sequence is known.

Of the 21 Arabidopsis Hsfs 15 belong to class A, 5 to class B and one to class C (see Table 1). So far only very few of them were studied experimentally in more detail (104, 105, 115–117). The question of how many Hsfs do we need, cannot be answered at present. At least, the overall complexity of Hsfs in tomato and other plants is comparable to Arabidopsis. Searching expressed sequence tags (EST) libraries, we found expression data for 15 additional tomato Hsfs with representatives in practically all groups and classes compiled in Table 1 (83).

Although the picture is far from complete, the experimental data obtained with the tomato Hsfs indicate that the multiplicity may be connected with distinct functions in the network of Hsfs. The data can be briefly summarized as follows (see Plate 1 for structural details):

- HsfA1 (527 aa residues) is the largest of the four tomato Hsfs. It is constitutively expressed in cell cultures as well as in different parts of the developing tomato plant. Probably HsfA1 plays a central role for the hs-inducible expression of HsfA2 and HsfB1. The C-terminal activator domain (CTAD) harbours two AHA motifs (Table 2) which are essential for the activator function (97).

- Synthesis of HsfA2 (351 aa residues) is strictly hs-dependent. HsfA2 accumulates to fairly high levels in tomato cell cultures and different plant tissues especially

Table 1 Survey of plant Hsfs

No.	Name (former name)	Chr. loc.	ORF (aa)	MW (kDa)	pI	Protein Accession Number	Ref.
Class A							
Group A1							
1	At-HsfA1a (Hsf1)	IV	495	55.7	4.94	CAB10555	104
2	At-HsfA 1b (Hsf3)	V	481	53.6	4.60	CAA74397	105
3	At-HsfA 1d	I	482	54.2	4.68	AAF81328.1	DS
4	At-HsfA 1e	III	468	52.0	4.59	AAF26960	DS
5	Lp-HsfA1	VIII	505	55.7	5.02	CAA47869	100, 101
Group A2							
6	At-HsfA2	II	345	39.1	4.95	AAC31222	106
7	Gm-HsfA2 (Hsf21)	–	P.C.	–	–	S59537	107
8	Lp-HsfA2	VIII	351	40.2	4.57	CAA47870	101
9	Ps-HsfA2 (HsfA)	–	334	38.1	4.63	CAA09300	108
Group A3							
10	At-HsfA3	V	412	46.4	5.15	CAB82937	DS
11	Lp-HsfA3	–	508	55.8	4.55	AAF74563	103
Group A4							
12	At-HsfA4a (Hsf21)	IV	401	46.2	5.18	CAA16745	DS
13	At-HsfA4c	V	345	39.6	5.51	BAB09213	DS
14	Ms-HsfA4	–	402	46.2	5.65	AAF37579	DS
15	Nt-HsfA4 (Hsf2)	–	408	46.4	4.96	BAA83711	109
16	Zm-HsfA4	–	308	35.4	6.50	CAA58117	110
Group A5							
17	At-HsfA5 (HsfA4b)	IV	466	52.4	6.15	CAB10177	DS
Group A6							
18	At-HsfA6a	V	282	33.2	5.39	BAB11313	DS
19	At-HsfA6b	III	406	46.7	4.74	BAB01258	111

during periods of repeated hs and recovery. Similar to HsfA1, HsfA2 is a strong activator protein with two AHA motifs in its CTAD. In the course of a hs regime HsfA2 exists in three different forms characterized by their intracellular distribution and modes of protein interaction:

(i) *Nuclear form*: due to the imbalance between the NLS-mediated nuclear import and the NES-mediated export, HsfA2 is found in the nucleus only as hetero-oligomer with HsfA1 (94, 95).

(ii) *Cytoplasmic insoluble form*: the ongoing accumulation of HsfA2 during long-term hs and its stability coincides with its interaction with Hsp17 class CII, one of the dominant forms of small Hsps in plants. During hs, both proteins are reversibly incorporated into high molecular weight cytoplasmic chaperone complexes built of the heat-stress granules (HSG).

Group A7							
20	At-HsfA7a	III	272	31.8	6.01	CAB41311	DS
21	At-HsfA7b	III	282	32.6	4.95	CAB86436	DS
Groups A8, A9							
22	At-HsfA8	I	374	42.6	4.64	AAF16564.1	DS
23	At-HsfA9	V	331	38.2	5.35	BAA97129	112
Class B							
Group B1							
24	At-HsfB1 (Hsf4)	IV	284	31.3	6.27	CAB16764	82,105
25	Gm-HsfB1 (Hsf34)	–	282	31.2	9.16	S59538	107
26	Lp-HsfB1	II	301	33.2	5.60	CAA39034	100,101
27	Nt-HsfB1 (Hsf1)	–	292	32.1	5.97	BAA83710	109
Group B2							
28	At-HsfB2a (Hsf6)	V	299	34.1	5.78	CAB63802	113
31	At-HsfB2b	IV	377	39.7	4.72	CAA39937	DS
Group B3							
34	At-HsfB3	II	244	28.3	5.11	AAB84350	DS
Group B4							
32	At-HsfB4	I	348	39.6	7.61	AAG34256.1	DS
33	Gm-HsfB4 (Hsf5)	–	370	42.1	7.69	S59539	107
Class C							
22	At-HsfC1	III	330	37.7	5.74	BAB02003	113
23	Os-HsfC1	I	339	36.9	6.22	BAB19067	DS

Hsfs from *Arabidopsis thaliana* (At), *Glycine max* (Gm), *Lycopersicon peruvianum* (Lp), *Medicago sativa* (Ms), *Nicotiana tabacum* (Nt), *Pisum sativum* (Ps), *Oryza sativa* (Os) and *Zea mays* (Zm) are listed with their new names based on their structure and evolutionary relation (83). The molecular weight (MW) and isoelectric points (pI) were calculated on the basis of the amino acid sequences using Clone Manager 5 software. The chromosomal localization (Chr. loc.) are indicated whenever possible. For identification, the accession numbers and references (Ref.) are given. DS, direct submission to the date base as part of the *Arabidopsis* sequencing project; P.C., partial clone (Gm-HsfA2).

(iii) *Cytoplasmic soluble form*: after dissociation of the HSG complexes in the recovery, HsfA2 and small Hsps are found in soluble oligomers in the cyto-plasm. The particular state of HsfA2 under these conditions and its DNA-binding capacity remain to be investigated.

• HsfA3 (508 aa residues) is the least studied of the four tomato Hsfs. It is found constitutively expressed in cell cultures but is barely detectable in plant tissues. It may represent a developmentally regulated Hsf. Four AHA motifs in the CTAD each with a central Trp residue contribute to the activator function (103).

• HsfB1 (301 aa residues) is the only B-type Hsf so far characterized in tomato. Its very low level, found in unstressed cell cultures or tissues, is transiently increased after hs. HsfB1 is relatively short-lived and always found in the nucleus. In contrast to the class A Hsfs, HsfB1 has no detectable activator function (93, 95, 100, 102, 118). This can be explained by the differences of the C-terminal domains

Table 2 Activator motifs (AHA motifs) of plant Hsfs

No.	Hsf (see Table 1)	Putative AHA motifs[a] (<u>AHA motifs</u> characterized)
1	At-HsfA1a	AHA (433) – FE**FLEEY**MPE–
2	At-HsfA1b	AHA (418) – DP**FWEQFF**SV–
3	At-HsfA1e	AHA1 (402) – DS**FWEQF**IGE– AHA2 (442) – SNV**W**SKN**Q**QM–
4	Lp-HsfA1	AHA1 (428) – <u>I**DW**QSGLLDE</u>– AHA2 (446) – <u>DP**FWEKFL**QS</u>–
5	At-HsfA2	AHA (324) – L**DW**DSQDLHD–
6	Lp-HsfA2	AHA1 (294) – <u>D**DIWEEL**LSE</u>– AHA2 (335) – <u>PE**WGEEL**QDL</u>–
7	Ps-HsfA2	AHA1 (277) – LS**DWEEL**LNQ– AHA2 (296) – EV**LIGDF**SQI–
8	At-HsfA3	AHA1 (277) – D**DWERLLM**YD– AHA2 (381) – DV**CWEQF**AAG–
9	Lp-HsfA3	AHA1 (429) – <u>EE**LWGMGFEA**</u>– AHA2 (447) – <u>PE**LWDSL**SSY</u>– AHA3 (467) – <u>SD**LWDIDPL**Q</u>– AHA4 (483) – <u>VD**KWPADGSP**</u>–
10	At-HsfA4a	AHA1 (256) – IA**IWENL**VSD– AHA2 (341) – DG**FWQQFF**SE–
11	At-HsfA4c	AHA1 (226) – LT**FWENL**VSE– AHA2 (289) – DD**FWEQC**LTE–
12	Ms-HsfA4	AHA1 (338) – DV**FWEQF**LTE– AHA2 (377) – GR**FWNM**RKS–
13	Nt-HsfA4	AHA1 (257) – LT**FWENV**LQD– AHA2 (344) – DI**FWEQF**LTE–
14	At-HsfA5	AHA (414) – DV**FWEQF**LTE–
15	At-HsfA6a	AHA (261) – EG**IW**KG**FV**LS–
16	At-HsfA6b	AHA (367) – EG**FW**EDLLNE–
17	At-HsfA7a	AHA (256) – DG**FWEEL**LSD–
18	At-HsfA7b	AHA (259) – DG**FWEEL**LMN–
19	At-HsfA8	AHA1 (258) – DGA**W**EKLLLL– AHA2 (330) – KS**YML**KLISE–

[a] Numbers in parentheses indicate position of the peptide motif in the C-terminal domain. Aromatic and large hydrophobic residues are printed in bold. Function of the underlined motifs were tested experimentally (see reference 97 for Lp-Hsfs A1 and A2 and reference 103 for Lp-HsfA3).

(CTD). The three class A Hsfs have an acidic CTD with the activator modules (AHA motifs), whereas the CTD of HsfB1 is basic. Typical AHA motifs are lacking. Interestingly, combinations of HsfB1 with class A Hsfs lead to strong synergistic effects in appropriate reporter assays. It is tempting to speculate that the adjacent positioning of the two Hsfs on a promoter creates a cooperative surrounding of the divergent CTDs facilitating recruitment of essential components of the transcriptional machinery.

3.3 Control of Hsf activity

3.3.1 The Hsf cycle

Based on *in vitro* and *in vivo* data obtained with vertebrate and *Drosophila* Hsfs (81, 119–124) the five-step activation process for Hsf shown in Fig. 1 can be defined. Upon heat stress, the inactive monomer in the cytoplasm (state 1) undergoes a conformational change (state 2) with subsequent oligomerization (state 3). The Hsf trimer, transported to the nucleus, is able to bind to the hs promoter elements (HSE, states 4, 5).

Although this model certainly reflects essential aspects of the Hsf cycle, hs-induced transcription activation in the reality of a given cell in a given situation may not strictly follow the sequence of steps given in the model.

- Treatment of yeast, *Drosophila*, or mammalian cells with salicylate or similar drugs results in Hsf activation, nuclear transport, and DNA-binding (state 4) but no hs gene transcription (125–127).

- Hsfs in yeasts (128, 129) and *Xenopus* oocytes (127, 130) are always nuclear and DNA-bound, and yet transcription of hs genes needs stress induction. Also, in vertebrate cells, most of the Hsf1 may be a nuclear protein which undergoes hs-induced changes with tight binding to nuclear structures (130, 131).

- As outlined above (Section 3.2), the situation in plants is even more complex because each of the Hsfs exhibits its own intracellular distribution pattern influenced not only by hs but also by the presence of and interaction with other Hsfs (94, 95).

3.3.2 Intrinsic elements of Hsf activity control

The two-dimensional presentation of Hsfs with their functional domains/modules (Plate 1) is convenient, and it reflects part of the reality derived from mutational analyses and testing of appropriate fusion proteins. But it is also misleading. It should be kept in mind that essential details of the Hsf function, and in particular of its activity control, evidently result from the precise molecular context of the functional modules within the still unknown three-dimensional structure of the Hsfs.

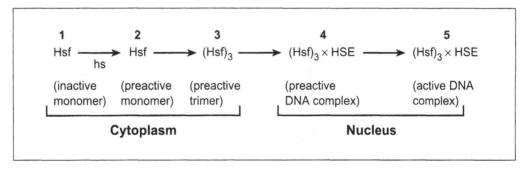

Fig. 1

It is interesting to notice that in some cases hybrid activator proteins built of heterologous DNA-binding domains (DBDs), e.g. of the bacterial LexA or the yeast Gal4 DBDs, and the C-terminal domains of the mammalian Hsf1 were fully active and hs-regulated (132, 133). Evidently, the CTAD contains all information necessary for the regulated phenotype.

It is now generally accepted that the major point of stress control is the maintenance of Hsf in an inactive state and/or the regeneration of this state after the stress activation. Depending on the cell type, organism, and Hsf investigated, this inactive state can be anything between states 1 and 4. Despite this evident flexibility in detail, there are a number of common features for all or at least for a group of Hsfs.

A frequently discussed model used to explain the inactive state of Hsfs involves a hypothetical intrinsic repressor domain, in particular the HR-C region for intramolecular interactions with the central HR-A/B/NLS region. This part of the proteins is required for oligomerization and nuclear transport, which are both steps of the activation process. Hence, intramolecular shielding may efficiently block Hsf function as transcription activator. In keeping with this, deletions or point mutations in the HR-C or HR-A/B regions were repeatedly reported to abolish hs regulation (132–135). Particularly interesting are results obtained with the human and the *Drosophila* Hsfs. Single point mutations in the HR-B part or in the HR-C region of human Hsf1 (132) or the NLS region of *Drosophila* Hsf (123) create constitutively active and multimeric factors. Using yeast as a test system for constitutive trimerization and function of human Hsf1, Liu and Thiele (136) also found that a small part of the linker region between DNA-binding domain and the HR-A/B region is required for the maintenance of the monomeric state. Unfortunately, physical interaction between the two hydrophobic heptad repeat regions was never demonstrated experimentally, e.g. in a yeast two-hybrid interaction test or in pull-down assays.

The special role of the central part of the Hsfs for their stress-regulated activity is supported by investigations on Hsf phosphorylation. Except for plants, change of the phosphorylation state accompanies Hsf activation in yeast, *Drosophila*, and vertebrates. In mammalian cells, Hsf phosphorylation is the result of the activity of at least four protein kinases, which may act independently or in cooperation. These are MAPK, PKC2, JNK and GSK-3β (see Section 2.3 and references 137–139). Phosphorylation of distinct serine residues is evidently not essential for the immediate activation process (135, 140), but rather for sustained Hsf activity under long-term stress (141, 142). At any rate, it is important for the shut-off of Hsf activity, i.e. the regeneration of the inactive state (139, 143–146). Deactivation coincides with dephosphorylation, and it is blocked by phosphatase inhibitors (142, 147, 148). An interesting peculiarity of Hsf phosphorylation in yeast was reported by Liu and Thiele (149). Both oxidative stress (os) and heat stress are connected with Hsf phosphorylation. But the phosphopeptide pattern and, as a consequence, the promoter specificity are different for Hsf-P(hs) and Hsf-P(os).

Despite all similarities in basic structure and regulatory behaviour, the situation with plant Hsfs in this respect is unclear. On the one hand, *in vitro* phosphorylation of the *Arabidopsis* HsfA1a (Hsf1) by a cyclin-dependent Cdc2a kinase was shown to influence its DNA-binding properties *in vitro* as detected by electrophoretic mobility shift assay (150), but so far there is no evidence for phosphorylation of plant Hsfs *in vivo* (Scharf *et al.*, unpublished). On the other hand, we discussed a model of intramolecular shielding of the HR-A/B-NLS region by the C-terminal HR-C region to explain the exclusive cytoplasmic localization of tomato HsfA2. As discussed above, this cytoplasmic retention can be relieved by hetero-oligomerization with HsfA1 or by deletion of the HR-C region (93). However, recent results indicate that, in fact, a strong nuclear export signal (NES) in the C-terminus of HsfA2 is responsible for the lacking accumulation in the nucleus (95). Thus, in this case the concept for a role of the HR-C region as intramolecular repressor domain needs modification.

3.3.3 Search for coregulators

Evidently, the concept of intramolecular repressor domains as discussed in Section 3.3.2 does not exclude the role of such domains as docking sites for coregulators. Indeed, more and more Hsf interacting proteins are reported connecting the activity of Hsfs with other parts of the regulatory network of yeast and mammals. Such coregulators can be other transcription activating factors, e.g. c-Myb, STAT1 (151–154) or repressors. Most remarkable is the detection of a small Hsf-binding protein (HSBP1) in mammals and nematodes, which specifically masks the HR-A/B region and thus helps to stabilize the monomeric form of Hsfs (155). Inactive forms of Hsfs likewise might act as repressors. Due to alternative splicing Hsfs 1, 2, and 4 of mammals can exist in an active (α form) and an inactive form (β form). At least for the human, Hsf1 and Hsf4 reporter assays demonstrate that the inactive form inhibits the activator function of the active form (156–159). For plants, there is experimental evidence that some of the class B Hsfs may have a similar repressor function (118).

Most important for the attenuation of the Hsf activity are components of the Hsp90 and Hsp70 chaperone complexes. The interaction of yeast, Drosophila and mammalian Hsf1 with Hsp70 as the first step of the deactivation and release from the DNA has been repeatedly documented (160–162). In yeast, large ATP-sensitive complexes of Hsf1 with Hsp70 were found under long-term hs and recovery conditions (162). In agreement with these earlier results, the human Hsf1 can be assembled *in vitro* into a multichaperone complex with all known components of the Hsp70 and Hsp90 machines (163–165). Moreover, geldanamycin, which specifically interferes with the Hsp90 function by blocking its ATP-binding site, was shown to promote Hsf activation (165). Over-expression of steroid receptors in COS cells induces Hsf activity, and this effect can be abolished by addition of the steroid ligand or by coexpression of the receptors together with Hsp70 (166). These results indicate that activity control of Hsfs and steroid receptors in vertebrates may require interaction with the same chaperone systems.

Although many details remain to be elaborated, the function of Hsps as coregulators of Hsf activity appears to be similar in plants as well.

- Lee and Schöffl (167) observed that transgenic *Arabidopsis* expressing a Hsp70 antisense RNA had strongly reduced Hsp70 levels and as a result impaired shut-off of Hsf activity.

- In transient reporter assays with tobacco protoplasts, coexpression of Hsfs with Hsp70 and/or Hsp90 chaperones markedly reduced the basal activity and thus improved the hs inducibility. Moreover, addition of geldanamycin as inhibitor of the Hsp90 function created a completely unregulated high Hsf activity (K.-D. Scharf, unpublished). Most intriguing in this context is the specific and reversible association of tomato HsfA2 with cytoplasmic chaperone complexes (see Section 3.2), which probably reflects its specific interaction with Hsp17 class CII. The significance of the HsfA2/Hsp17 complex for the function of HsfA2 in the hs response is not yet clear.

3.3.4 Signal transduction and Hsf activity

The intriguing observations about the role of chaperones for the control of Hsf activity help to understand earlier reports that the stress-sensitive system of cells is protein homeostasis, i.e. the balance between new synthesis, folding, intracellular distribution, assembly to multimeric complexes and degradation of proteins. Our present concept is summarized in Fig. 2. Depending on the particular protein, some or all parts of its life cycle need chaperones. Stress-induced denaturation of mature proteins or of intermediates of the maturation process creates a cellular situation with an increased requirement of chaperones, and thus leads to a derepression of Hsf activity. In keeping with this, Hsp70 or better the pool of free Hsp70 was considered as the cellular thermometer (168), and accumulation of abnormal proteins or inhibition of the proteasome function were repeatedly found to induce Hsp synthesis (142, 169–173). The same holds true for reactive oxygen species as stressors creating proteins with non-native disulfide bonds or oxidized side chains (174–177).

Although our interest in this part of the review was focused on Hsfs and their regulated function in the hs response, it is important to recall that other transcription factors respond to heat and oxidative stress as well (178). Due to the complexity of stress-responsive promoters, expression of hs genes may result from activation of two or more transcription factors. The following examples may illustrate this point.

- In yeast, the Msn2p/Msn4p transcription factors bind to the general stress-responsive elements (STRE), which are not recognized by Hsf. Phosphorylation and activation of Msn2p/Msn4p is induced by hs (179).

- The immediate link between heat and oxidative stress in yeast is brought about by physical interaction of Hsf1 with Skn7. The latter is hybrid transcription factor with a C-terminal domain sensitive to oxidative stress and a DBD with high degree of sequence and structural homology to the DBD of Hsf1 (82, 180).

- In mammalian cells, interaction of STAT1 with Hsf1 mediates the cytokinine-activated transcription of Hsp70 and Hsp90 (153, 154).

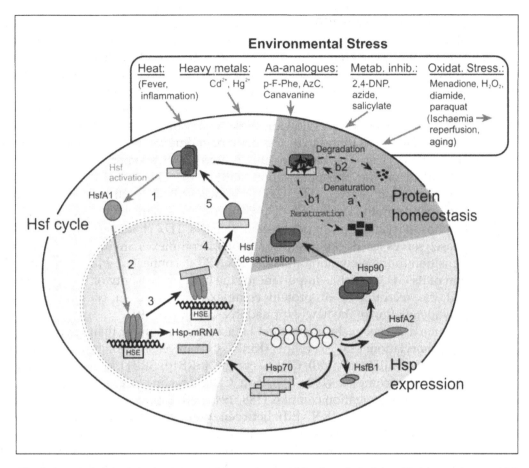

Fig. 2 Autoregulation of the heat-stress response in a model eukaryotic cell (modified from reference 114). Besides heat stress, a considerable number of chemical stressors (see examples given above) are able to trigger transcription of hs genes. Very likely, they all cause accumulation of denatured or immature proteins, which tend to form aggregates and bind molecular chaperones of the Hsp70 and Hsp90 family. Thus, the transient disruption of protein homeostasis results in a deficiency of free chaperones in the cytoplasm with subsequent activation (1) and nuclear import (2) of Hsf. In a type of autoregulatory circuit, hs gene expression is assumed to restore the initial state including inactivation of Hsf (steps 3–5 of the Hsf cycle) and protein renaturation (b1) or degradation (b2). A peculiarity of plants is the synthesis of new Hsfs, e.g. Hsfs A2 and B1 in tomato, as a result of the hs response (see text).

- Activation of mammalian AP1 (yeast yAP1) by heat or oxidative stress may be the result of direct oxidation of Cys residues in the AP1 subunits, which are considered as a redox-sensitive sulfhydryl switch (181, 182).
- In vertebrates hetero-oligomers of Hsf3 with the c-Myb proto-oncogene are crucial for the cell-cycle-controlled synthesis of Hsps. Induction of p53 activity by genotoxic agents disrupts the heterooligomer and down-regulates Hsp synthesis (151, 183–185).

3.3.5 CTD phosphorylation

More than 15 years ago, investigations by J. Lis and co-workers on the architecture of *Drosophila* hs promoters resulted in an unexpected finding. Even in uninduced cells, RNA polymerase II was bound to the hs promoters and started transcription. However, elongation was blocked after synthesis of about 25 nucleotides of RNA, and this block was released by binding activated Hsf (for references to early literature see 186, 187). Following these observations, the existence of stalled RNA polymerase II was found to be a general phenomenon not restricted to hs genes and not to *Drosophila*. The basis for this is two activity states of RNA polymerase II dependent on the phosphorylation state of the C-terminal domain (CTD) of the largest subunit (Fig. 3). The CTD is composed of 25–60 repeats of a conserved heptapeptide motif (Tyr-Ser-Pro-Thr-Ser-Pro-Ser) recognized by specific CTD kinases. Unphosphorylated or hypophosphorylated CTD, characteristic of the initiation phase, serves as docking site for protein complexes involved in initiation. The transition to the elongation mode of RNAPII is connected with hyperphosphorylation of its CTD, which is important for the high processivity of the enzyme complex and the recruitment of protein complexes involved in capping, splicing, and polyadenylation of the newly formed RNA (186, 188–191).

Three different cyclin-dependent kinases are known for their specific role in CTD phosphorylation: CDK7/cyclinH (Kin28/Ccl1 in yeast) are subunits of the transcription factor TFIIH (187, 190); CDK8/cyclinC (SRB10/SRB11 in yeast) are components of mediator complex (192, 193), and CDK9/cyclinT are subunits of the P-TEFb transcription elongation complex (188, 194, 195). Essential for the rapid transition of RNA polymerase II (RNAPII) between the two states (Fig. 3) is the dephosphorylation of the CTD after transcription termination. So far only one CTD-specific phosphatase was identified in human cells and yeast. Its activity is modulated by general transcription factors TFIIB and TFIIF (196). The sequence of the heptapeptide of CTD of RNAPII as well as the kinases associated with it are conserved between animals, yeast, and plants. Unfortunately, there are only two reports from plants on the partial purification of a CTD kinase from wheat germ extract (197) and on the characterization of a kinase R2 from rice, which was shown by functional complementation to be homologous to the yeast CDK7 (198).

With our increasing knowledge about the multiple functions of CTD phosphorylation for RNA transcription and processing, it became apparent that not only the phosphorylation level but also the particular pattern are important for the function of RNAPII. A series of monoclonal antibodies and specific Cdk inhibitors helped to discriminate between different states (188, 199, 200). In yeast and mammalian cells, there is evidence that the CTD phosphorylation pattern changes during heat stress, probably as a result of reduced activity of Cdk7 and Cdk8, and this contributes to the selective transcription of hs genes (187, 188, 201, 202). Although many details of this reprogramming of transcription due the altered CTD phosporylation pattern are unclear, it is important to notice that the activity state of the P-TEFb-connected Cdk9 needs interaction with the Hsp90 chaperone system (195)

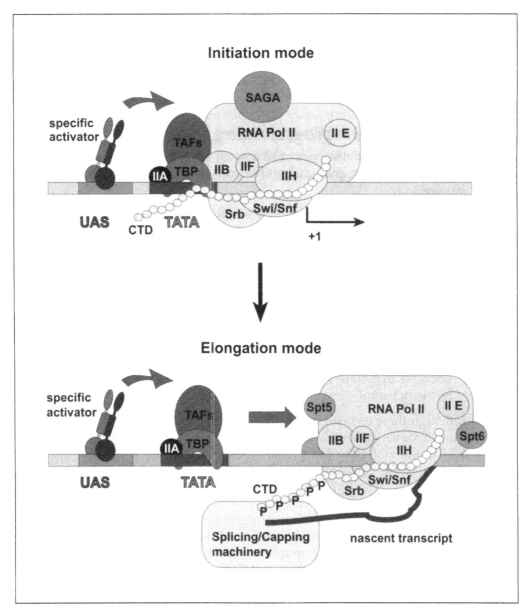

Fig. 3 Two activity states of eukaryotic RNA polymerase II depend on the phosphorylation level of the C-terminal domain (CTD) of the largest subunit (figure kindly provided by P. Döring). (A) In the initiation mode (low phosphorylation of CTD) the CTD helps in the recruitment of transcription initiation factors, e.g. of TFIIA and TFIIB, to the transcription start point downstream of the TFIID complex formed by the TATA-box binding protein (TBP) and about 10 associated factors (TAFs). SAGA, Srb, and Swi/Snf designate additional multiprotein complexes involved in histone acetylation, CTD phosphorylation, and chromatin remodelling. (B) The transition to the elongation mode is triggered by the contacts to an activator protein, e.g. Hsf. The RNA polymerase II machinery is released from the initiation site and starts or continues polymerization. New elongation-specific proteins are bound (Spt5 and Spt6), and the highly phosphorylated CTD serves as attachment site for protein complexes necessary for processing of the newly formed RNA [capping, splicing, termination, and poly(A) addition].

and that in mammalian cells the CTD phosphatase is inactivated during hs by binding to the nuclear matrix (196).

4. Complex cellular programmes influenced by heat stress

4.1 Translational reprogramming

The selectivity and efficiency of Hsp synthesis is intimately connected with the rapid and reversible reprogramming of the translation machinery. The following steps are involved:

- Modification of initiation factors results in an immediate halt of translation initiation followed by a decay of the polysomal population within 5–10 minutes after the onset of the hs.
- More delayed (30–60 min.) is the dephosphorylation of ribosomal protein S6 localized in the initiation centre of the 40S ribosomal subunit essential for mRNA binding.
- New synthesis of hs mRNAs allows regeneration of polysomes, which are almost exclusively engaged in Hsp synthesis.
- Removal of housekeeping mRNAs from the polysomes is not necessarily connected with their degradation, but they may be sequestered and reactivated in the recovery period.

Although these elements of the translational reprogramming have been well documented for vertebrates, insects (*Drosophila*), and plants for many years (see reviews 5, 6, 203), the molecular basis is still mostly unknown. No doubt, the complexity of changes at the translational level and the ready reversibility involve the activity of protein kinases and phosphatases as well as Hsps. We briefly summarize here the present state of our knowledge.

Because of the efficient discrimination between housekeeping and hs mRNAs in the initiation process, modification of initiation factors were always considered as central parts of the reprogramming. Indeed, in heat-stressed mammalian cells phosphorylation of the eIF2α subunit inhibits preinitiation complex formation, whereas dephosphorylation of eIF4B and inactivation of eIF4E by complex formation with an inhibitory subunit (4E-BP1) precludes assembly of the cap complex at the 5′ end of the mRNA (204–209). The eIF2α kinase is kept in an inactive but activation competent complex with Hsp70 and Hsp90. Similar to the activation of Hsfs (see Section 3.3.3), the depletion of the pool of free Hsp70 as a result of the interaction with denatured proteins might trigger the block of translation by activation of eIF2α kinase (210, 211). In plants the hs induced polysomal decay (5) and reduction of cap-dependent translation (212) is evidently not connected with modification of eIF2α nor with changes of eIF4E function but with the rapid dephosphorylation of the 59 kDa eIF4B (213).

The role of Hsps for different aspects of the translational reprogramming has been repeatedly documented. Well known and probably unique for plants is the assembly of large cytoplasmic multichaperone complexes (214–216). They are built of hundreds of 40 nm particles (heat-stress granules, HSG) containing class CI and class CII small cytosolic Hsps, Hsp70, and the heat-stress transcription factor HsfA2 (94, 217, 218). HSG are assembled in a stress-dependent manner without the necessity of ongoing protein synthesis (215). Cell fractionation combined with *in vitro* translation as well as electron microscopic investigations indicate that HSG complexes may also serve as transient storage sites for housekeeping mRNAs not contained in polysomes during the stress period (215).

Earlier findings about electron-dense aggregates containing small Hsps also in vertebrate cells (5) were recently extended by the analysis of stress granules in mammalian cells. They contain poly(A) RNA, RNA-binding proteins, initiation factors, and, in heat-stressed cells, also Hsp27 (219, 220). An intriguing observation in this context is the formation of insoluble aggregates (HSG) of Hsp27 with the adapter subunit eIF4G of the cap recognition complex (221). The inhibition of cap-dependent translation by entrapping eIF4G can be relieved by Hsp70. In fact, the eIF4G deficiency together with the hs-induced overload of the proteasome pathway delays the turnover of short-lived mRNAs with an AU-rich sequence in their 3'-UTR (222), and, together with the AU-specific binding proteins (AUF1/HuR), these mRNPs are probably integrated into the stress granules (223).

Impairment of the cap-dependent mechanism of translation initiation and sequestration of housekeeping and/or developmental mRNPs evidently contribute to the striking shift in the translation specificity during stress. Efficient recognition of internal ribosomal entry sites (IRES) in the 5' untranslated leader sequences of hs and other mRNAs allow recognition by base pairing with the 18S rRNA. They are the basis for the internal initiation as described for the Hsp70 mRNAs in yeast, *Drosophila*, and mammalian cells (224–226). Similar IRES motifs were also identified in hs mRNAs of plants (227). In addition, Hsp101 was reported to be an RNA-binding protein with high affinity to a translation enhancer motif (omega element) found in the 5'-UTRs of tobacco mosaic virus derived mRNA and the feredoxin (FED1) mRNA (228, 229). As a result, translation of both mRNAs is not repressed but rather enhanced under hs conditions.

A remarkable consequence of ribosome modification by dephosphorylation of small subunit protein S6, observed in animals and plants [see summary by Nover (5)] is the block of cell cycling under hs conditions because of the failure to translate a unique family of cell-cycle-controlled mRNAs. They are characterized by oligopyrimidine tracts in their leader sequences and encode ribosomal proteins, translation factors, and similar proteins. Translation of other mRNAs is not affected by S6 dephosphorylation. The signalling pathway required for maintenance of the proper phosphorylation state of the small ribosomal subunit involves the p70/p85^{S6K} kinase, whose activity is controlled by a rapamycin-sensitive upstream kinase complex (FRAP/TOR) and the phosphoinositol-dependent kinase PI3K (230–232).

4.2   Cell cycle

Control of the cell cycle requires stage-specific activation of cyclin-dependent protein kinases (Cdks) by phosphorylation/dephosphorylation and their association with transiently formed cyclins (Clns). Basic mechanisms and regulatory components of the cell cycle are conserved between plants, animals, and yeast (233–237). Because the complete reproduction of all cellular components within the cell cycle is a dominant housekeeping function, it was not surprising that heat stress was found to reversibly block cell-cycle progression. Earlier, temperature-shift programmes with repeated hs and recovery periods were used for synchronization of proliferating cells of *Tetrahymena*, *Schizosaccharomyces pombe* and plant cells (5, 238).

As expected, the effects of hs on cell cycle are complex and stage-specific with many steps or structures influenced by the activity of chaperones (234, 239–241), and, vice versa, expression of chaperones of the Hsp70 and Hsp90 families is also under cell-cycle control (see Section 3.3.4). The central role of the cytoskeletal systems, in particular of the tubulin system, for the ordered redistribution of chromosomes and other cellular components, is immediately affected. Restoration of cytoskeletal functions is dependent or at least markedly improved by chaperones of the Hsp70 and Hsp90 families (5, 242, 243). Moreover, the Hsp90 chaperone machinery is an integral component of the centromer complex (244). Long-lasting and more specific effects of hs result in G1/S and G2/M arrest [see the excellent summary of Kühl and Rensing (234)]. The lack of synthesis of stage-specific cyclins (245–247) or increased formation of repressors, e.g. the p53-dependent formation of the p21 Cdk inhibitor in mammalian cells (183, 248), may be responsible for these effects. It is important to notice in this context that Hsp90 is an integral component of many mammalian protein kinase signalling pathways involved in cell-cycle control, e.g. the MAP kinase cascade, the Raf-1 cascade, Cdk4, Cdk9, Cdc28, and the Wee-1 tyrosine kinase (195, 249–252). Down-regulation of these signalling pathways during hs may contribute to the cell-cycle block (234, 253).

In all organisms investigated in this respect, a central control point of cell cycling is the rapamycin-sensitive TOR protein kinase complex. It controls the activity state of the ribosomal protein S6 kinase (p70^{S6K}) and thus the phosphorylation state of the ribosomal small subunit (230, 231). Dephosphorylation of S6 under conditions of heat stress or amino acid deprivation blocks the cell cycle because of the lack of synthesis of translation initiation factors and distinct ribosomal proteins (see Section 4.1).

4.3   Ribosome biosynthesis

Ribosome biosynthesis in yeast, plants, and animals, i.e. the transcription of rRNA precursors (pre-rRNAs) by RNA polymerases I and III, the assembly of pre-ribosomal particles (pre-rRNP) in the nucleolus as well as their processing and nuclear export as ribosomal subunits, is a dominant housekeeping function tightly connected to the cell cycle. For rapidly growing yeast, it was calculated that 60 % of total transcription and 90 % of total splicing capacity is needed for ribosomal synthesis involving hundreds of nucleolar helper RNAs and proteins (254–256). Under heat-stress conditions, ribosome biosynthesis is shut down in a multistep process (5, 257, 258).

(1) Immediately after the onset of the hs, processing of pre-rRNP is blocked resulting in an increasing accumulation of rRNP granules in the nucleolus.

(2) Although translation of ribosomal proteins in general is not impaired, the increasing dephosphorylation of the ribosomal protein S6, which is complete about 30 minutes after the onset of hs, affects translation of mRNAs encoding distinct ribosomal proteins as well as translation factors. As a consequence, defective pre-ribosomal particles are formed.

(3) With ongoing hs, the accumulation of pre-rRNPs leads to a remarkable disintegration of the nucleolus and, finally, cessation of rRNA transcription. Newly synthesized Hsp70 is found tightly connected with the nucleosomal pre-rRNP.

(4) In the recovery period, the nucleolar structure and function rapidly normalize. The pre-rRNP granules disappear. Some of them can be processed to mature ribosomal subunits, most of them are defective and degraded. The recovery process is markedly enhanced in thermotolerant cells, i.e. in the presence of nucleolar Hsp70.

Signalling pathways involved in the four stages are barely characterized. The immediate but readily reversible shutdown of an early step of pre-rRNP processing is particularly striking. Availability or function of one or several snoRNP necessary for the strictly controlled sequence of site-specific modification and / or cleavage of pre-rRNA may be impaired (255). For the delayed stages (2) and (3), the protein kinase / phosphatase network controlling the S6 phosphorylation level (see point 4) and the feedback inhibition of RNA polymerase I by the accumulating pre-rRNP in the nucleolus are responsible. The complexity of the gene expression programme leading to the assembly and maturation of ribosomal precursor particles and the sheer mass of material required (254) necessitates a perfect coordination with other cellular activities. Although not fully understood in detail, it is not surprising that ribosome synthesis is tuned to membrane functions and to cell-cycle control (254, 255). On the other hand, the nucleolar structure and function are central not only for ribosome synthesis but also for mRNA export, nuclear protein import, processing of tRNA precursors, and other cellular processes (259, 260).

4.4 Photosynthesis

The capability for rapid adaptation of the energy conversion by photosystems I and II is a prerequisite for plant survival under different light and environmental conditions (261, 262). Two proteins (D1 and D2) in the reaction centre of PSII are particularly important for the activity control, because protein D1 has a fairly rapid turnover, i.e. in contrast to all other components of PSII, it needs permanent new synthesis and incorporation into the PSII complex. Prolonged high irradiance of leaves caused high levels of D1 phosphorylation and decreased activity of the PSII complex (photoinhibition). D1-P can evidently be reactivated *in situ* or degraded after dephosphorylation (263). Similar to many other unfavourable environmental conditions, hs also causes a decrease of photosynthetic activities connected with changes of the membrane properties, composition of PSII, and coupling between PSI and PSII (264–266), and Hsps protect plants against damage of the photosynthetic apparatus (34–37). Most remarkable is the dramatically increasing dephosphorylation of the D1, D2, and the chlorophyll *a* binding protein CP43. Evidently, an intrinsic membrane protein phosphatase is activated by hs-induced release of an inhibitory subunit (TLP40), which has peptidyl-prolyl *cis,trans*-isomerase activity (267). Enhanced turnover of D1 may be essential for the rapid repair of damaged PSII in the recovery.

4.5 Apoptosis

Apoptosis or programmed cell death is characterized by condensation of nuclear chromatin, cytoplasmic shrinkage, membrane blebbing, nuclear fragmentation, and encapsulation of these fragments into so-called 'apoptotic bodies' (268–274). Apoptosis is known to be induced by several developmental and pathological signals as well as stress, which includes heat and oxidative stress, UV irradiation, and others. In addition to the induction of apoptosis, heat stress also causes the induction of cellular heat-stress response, i.e. new synthesis of heat-shock proteins. But clearly the two processes have different, cell-specific temperature thresholds. Treatment of mammalian cells at 42°C induces Hsp synthesis and protects them from apoptosis. On the other hand, apoptosis is triggered at 43–44°C when Hsp synthesis is delayed or even lacking. Further increase of the temperature to 46°C leads to necrosis (142, 275, 276).

Despite the different temperature thresholds for triggering Hsp synthesis and apoptosis, respectively, common signalling components may be involved. One of these signals is the accumulation of partially denatured or abnormal proteins upon heat stress, oxidative stress or treatment of cells with inhibitors of proteasome function (275–278). One of the key regulators of heat-stress induced apoptosis in mammals is ceramide (see Section 2.1) reported to cause increased expression of c-Jun and activation of caspase-3 (279–281). Other elements of the apoptotic signalling are protein kinases such as c-Jun N-terminal kinase (JNK) and glycogen synthase kinase-3β (GSK3β). Both kinases are activated by heat stress or accumulation of abnormal proteins (142, 282–284). Interestingly, the threshold phenomenon described above is also observed at the level of JNK activity. Transient JNK activation by mild hs coincides with Hsp synthesis and protection from apoptosis, whereas severe stress leads to permanent JNK activation, block of Hsp synthesis and apoptosis (282, 285). The same antagonism was found for the apoptotic signalling pathway triggered by the membrane-bound FAS receptor. Activation of Hsf1 and as a consequence Hsp70 synthesis were inhibited, whereas preinduction of the Hsf1 signalling by mild hs precludes progress of the apoptotic pathway (285, 286).

The protection from apoptosis in thermotolerant cells is tightly connected with

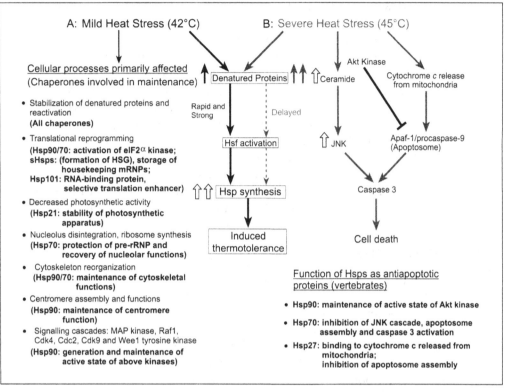

Fig. 4 Summary of temperature effects on the triggering of the hs response (A, left) or apoptosis (B, right) and the role of Hsps in the two pathways. The multiplicity of functions of Hsps for thermotolerance effects (left) and as antiapoptotic proteins (right) are indicated. The decision between the hs or apoptotic response is dependent on the extend of Hsp synthesis, which is rapid and strong under mild but delayed and reduced under severe temperature stress.

synthesis of Hsps, in particular of Hsp90, Hsp70, and Hsp27, which act as anti-apoptotic proteins (see reviews see 287–289) (Fig. 4). Different parts of the cell-death programme are inhibited. The antiapoptotic effect of Hsp90 results from the stabilization of the active state of the Akt Ser/Thr-specific protein kinase (289). Hsp27 binds to cytochrome c released from the mitochondria and thus prevents formation of the Apaf-1/procaspase-9 apoptosome complex (290–293). A similar block of apoptosome formation was observed for Hsp70 because of its interaction with Apaf-1 (294–296). As a consequence, depletion of Hsp70 in human tumour cell lines by antisense techniques or disruption of the Hsf gene enhanced the apoptotic response (286, 297–299), whereas Hsp70 overexpression had a protective effect (300–303). Hsp70 has multiple points of interaction with the apoptotic programme. Using mammalian cells expressing wild-type and mutant forms of Hsp70, Mosser *et al.* (304) demonstrated that the inhibition of the procaspase-9 activation requires the full-length and active chaperone with N-terminal ATP-binding and C-terminal peptide-binding domains. In contrast to this, inhibition of JNK activation was also observed with a deletion form of Hsp70 lacking its ATP-binding domain. Finally, protection from cytochrome c release required full-length Hsp70 but not its functional peptide-binding domain.

Similar to mammalian systems, apoptosis in plants is triggered by pathogen inter-action (305, 306), oxidative stress (307, 308), and heat stress (309–311). The typical morphological changes with DNA fragmentation, cytoplasmic condensation or membrane blebbing (273, 306, 309) are also connected with the well-known molecular characteristics. These are release of cytochrome c (309, 312), activation of a caspase-3-like proteinase (274, 310, 313), and subsequent cleavage of lamin-like proteins and poly(ADP-ribose) polymerase (308, 310–312). New synthesis and/or activation of additional proteases were also reported (314, 315). Finally, there is evidence for a role of a membrane-associated Bax protein (Bcl-2 family) as a death-promoting factor and the existence of corresponding Bax inhibitor proteins in plants (274, 316).

5. Conclusions

The rapid advances in plant molecular biology, which will dramatically accelerate in the forthcoming years due to the availability of the sequence information of the *Arabidopsis* genome, have finally opened the view to the whole complexity of stress-response systems. The fragmentary picture with many pieces derived from the yeast and vertebrate literature, outlined in this review, will soon be completed. This development will certainly facilitate research on the dynamic changes in the course of a stress response and on the cross-talks between different stress-response systems characteristic of the stressful real life of plants.

References

1. Ritossa, F. (1962) A new puffing pattern induced by heat shock and DNP in *Drosophila. Experientia*, **18**, 571–573.
2. Tissieres, A., Mitchell, H. K. and Tracy, U. M. (1974) Protein synthesis in salivary glands of *D. melanogaster*. Relation to chromosome puffs. *J. Mol. Biol.*, **84**, 389–398.
3. McKenzie, S. L. and Meselson, M. (1977) Translation in vitro of *Drosophila* heat-shock messages. *J. Mol. Biol.*, **117**, 279–283.
4. Ashburner, M. and Bonner, J. J. (1979) The induction of gene activity in *Drosophila* by heat shock. *Cell*, **17**, 241–254.
5. Nover, L. (ed.) (1991) *Heat Shock Response*. CRC Press, Boca Raton, FL.
6. Nover, L., Neumann, D. and Scharf, K.-D. (1989) Heat shock and other stress response systems of plants. *Res. Problems Cell Diff.*, Springer, Berlin.
7. Nover, L. and Miernyk, J. (2001) A genomic approach to the chaperone system of *Arabidospis thaliana. Cell Stress Chap.*, **6**, 175–176 and following pages.
8. Forreiter, C. and Nover, L. (1998) Heat stress-induced proteins and the concept of molecular chaperones. *J. Biosci*, **23**, 287–302.
9. Ellis, R. J. (2000) Chaperone substrates inside the cell. *Trends Biochem. Sci.*, **25**, 210–212.
10. Richter, K. and Buchner, J. (2001) Hsp90: chaperoning signal transduction. *J. Cell Physiol.*, **188**, 281–290.

11. MacRae, T. H. (2000) Structure and function of small heat shock/alpha-crystallin proteins: established concepts and emerging ideas. *Cell Mol. Life Sci.*, **57**, 899–913.
12. Slavotinek, A. M. and Biesecker, L. G. (2001) Unfolding the role of chaperones and chaperonins in human disease. *Trends Genet.*, **17**, 528–535.
13. Thirumalai, D. and Lorimer, G. H. (2001) Chaperonin-mediated protein folding. *Annu. Rev. Biophys. Biomolec. Struct.*, **30**, 245–269.
14. Schleiff, E. and Soll, J. (2000) Travelling of proteins through membranes: translocation into chloroplasts. *Planta*, **211**, 449–456.
15. German, J. (1984) Embryogenic stress hypothesis of teratogenesis. *Am. J. Med.*, **76**, 293–301.
16. Mitchell, H. K., Moller, G., Petersen, N. S. and Lipps-Sarmiento, L. (1979) Specific protection from phenocopy induction by heat shock. *Dev. Genet.*, **1**, 181–192.
17. Vidal, M. (2001) A biological atlas of functional maps. *Cell*, **104**, 333–339.
18. Zhu, H., Bilgin, M., Bangham, R., Hall, D., Casamayor, A., Bertone, P., Lan, N., Jansen, R., Bidlingmaier, S., Houfek, T., Mitchell, T., Miller, P., Dean, R. A., Gerstein, M. and Snyder, M. (2001) Global analysis of protein activities using proteome chips. *Science*, **293**, 2101–2105.
19. Seki, M., Narusaka, M., Abe, H., Kasuga, M., Yamaguchi-Shinozaki, K., Carninci, P., Hayashizaki, Y. and Shinozaki, K. (2001) Monitoring the expression pattern of 1300 *Arabidopsis* genes under drought and cold stresses by using full-length cDNA microarray. *Plant Cell*, **13**, 61–72.
20. Petersohn, A., Brigulla, M., Haas, S., Hoheisel, J. D., Völker, U. and Hecker, M. (2001) Global analysis of the general stress response of *Bacillus subtilis*. *J. Bacteriol.*, **183**, 5617–5631.
21. Kamada, Y., Jung, U. S., Piotrowski, J. and Levin, D. E. (1995) The protein kinase C-activated MAP kinase pathway of *Saccharomyces cerevisiae* mediates a novel aspect of the heat shock response. *Genes Dev.*, **9**, 1559–1571.
22. Brown, D. A. and London, E. (1998) Functions of lipid rafts in biologial membranes. *Annu. Rev. Cell Dev. Biol.*, **14**, 111–136.
23. Smart, E. J., Graf, G. A., McNiven, M. A., Sessa, W. C., Engelman, J. A., Scherer, P. E., Okamoto, T. and Lisanti, M. P. (1999) Caveolins, liquid-ordered domains, and signal transduction. *Mol. Cell. Biol.*, **19**, 7289–7304.
24. de Kruiff, B. (1997) Biomembranes. Lipids beyond the bilayer. *Nature*, **386**, 129–130.
25. Slater, S. J., Kelly, M. B., Taddeo, F. J., Rubin, E. and Stubbs, C. D. (1994) The modulation of protein kinase C activity by membrane lipid bilayer structure. *J. Biol. Chem.*, **269**, 4866–4871.
26. Simons, K. and Ikonen, E. (1997) Functional rafts in cell membranes. *Nature*, **387**, 569–572.
27. Vigh, L., Maresca, B. and Harwood, J. L. (1998) Does the membrane's physical state control the expression of heat shock and other genes?. *Trends Biochem. Sci.*, **23**, 369–374.
28. Carratu, L., Franceschelli, S., Pardini, C. L., Kobayashi, G. S., Horvath, I., Vigh, L. and Maresca, B. (1996) Membrane lipid perturbation modifies the set point of the temperature of heat shock response in yeast. *Proc. Natl Acad. Sci. USA*, **93**, 3870–3875.
29. Gombos, Z., Wada, H., Hideg, E. and Murata, N. (1994) The unsaturation of membrane lipids stabilizes photosynthesis against heat stress. *Plant Physiol.*, **104**, 563–567.
30. Gombos, Z., Wada, H. and Murata, N. (1994) The recovery of photosynthesis from low-temperature photoinhibition is accelerated by the unsaturation of membrane lipids: a mechanism for chilling tolerance. *Proc. Natl Acad Sci. USA*, **91**, 8787–8791.
31. Grindstaff, K. K., Fielding, L. A. and Brodl, M. R. (1996) Effect of gibberellin and heat shock on the lipid composition of endoplasmic reticulum in barley aleurone layers. *Plant Physiol.*, **110**, 571–581.

32. Mariamma, M., Muthukumar, B. and Gnanam, A. (1997) Thermotolerance and effect of heat shock on the stability of the ATPase enzyme in rice. *J. Plant Physiol.*, **150**, 739–742.

33. Chatterjee, M. T., Khalawan, S. A. and Curran, B. P. (2000) Cellular lipid composition influences stress activation of the yeast general stress response element (STRE). *Microbiology*, **146**, 877–884.

34. Török, Z., Goloubinoff, P., Horvath, I., Tsvetkova, N. M., Glatz, A., Balogh, G., Varvasovszki, V., Los, D. A., Vierling, E., Crowe, J. H. and Vigh, L. (2001) *Synechocystis* HSP17 is an amphitropic protein that stabilizes heat-stressed membranes and binds denatured proteins for subsequent chaperone-mediated refolding. *Proc. Natl Acad. Sci. USA*, **98**, 3098–3103.

35. Glatz, A., Vass, I., Los, D. A. and Vigh, L. (1999) The *Synechocystis* model of stress: From molecular chaperones to membranes. *Plant Physiol. Biochem.*, **37**, 1–12.

36. Gustavsson, N., Kokke, B. P. A., Anzelius, B., Boelens, W. C. and Sundby, C. (2001) Substitution of conserved methionines by leucines in chloroplast small heat shock protein results in loss of redox-response but retained chaperone-like activity. *Protein Sci.*, **10**, 1785–1793.

37. Schroda, M., Kropat, J., Oster, U., Rüdiger, W., Vallon, O., Wollman, F. A. and Beck, C. F. (2001) Possible role for molecular chaperones in assembly and repair of photosystem II. *Biochem. Soc. Trans.*, **29**, 413–418.

38. Schneiter, R. (1999) Brave little yeast, please guide us to Thebes: sphingolipid function in *S. cerevisiae*. *BioEssays*, **21**, 1004–1010.

39. Spiegel, S. and Milstien, S. (2000) Sphongosine-1-phosphate: signaling inside and out. *FEBS Lett.*, **476**, 55–57.

40. Dickson, R. C. (1998) Sphingolipid functions in *Saccharomyces cerevisiae*: comparison to mammals. *Annu. Rev. Biochem.*, **67**, 27–48.

41. Dickson, R. C., Nagiec, E. E., Skrzypek, M., Tillman, P., Wells, G. B. and Lester, R. L. (1997) Sphingolipids are potential heat stress signals in *Saccharomyces*. *J. Biol. Chem.*, **272**, 30196–30200.

42. Chung, N., Jenkins, G., Hannun, Y. A., Heitman, J. and Obeid, L. M. (2000) Sphingolipids signal heat stress-induced ubiquitin-dependent proteolysis. *J. Biol. Chem.*, **275**, 17229–17232.

43. Wells, G. B., Dickson, R. C. and Lester, R. L. (1998) Heat-induced elevation of ceramide in *Saccharomyces cerevisiae* via de novo synthesis. *J. Biol. Chem.*, **273**, 7235–7243.

44. Jenkins, G. M., Richards, A., Wahl, T., Mao, C., Obeid, L. and Hannun, Y. (1997) Involvement of yeast sphingolipids in the heat stress response of *Saccharomyces cerevisiae*. *J. Biol. Chem.*, **272**, 32566–32572.

45. Simola, M., Hanninen, A. L., Stranius, S. M. and Makarov, M. (2000) Trehalose is required for conformational repair of heat-denatured proteins in the yeast endoplasmic reticulum but not for maintenance of membrane traffic functions after severe heat stress. *Mol. Microbiol.*, **37**, 42–53.

46. Zielinski, R. E. (1998) Calmodulin and calmodulin-binding proteins in plants. *Annu. Rev. Plant Physiol. Plant Mol. Biol.*, **49**, 697–725.

47. Trewavas, A. (1999) How plants learn. *Proc. Natl Acad. Sci. USA*, **96**, 4216–4218.

48. Harmon, A. C., Gribskov, M. and Harper, J. F. (2000) CDPKs - a kinase for every Ca^{2+} signal? *Trends Plant Sci.*, **5**, 154–159.

49. Knight, H. and Knight, M. R. (2000) Imaging spatial and cellular characteristics of low temperature calcium signature after cold acclimation in *Arabidopsis*. *J. Exp. Bot.*, **51**, 1679–1686.

50. Gong, M., Li, Y. J., Dai, X., Tian, M. and Li, Z. G. (1997) Involvement of calcium and calmodulin in the acquisition of heat-shock induced thermotolerance in maize seedlings. *J. Plant Physiol.*, **150**, 615–621.

51. Torrecilla, I., Leganes, F., Bonilla, I. and Fernandez-Pinas, F., (2000) Use of recombinant aequorin to study calcium homeostasis and monitor calcium transients in response to heat and cold shock in cyanobacteria. *Plant Physiol.*, **123**, 161–175.

52. Kiegle, E., Moore, C. A., Haseloff, J., Tester, M. A. and Knight, M. R. (2000) Cell-type-specific calcium responses to drought, salt and cold in the *Arabidopsis* root. *Plant J.*, **23**, 267–278.

53. Iida, H., Ohya, Y. and Anraku, Y. (1995) Calmodulin dependent protein kinase II and calmodulin are required for induced thermotolerance in *Saccharomyces cerevisiae*. *Curr. Genet.*, **27**, 190–193.

54. Stevenson, M. A. and Calderwood, S. K. (1990) Members of the 70-kilodalton heat shock protein family contain a highly conserved calmodulin-binding domain. *Mol. Cell. Biol.*, **10**, 1234–1238.

55. Sun, X. T., Li, B., Zhou, G. M., Tang, W. Q., Bai, J., Sun, D. Y. and Zhou, R. G. (2000) Binding of the maize cytosolic Hsp70 to calmodulin, and identification of calmodulin-binding site in Hsp70. *Plant Cell Physiol.*, **41**, 804–810.

56. Yang, T. B. and Poovaiah, B. W. (2000) *Arabidopsis* chloroplast chaperonin 10 is a calmodulin-binding protein. *Biochem. Biophys. Res. Commun.*, **275**, 601–607.

57. Braam, J. (1992) Regulated expression of the calmodulin-related TCH genes in cultured *Arabidopsis* cells: induction by calcium and heat shock. *Proc. Natl Acad. Sci. USA*, **89**, 3213–3216.

58. Gong, M., van der Luit, A. H., Knight, M. R. and Trewavas, A. J. (1998) Heat-shock-induced changes in intracellular Ca^{2+} level in tobacco seedlings in relation to thermo-tolerance. *Plant Physiol.*, **116**, 429–437.

59. Pouyssegur, J. (2000) An arresting start for MAPK. *Science*, **290**, 1515–1518.

60. Kovtun, Y., Chiu, W. L., Tena, G. and Sheen, J. (2000) Functional analysis of oxidative stress-activated mitogen-activated protein kinase cascade in plants. *Proc. Natl Acad. Sci. USA*, **97**, 2940–2945.

61. Hardie, D. G. (1999) Plant protein serine/threonine kinases: classification and functions. *Annu. Rev. Plant Physiol. Plant Mol. Biol.*, **50**, 97–131.

62. Meskiene, I. and Hirt, H. (2000) MAP kinase pathways: molecular plug-and-play chips for the cell. *Plant Mol. Biol.*, **42**, 791–806.

63. Bögre, L., Meskiene, I., Heberle-Bors, E. and Hirt, H. (2000) Stressing the role of MAP kinases in mitogenic stimulation. *Plant Mol. Biol.*, **43**, 707–720.

64. Lee, Y. L. and Corry, P. M. (1998) Metabolic oxidative stress-induced Hsp70 gene expression is mediated through SAPK pathway. *J. Biol. Chem.*, **273**, 29857–29863.

65. Dorion, S., Berube, J., Huot, J. and Landry, J. (1999) A short-lived protein involved in the heat shock sensing mechanism responsible for stress-activated protein kinase 2 (SAPK2/p38) activation. *J. Biol. Chem.*, **274**, 37591–37597.

66. Shiozaki, K., Shiozaki, M. and Russell, P. (1998) Heat stress activates fission yeast Spc1/StyI MAPK by a MEKK-independent mechanism. *Mol. Cell. Biol.*, **9**, 1339–1349.

67. Nguyen, A. N. and Shiozaki, K. (1999) Heat shock-induced activation of stress MAP kinase is regulated by threonine- and tyrosine-specific phosphatases. *Genes Dev.*, **13**, 1653–1663.

68. Mattison, C. P. and Ota, I. M. (2000) Two protein tyrosine phosphatases, Ptp2 and Ptp3, modulate the subcellular localization of the Hog1 MAP kinase in yeast. *Genes Dev.*, **14**, 1229–1235.

69. Yamboliev, I. A., Hedges, J. C., Mutnick, J. L. M., Adam, L. P. and Gerthoffer, W. T. (2000) Evidence for modulation of smooth muscle force by the p38 MAP kinase/HSP27 pathway. *Am. J. Physiol.-Heart Circul. Physiol.*, **278**, H1899–H1907.

70. Rouse, J., Cohen, P., Trigon, S., Morange, M., Alouso-Liamazares, A., Zamanillo, D., Hunt, T. and Nebreda, A. R. (1994) A novel kinase cascade triggered by stress and heat shock that stimulates MAPKAP kinase-2 and phosphorylation of the small heat shock protein. *Cell*, **78**, 1027–1037.

71. Engel, K., Ahlers, A., Brach, M. A., Herrmann, F. and Gaestel, M. (1995) MAPKAP kinase 2 is activated by heat shock and TNF-alpha - in vivo phosphorylation of small heat shock protein results from stimulation of the MAP kinase cascade. *J. Cell Biochem.*, **57**, 321–330.

72. Kato, K., Ito, H., Iwamoto, I., Iida, K. and Inaguma, Y. (2001) Protein kinase inhibitors can suppress stress-induced dissociation of Hsp27. *Cell Stress Chap.*, **6**, 16–20.

73. Nover, L. and Scharf, K.-D. (1984) Synthesis, modification and structural binding of heat shock proteins in tomato cell cultures. *Eur. J. Biochem.*, **139**, 303–313.

74. Heider, H., Boscheinen, O. and Scharf, K.-D. (1998) A heat stress pulse inactivates a 50 kDa myelin basic protein kinase in tomato. *Botan. Acta,***111**, 398–401.

75. Kamada, Y., Jung, U. S., Piotrowski, J. and Levin, D. E. (1995) The protein kinase C-activated MAP kinase pathway of *Saccharomyces cerevisiae* mediates a novel aspect of the heat shock response. *Genes Dev.*, **9**, 1559–1571.

76. Gerber, D. A., Souquere-Besse, S., Puvion, F., Dubois, M.-F., Bensaude, O. and Cochet, C. (2000) Heat-induced relocalization of protein kinase CK2: implication of CK2 in the context of cellular stress. *J. Biol. Chem.*, **275**, 23919–23926.

77. Kim, D. K., Yamaguchi, Y., Wada, T. and Handa, H. (2001) The regulation of elongation by eukaryotic RNA polymerase II: a recent view. *Mol. Cells*, **11**, 267–274.

78. Carey, M. (1998) The enhanceosome and transcriptional synergy. *Cell*, **92**, 5–8.

79. Glass, C. K. and Rosenfeld, M. G. (2000) The coregulator exchange in transcriptional functions of nuclear receptors. *Genes Dev.*, **14**, 121–141.

80. Tansey, W. P. (2001) Transcriptional activation: risky business. *Genes Dev.*, **15**, 1045–1050.

81. Wu, C. (1995) Heat stress transcription factors. *Annu. Rev. Cell Biol.*, **11**, 441–469.

82. Nover, L., Scharf, K.- D., Gagliardi, D., Vergne, P., Czarnecka-Verner, E. and Gurley, W. B. (1996) The Hsf world: classification and properties of plant heat stress transcription factors. *Cell Stress Chap.*, **1**, 215–223.

83. Nover, L., Bharti, K., Döring, P., Ganguli, A. and Scharf, K.-D. (2001) *Arabidopsis* and the Hsf world: how many heat stress transcription factors do we need? *Cell Stress Chap.*, **6**, 77–189.

84. Morimoto, R. I. (1998) Regulation of the heat shock transcriptional response: cross talk between family of heat shock factors, molecular chaperones, and negative regulators. *Genes Dev.*, **12**, 3788–3796.

85. Nakai, A. (1999) New aspects in the vertebrate heat shock factor system: HsfA3 and HsfA4. *Cell Stress Chap.*, **4**, 86–93.

86. Harrison, C. J., Bohm, A. A. and Nelson, H. C. M. (1994) Crystal structure of the DNA binding domain of the heat shock transcription factor. *Science*, **263** , 224–227.

87. Schultheiss, J., Kunert, O., Gase, U., Scharf, K.-D., Nover, L. and Rüterjans, H. (1996) Solution structure of the DNA-binding domain of the tomato heat stress transcription factor HSF24. *Eur. J. Biochem.*, **236**, 911–921.

88. Nover, L. (1987) Expression of heat shock genes in homologous and heterologous systems. *Enzyme Microb. Technol.*, **9**, 130–144.

89. Littlefield, O. and Nelson, H. C. M. (1999) A new use for the 'wing' of the 'winged' helix-turn-helix motif in the HSF-DNA cocrystal. *Nat. Struct. Biol.*, **6**, 464–470.

90. Cicero, M. P., Hubl, S. T., Harrison, C. J., Littlefield, O., Hardy, J. A. and Nelson, H. C. M. (2001) The wing in yeast heat shock transcription factor (HSF) DNA-binding domain is required for full activity. *Nucleic Acids Res.*, **29**, 1715–1723.

91. Ahn, S. G., Liu, P. C. C., Klyachko, K., Morimoto, R. I. and Thiele, D. J. (2001) The loop domain of heat shock transcription factor 1 dictates DNA-binding specificity and responses to heat stress. *Genes Dev.*, **15**, 2134–2145.

92. Peteranderl, R., Rabenstein, M. Shin, Y., Liu, C. W., Wemmer, D. E., King, D. S. and Nelson, H. C. M. (1999) Biochemical and biophysical characterization of the trimerization domain from the heat shock transcription factor. *Biochemistry*, **38**, 3559–3569.

93. Lyck, R., Harmening, U., Höhfeld, I., Treuter, E., Scharf, K.- D. and Nover, L. (1997) Intracellular distribution and identification of the nuclear localization signals of two plant heat-stress transcription factors. *Planta*, **202**, 117–125.

94. Scharf, K.-D., Heider, H., Höhfeld, I., Lyck, R., Schmidt, E. and Nover, L. (1998) The tomato Hsf system: HsfA2 needs interaction with HsfA1 for efficient nuclear import and may be localized in cytoplasmic heat stress granules. *Mol. Cell. Biol.*, **18**, 2240–2251.

95. Heerklotz, D., Döring, P., Bonzelius, F., Winkelhaus, S. and Nover, L. (2001) The balance of nuclear import and export determines the intracellular distribution and function of tomato heat stress transcription factor HsfA2. *Mol. Cell. Biol.*, **21**, 1759–1768.

96. Nover, L. and Scharf, K.-D. (1997) Heat stress proteins and transcription factors. *Cell. Mol. Life Sci.*, **53**, 80–103.

97. Döring, P., Treuter, E., Kistner, C., Lyck, R., Chen, A. and Nover, L. (2000) Role of AHA motifs for the activator function of tomato heat stress transcription factors HsfA1 and HsfA2. *Plant Cell*, **12**, 265–278.

98. Mason, P. B. and Lis, J. T. (1997) Cooperative and competitive protein interactions at the *hsp70* promoter. *J. Biol. Chem.*, **272**, 33227–33233.

99. Yuan, C. X. and Gurley, W. B. (2000) Potential targets for HSF1 within the preinitiation complex. *Cell Stress Chap.*, **5**, 229–242.

100. Scharf, K.-D., Rose, S., Zott, W., Schöffl, F. and Nover, L. (1990) Three tomato genes code for heat stress transcription factors with a region of remarkable homology to the DNA-binding domain of the yeast HSF. *EMBO J.*, **9**, 4495–4501.

101. Scharf, K.-D., Rose, S., Thierfelder, J. and Nover, L. (1993) Two cDNAs for tomato heat stress transcription factors. *Plant Physiol.*, **102**, 1355–1356.

102. Treuter, E., Nover, L., Ohme, K. and Scharf, K.-D. (1993) Promoter specificity and deletion analysis of three heat stress transcription factors of tomato. *Mol. Gen. Genet.*, **240**, 113–125.

103. Bharti, K., Schmidt, E., Lyck, R., Bublak, D. and Scharf, K.-D. (2000) Isolation and characterization of HsfA3, a new heat stress transcription factor of *Lycopersicon peruvianum*. *Plant. J.*, **22**, 355–365.

104. Hübel, A. and Schöffl, F. (1994) *Arabidopsis* heat shock factor: isolation and characterization of the gene and the recombinant protein. *Plant Mol. Biol.*, **26**, 353–362.

105. Prändl, R., Hinderhofer, K., Eggers-Schumacher, G. and Schöffl, F. (1998) Hsf3, a new heat shock factor from *Arabidopsis thaliana*, derepresses the heat shock response and confers thermotolerance when overexpressed in transgenic plants. *Mol. Gen. Genet.*, **258**, 269–278.

106. Lin, X., Kaul, S., Rounsley, S. D., Shea, T. P., Benito, M.- I., Town, C. D., Fujii, C. Y., Mason, T. M., Bowman, C. L., Barnstead, M. E., Feldblyum, T. V., Buell, C. R., Ketchum, K. A., Lee, J. J., Ronning, C. M., Koo, H., Moffat, K. S., Cronin, L. A., Shen, M., VanAken,

S. E., Umayam, L., Tallon, L. J., Gill, J. E., Adams, M. D., Carrera, A. J., Creasy, T. H., Goodman, H. M., Somerville, C. R., Copenhaver, G. P., Preuss, D., Nierman, W. C., White, O., Eisen, J. A., Salzberg, S. L., Fraser, C. M. and Venter, J. C. (1999) Sequence and analysis of chromosome 2 of the plant *Arabidopsis thaliana*. *Nature*, **402**, 761–768.

107. Czarnecka-Verner, E., Yuan, C. X., Fox, P. C. and Gurley, W. B. (1995) Isolation and characterization of six heat shock transcription factor cDNA clones from syobean. *Plant Mol. Biol.*, **29** , 37–51.

108. Aranda, M. A., Escaler, M., Thomas, C. L. and Maule, A. J. (1999) A heat shock transcription factor in pea is differentially controlled by heat and virus replication. *Plant J.*, **20**, 153–161.

109. Shoji, T., Kato, K., Sekine, M., Yoshida, K., Shinmyo, A. (2000) Two types of heat shock factors in cultured tobacco cells. *Plant Cell Reports*, **19**, 414–420.

110. Gagliardi, D., Breton, C., Chaboud, A., Vergne, P. and Dumas, C. (1995) Expression of heat shock factor and heat shock protein 70 genes during maize pollen development. *Plant Mol. Biol.*, **29** , 841–856.

111. Sato, S., Nakamura, Y., Kaneko, T., Katoh, T., Asamizu, E. and Tabata, S. (2000) Structural analysis of *Arabidopsis thaliana* chromosome 3. I. Sequence features of the regions of 4,504,864 bp covered by sixty P1 and TAC clones. *DNA Res.*, **7**, 131–135.

112. Sato, S., Nakamura, Y., Kaneko, T., Katoh, T., Asamizu, E., Kotani, H. and Tabata, S. (2000) Structural analysis of *Arabidopsis thaliana* chromosome 5. X. Sequence features of the regions of 3,076,755 bp covered by sixty P1 and TAC clones. *DNA Res.*, **7**, 31–63.

113. Kaneko, T., Katoh, T., Sato, S., Nakamura, A., Asamizu, E. and Tabata, S. (2000) Structural analysis of *Arabidopsis thaliana* chromosome 3. II. Sequence features of the 4,251,695 bp regions covered by 90 P1, TAC and BAC clones. *DNA Res.*, **7**, 217–221.

114. Scharf, K.-D., Höhfeld, I. and Nover, L. (1998) Heat stress response and heat stress transcription factors. *J. Biosci.*, **23**, 313–329.

115. Hübel, A., Lee, J. H., Wu, C. and Schöffl, F. (1995) *Arabidopsis* heat shock factor is constitutively active in *Drosophila* and human cells. *Mol. Gen. Genet.*, **248**, 136–141.

116. Lee, J. H., Hübel, A. and Schöffl, F. (1995) Derepression of the activity of genetically engineered heat shock factor causes constitutive synthesis of heat shock proteins and increased thermotolerance in transgenic *Arabidopsis*. *Plant J.*, **8**, 603–612.

117. Schöffl, F., Prändl, R. and Reindl, A. (1998) Regulation of the heat shock response. *Plant Physiol.*, **117**, 1135–1141.

118. Czarnecka-Verner, E., Yuan, C. X., Scharf, K.-D., Englich, G. and Gurley, W. B. (2000) Plants contain a novel multi-member class of heat shock factors without transcriptional activator potential. *Plant Mol. Biol.*, **43**, 459–471.

119. Zhong, M., Kim, S. J. and Wu, C. (1999) Sensitivity of *Drosophila* heat shock transcription factor to low pH. *J. Biol. Chem.*, **274**, 3135–3140.

120. Baler, R., Dahl, G. and Voellmy, R. (1993) Activation of human heat shock-genes is accompanied by oligomerisation, modification, and rapid translocation of heat shock transcription factor. *Mol. Cell. Biol.*, **13**, 2486–2496.

121. Rabindran, S. K., Haroun, R. I., Clos, J., Wisniewski, J. and Wu, C. (1993) Regulation of heat shock factor trimerization: Role of a conserved leucine zipper. *Science*, **259**, 230–234.

122. Zuo, J., Runger, D. and Voellmy, R. (1995) Multiple layers of regulation of human heat shock transcription factor 1. *Mol. Cell. Biol.*, **15**, 4319–4330.

123. Zandi, E., Tran, T. N. T., Chamberlain, W. and Parker, C. S. (1997) Nuclear entry, oligomerization, and DNA binding of the *Drosophila* heat shock transcription factor are regulated by a unique nuclear localization sequence. *Genes Dev.*, **11**, 1299–1314.

124. Farkas, T., Kutskova, Y. A. and Zimarino, V. (1998) Intramolecular repression of mouse heat shock factor 1. *Mol. Cell. Biol.*, **18**, 906–918.

125. Cotto, J. J., Kline, M. and Morimoto, R. I. (1996) Activation of heat shock factor 1 DNA-binding precedes stress-induced serine phosphorylation. *J. Biol. Chem.*, **271**, 3335–3358.

126. Winegarden, N. A., Wong, K. S., Sopta, M. and Westwood, J. T. (1996) Sodium salicylate decreases intracellular ATP, induces both heat shock factor binding and chromosomal puffing, but does not induce hsp 70 gene transcription in *Drosophila*. *J. Biol. Chem.*, **271**, 26971–26980.

127. Bharadwaj. S., Hnatov. A., Ali. A. and Ovsenek, N. (1998) Induction of the DNA-binding and transcriptional activities of heat shock factor 1 is uncoupled in *Xenopus* oocytes. *BBA Mol. Cell Res.*, **1402**, 79–85.

128. Sorger, P. K. and Pelham, H. R. B. (1988) Yeast heat shock factor is an essential DNA-binding protein that exhibits temperature-dependent phosphorylation. *Cell*, **54**, 855–864.

129. Wiederrecht, G., Seto, D. and Parker, C. S. (1988) Isolation of the gene encoding the *S. cerevisiae* heat shock transcription factor. *Cell*, **54**, 841–853.

130. Mercier, P. A., Winegarden, N. A. and Westwood, J. T. (1999) Human heat shock factor 1 is predominantly a nuclear protein before and after heat stress. *J. Cell. Sci.*, **112**, 2765–2774.

131. Jolly, C., Usson, Y. and Morimoto, R. I. (1999) Rapid and reversible relocalization of heat shock factor 1 within seconds to nuclear stress granules. *Proc. Natl Acad. Sci. USA*, **96**, 6769–6774.

132. Zuo, J., Baler, R., Dahl, G. and Voellmy, R. (1994) Activation of the DNA-binding ability of human heat shock transcription factor 1 may involve the transition from an intra-molecular to an intermolecular triple-stranded coiled-coil structure. *Mol. Cell. Biol.*, **14**, 7557–7568.

133. Shi, Y., Kroeger, P. E. and Morimoto, R. I. (1995) The carboxyl-terminal transactivation domain of heat shock factor 1 is negatively regulated and stress responsive. *Mol. Cell. Biol.*, **15**, 4309–4318.

134. Boscheinen, O., Lyck, R., Queitsch, C., Treuter, E., Zimarino, V. and Scharf, K.-D. (1997) Heat stress transcription factors from tomato can functionally replace HSF1 in the yeast *Saccharomyces cerevisiae*. *Mol. Gen. Genet.*, **255**, 322–331.

135. Chen, Y. Q., Barlev, N. A., Westergaard, O. and Jakobsen, B. K. (1993) Identification of the C-terminal activator domain in yeast heat shock factor - Independent control of transient and sustained transcriptional activity. *EMBO J.*, **12**, 5007–5018.

136. Liu, P. C. C. and Thiele, D. J. (1999) Modulation of human heat shock factor trimerization by the linker domain. *J. Biol. Chem.*, **274**, 17219–17225.

137. Chu, B., Zhong, R., Soncin, F., Stevenson, M. A. and Calderwood, S. K. (1998) Transcriptional activity of heat shock factor 1 at 37°C is repressed through phosphorylation on two distinct serine residues by glycogen synthase kinase 3 alpha and protein kinases Cα and Cζ. *J. Biol. Chem.*, **273**, 18640–18646.

138. Dai, R. J., Frejtag, W., He, B., Zhang, Y. and Mivechi, N. F. (2000) c-Jun NH2-terminal kinase targeting and phosphorylation of heat shock factor-1 suppress its transcriptional activity. *J. Biol. Chem.*, **275**, 18210–18218.

139. Baek, S. H., Lee, U. Y., Park, E. M., Han, M. Y., Lee, Y. S. and Park, Y. M. (2001). Role of protein kinase C delta in transmitting hypoxia signal to HSF and HIF-1. *J. Cell Physiol.*, **188**, 223–235.

140. Fritsch, N. and Wu, C. (1999) Phosphorylation of *Drosophila* heat shock transcription factor. *Cell Stress Chap.*, **4**, 102–117.

141. Ohnishi, K., Wang, X., Takahashi, A., Matsumoto, H. and Ohnishi, T. (1999) The protein kinase inhibitor, H-7, suppresses heat-induced activation of heat shock transcription factor 1. *Mol. Cell. Biochem.*, **197**, 129–135.

142. Park, J. H. and Liu, A. Y. C. (2001) JNK phosphorylates the HSF1 transcriptional activation domain: role of JNK in the regulation of the heat shock response. *J. Cell Biochem.*, **82**, 326–338.

143. Kline, M. P. and Morimoto, R. I. (1997) Repression of the heat shock factor 1 transcriptional activation domain is modulated by constitutive phosphorylation. *Mol. Cell. Biol.*, **17**, 2107–2115.

144. Xia, W. L., Guo, Y. L., Vilaboa, N., Zuo, J. R. and Voellmy, R. (1998) Transcriptional activation of heat shock factor HSF1 probed by phosphopeptide analysis of factor P-32-labeled in vivo. *J. Biol. Chem.*, **273**, 8749–8755.

145. Kim, D. and Li, G. C. (1999) Proteasome inhibitors lactacystin and MG132 inhibit the dephosphorylation of HSF1 after heat shock and suppress thermal induction of heat shock gene expression. *Biochem. Biophys. Res. Commun.*, **264**, 352–358.

146. Xavier, I. J., Mercier, P. A., McLoughlin, C. M., Ali, A., Woodgett, J. R. and Ovsenek, N. (2000) Glycogen synthase kinase 3 beta negatively regulates both DNA-binding and transcriptional activities of heat shock factor 1. *J. Biol. Chem.*, **275**, 29147–29152.

147. Rossi, A., Elia, G. and Santoro, M. G. (1998) Activation of the heat shock factor 1 by serine protease inhibitors. *J. Biol. Chem.*, **273**, 16446–16452.

148. Joyeux, M., Arnaud, C., Richard, M. J., Yellon, D. M., Demenge, P. and Ribuot, C. (2000) Effect of okadaic acid, a protein phosphatase inhibitor on heat stress-induced HSP72 synthesis and thermotolerance. *Cardiovasc. Drugs Ther.*, **14**, 441–446.

149. Liu, H. D. and Thiele, D. J. (1996) Oxidative stress induces heat shock factor phosphorylation and Hsf-dependent activation of yeast metallothionein gene transcription. *Genes Dev.*, **10**, 592–603.

150. Reindl, A., Schöffl, F., Schell, J., Koncz, C. and Bako, L. (1997) Phosphorylation by a cyclin-dependent kinase modulates DNA binding of the *Arabidopsis* heat-shock transcription factor HSF1 in vitro. *Plant Physiol.*, **115**, 93–100.

151. Kanei-Ishii, C., Tanikawa, J., Nakai, A., Morimoto, R. I. and Ishii, S. (1997) Activation of heat shock transcription factor 3 by c-Myb in the absence of cellular stress. *Science*, **277**, 246–248.

152. Lin, J. T. and Lis, J. T. (1999) Glycogen synthase phosphatase interacts with heat shock factor to activate CUP1 gene transcription in *Saccharomyces cerevisiae*. *Mol. Cell. Biol.*, **19**, 3237–3245.

153. Stephanou, A., Isenberg, D. A., Akira, S., Kishimoto, T. and Latchman, D. S. (1998) The nuclear factor interleukin-6 (NF-IL6) and signal transducer and activator of transcription-3 (STAT-3) signalling pathways co-operate to mediate the activation of the hsp90 beta gene by interleukin-6 but have opposite effects an its inducibility by heat shock. *Biochem. J.*, **330**, 189–195.

154. Stephanou, A., Isenberg, D. A., Nakajima, K. and Latchman, D. S. (1999) Signal transducer and activator of transcription-1 and heat shock factor-1 interact and activate the transcription of the Hsp-70 and Hsp-90 beta gene promoters. *J. Biol. Chem.*, **274**, 1723–1728.

155. Satyal, S. H., Chen, D., Fox, S. G., Kramer, J. M. and Morimoto, R. I. (1998) Negative regulation of the heat shock transcriptional response by HSBP1. *Genes Dev.*, **12**, 1962–1974.

156. Frejtag, W., Zhang, Y., Dai, R. J., Anderson, M. G. and Mivechi, N. F. (2001) Heat shock factor-4 (HSF-4a) represses basal transcription through interaction with TFIIF. *J. Biol. Chem.*, **276**, 14685–14694.

157. Nakai, A., Tanabe, M., Kawazoe, Y., Inazawa, J., Morimoto, R. I. and Nagata, K. (1997) HSF4, a new member of the human heat shock factor family which lacks properties of a transcriptional activator. *Mol. Cell. Biol.*, **17**, 469–481.

158. Tanabe, M., Sasai, N., Nagata, K., Liu, X. D., Liu, P. C. C., Thiele, D. J. and Nakai, A. (1999) The mammalian HSF4 gene generates both an activator and a repressor of heat shock genes by alternative splicing. *J. Biol. Chem.*, **274**, 27845–27856.

159. Goodson, M. L., Park-Sarge, O. and Sarge, K. D. (1995) Tissue-dependent expression of heat shock factor 2 isoforms with distinct transcriptional activities. *Mol. Cell. Biol.* **15**, 5288–5293.

160. Shi, Y. H., Mosser, D. D. and Morimoto, R. I. (1998) Molecular chaperones as HSF1-specific transcriptional repressors. *Genes Dev.*, **12**, 654–666.

161. Marchler, G. and Wu, C. (2001) Modulation of *Drosophila* heat shock transcription factor activity by the molecular chaperone DROJ1. *EMBO J.*, **20**, 499–509.

162. Bonner, J. J., Carlson, T., Fackenthal, D. L., Paddock, D., Storey, K. and Lea, K. (2000) Complex regulation of the yeast heat shock transcription factor. *Mol. Biol. Cell*, **11**, 1739–1751.

163. Bharadwaj, S., Ali, A. and Ovsenek, N. (1999) Multiple components of the HSP90 chaperone complex function in regulation of heat shock factor 1 in vivo. *Mol. Cell. Biol.*, **19**, 8033–8041.

164. Nair, S. C., Toran, E. J., Rimerman, R. A., Hyermstad, S., Smithgall, T. E. and Smith, D. F. (1996) A pathway of multi-chaperone interactions common to diverse regulatory proteins: estrogen receptor, Fes tyrosine kinase, heat shock transcription factor Hsf 1, and the aryl hydrocarbon receptor. *Cell Stress Chap.*, **1**, 237–250.

165. Ali, A, Bharadwaj, S., O'Carroll, R. and Ovsenek, N. (1998) HSP90 interacts with and regulates the activity of heat shock factor 1 in *Xenopus* oocytes. *Mol. Cell. Biol.*, **18**, 4949–4960.

166. Xiao, N. Q. and DeFranco, D. B. (1997) Overexpression of unliganded steroid receptors activates endogenous heat shock factor. *Mol. Endocrinol.*, **11**, 1365–1374.

167. Lee, J. H. and Schöffl, F. (1996) An hsp70 antisense gene affects the expression of Hsp70/Hsc70, the regulation of Hsf, and the acquisition of thermotolerance in transgenic *Arabidopsis thaliana*. *Mol. Gen. Genet.*, **252**, 11–19.

168. Craig, E. A. and Gross, C. A. (1991) Is hsp70 the cellular thermometer? *Trends Biochem. Sci.*, **16**, 135–140.

169. Bush, K. T., Goldberg, A. L. and Nigam, S. K. (1997) Proteasome inhibition leads to a heat-shock response, induction of endoplasmic reticulum chaperones, and thermotolerance. *J. Biol. Chem.*, **272**, 9086–9092.

170. Lee, D. H. and Goldberg, A. L. (1998) Proteasome inhibitors cause induction of heat shock proteins and trehalose, which together confer thermotolerance in *Saccharomyces cerevisiae*. *Mol. Cell. Biol.*, **18**, 30–38.

171. Mathew, A., Mathur, S. K. and Morimoto, R. I. (1998) Heat shock response and protein degradation: regulation of Hsf2 by the ubiquitin-proteasome pathway. *Mol. Cell. Biol.*, **18**, 5091–5098.

172. Kim, D., Kim, S. H. and Li, G. C. (1999) Proteasome inhibitors MG132 and lactacystin hyperphosphorylate HSF1 and induce hsp70 and hsp27 expression. *Biochem. Biophys. Res. Commun.*, **254**, 264–268.

173. Pirkkala, L., Alastalo, T. P., Zuo, X. X., Benjamin, I. J. and Sistonen, L. (2000). Disruption of heat shock factor 1 reveals an essential role in the ubiquitin proteolytic pathway. *Mol. Cell. Biol.*, **20**, 2670–2675.

174. McDuffee, A. T., Senisterra, G., Huntley, S., Lepock, J. R., Sekhar, K. R., Meredith, M. J., Borrelli, M. J., Morrow, J. D. and Freeman, M. L. (1997) Proteins containing non-native disulfide bonds generated by oxidative stress can act as signals for the induction of the heat shock response. *J. Cell. Physiol.*, **171**, 143–151.

175. Nishizawa, J., Nakai, A., Matsuda, K., Komeda, M., Ban, T. and Nagata, K. (1999) Reactive oxygen species play an important role in the activation of heat shock factor 1 in ischemic-reperfused heart. *Circulation*, **99**, 934–941.

176. Ito, H., Okamoto, K. and Kato, K. (1998) Enhancement of expression of stress proteins by agents that lower the levels of glutathione in cells. *BBA Gene Struct. Express.*, **1397**, 223–230.

177. Buchczyk, D. P., Briviba, K., Hartl, F. U. and Sies, H., (2000) Responses to peroxynitrite in yeast: glyceraldehyde-3-phosphate dehydrogenase (GAPDH) as a sensitive intracellular target for nitration and enhancement of chaperone expression and ubiquitination. *Biol. Chem.*, **381**, 121–126.

178. Santoro, M. G. (2000) Heat shock factors and the control of the stress response. *Biochem. Pharmacol.*, **59**, 55–63.

179. Garreau, H., Hasan, R. N., Renault, G., Estruch, F., Boy-Marcotte, E. and Jacquet, M. (2000) Hyperphosphorylation of Msn2p and Msn4p in response to heat shock and the diauxic shift is inhibited by cAMP in *Saccharomyces cerevisiae*. *Microbiology*, **146**, 2113–2120.

180. Raitt, D. C., Johnson, A. L., Erkine, A. M., Makino, K., Morgan, B., Gross, D. S. and Johnston, L. H. (2000) The Skn7 response regulator of *Saccharomyces cerevisiae* interacts with Hsf1 in vivo and is required for the induction of heat shock genes by oxidative stress. *Mol. Biol. Cell.*, **11**, 2335–2347.

181. Sugiyama, K., Izawa, S. and Inoue, Y. (2000). The Yap1p-dependent induction of glutathione synthesis in heat shock response of *Saccharomyces cerevisiae*. *J. Biol. Chem.* **275**, 15535–15540.

182. Diamond, D. A., Parsian, A., Hunt, C. R., Lofgren, S., Spitz, D. R., Goswami, P. C. and Gius, D. (1999) Redox factor-1 (Ref-1) mediates the activation of AP-1 in HeLa and NIH3T3 cells in response to heat shock. *J. Biol. Chem.*, **274**, 16959–16964.

183. Ohnishi, K. and Ohnishi, T. (2001) Heat-induced p53-dependent signal transduction and its role in hyperthermic cancer therapy. *Int. J. Hyperthermia*, **17**, 415–427.

184. Tanikawa, J., Ichikawa-Iwata, E., Kanei-Ishii, C., Nakai, A., Matsuzawa, S., Reed J. C. and Ishii, S. (2000) p53 suppresses the c-myb induced activation of heat shock transcription factor 3. *J. Biol. Chem.*, **275**, 15578–15585.

185. Kamano, H. and Klempnauer, K. H. (1997) B-Myb and cyclin D1 mediate heat shock element dependent activation of the human hsp70 promoter. *Oncogene*, **14**, 1223–1229.

186. Andrulis, E. D., Guzman, E., Döring, P., Werner, J. and Lis J. T. (2000). High resolution localization of *Drosophila* Spt5 and Spt6 at heat shock genes in vivo: roles in promoter proximal pausing and transcription elongation. *Genes Dev.*, **14**, 2535–2649.

187. Lee, D. and Lis, J. T. (1998) Transcriptional activation independent of TFIIH kinase and the RNA polymerase II mediator in vivo. *Nature*, **393**,389–392.

188. Dubois, M. F. and Bensaude, O. (1998) Phosphorylation of RNA polymerase II C-terminal domain (CTD): a new control for heat shock gene expression? *Cell Stress Chap.*, **3**, 147–51.

189. Bensaude, O., Bonnet, F., Casse, C., Dubois, M. F., Nguyen, V. T. and Palancade, B. (1999). Regulated phosphorylation of the RNA polymerase IIC-terminal domain (CTD). *Biochem. Cell Biol.*, **77**, 249–255.

190. Rodriguez, C. R., Cho, E. J., Keogh, M.C, Moore, C. L., Greenleaf, A. L. and Buratowski, S. (2000). Kin28, the TFIIH-associated carboxy-terminal domain kinase, facilitates the recruitment of mRNA processing machinery to RNA polymerase II. *Mol. Cell. Biol.,* **20**, 104–112.

191. Proudfoot, N. (2000). Connecting transcription to messenger RNA processing. *Trends Biochem. Sci.,* **25**, 290–293.

192. Ohkuni, K. and Yamashita, I. (2000). A transcriptional autoregulatory loop for KIN28-CCL1 and SRB10-SRB11, each encoding RNA polymerase IICTD kinase-cyclin pair, stimulates the meiotic development of S-cerevisiae. *Yeast* **16**, 829–846.

193. Akoulitchev, S., Chuikov, S. and Reinberg, D. (2000). TFIIH is negatively regulated by cdk8-containing mediator complexes. *Nature,* **407**, 102–106.

194. Napolitano, G., Majello, B., Licciardo, P., Giordano, A. and Lania, L. (2000). Transcriptional activity of P-TEFb kinase in vivo requires the C-terminal domain of RNA polymerase II. *Gene,* **254**, 139–145.

195. O'Keeffe, B., Fong, Y., Chen, D., Zhou, S. and Zhou, Q. (2000). Requirement for a kinase-specific chaperone pathway in the production of a Cdk9/cyclin T1 heterodimer responsible for P-TEFb-mediated tat stimulation of HIV-1 transcription. *J. Biol. Chem.,* **275**, 279–287.

196. Dubois, M. F., Marshall, N. F., Nguyen, V. T., Dahmus, G. K., Bonnet, F., Dahmus, M. E. and Bensaude, O. (1999). Heat shock of HeLa cells inactivates a nuclear protein phosphatase specific for dephosphorylation of the C-terminal domain of RNA polymerase II. *Nucleic Acids Res.,* **27**, 1338–1344.

197. Guilfoyle, T. J. (1989) A protein kinase from wheat germ that phosphorylates the largest subunit of RNA polymerase II. *Plant Cell,* **1**, 827–36.

198. Yamaguchi, M., Umeda, M. and Uchimiya, H. (1998). A rice homolog of Cdk7/MO15 phosphorylates both cyclin-dependent protein kinases and the carboxy-terminal domain of RNA polymerase II. *Plant J.,* **16**, 613–619.

199. Bonnet, F., Vigneron, M., Bensaude, O. and Dubois, M. F. (1999). Transcription-independent phosphorylation of the RNA polymerase IIC-terminal domain (CTD) involves ERK kinases (MEK1/2). *Nucleic Acids Res.,* **27**, 4399–4404.

200. Lavoie, S. B., Albert, A. L., Thibodeau, A. and Vincent, M. (1999). Heat shock-induced alterations in phosphorylation of the largest subunit of RNA polymerase II as revealed by monoclonal antibodies CC-3 and MPM-2. *Biochem. Cell Biol.,* **77**, 367–374.

201. Cooper, K. F., Mallory, M. J., Smith, J. B. and Strich, R. (1997) Stress and developmental regulation of the yeast C-type cyclin Ume3p (Srb11p/Ssn8p). *EMBO J.,* **16**, 4665–75.

202. Patturajan, M., Conrad, N. K., Bregman, D. B. and Corden, J. L. (1999). Yeast carboxyl-terminal domain kinase I positively and negatively regulates RNA polymerase II carboxyl-terminal domain phosphorylation. *J. Biol. Chem.,* **274**, 27823–27828.

203. Duncan, R. F. and Hershey, J. W. (1989) Protein synthesis and protein phosphorylation during heat stress, recovery, and adaptation. *J. Cell Biol.,* **109**, 1467–1481.

204. Vries, R. G. J., Flynn, A., Patel, J. C., Wang, X. M., Denton, R. M. and Proud, C. G. (1997) Heat shock increases the association of binding protein-1 with initiation factor 4E. *J. Biol. Chem.,* **272**, 32779–32784.

205. Holcik, M., Sonenberg, N. and Korneluk, R. G. (2000) Internal ribosome initiation of translation and the control of cell death. *Trends Genet.,* **16**, 469–473.

206. Dever, T. E. (1999) Translation initiation: adept at adapting. *Trends Biochem. Sci.,* **24**, 398–403.

207. Waskiewicz, A. J., Johnson, J. C., Penn, B., Mahalingam, M., Kimball, S. R. and Cooper, J. A. (1999) Phosphorylation of the cap-binding protein eukaryotic translation initiation factor 4E by protein kinase Mnk1 in vivo. *Mol. Cell. Biol.,* **19**, 1871–1880.

208. Scheper, G. C., Mulder, J., Kleijn, M., Voorma, H. O., Thomas, A. A. M. and van Wijk, R. (1997) Inactivation of eIF2B and phopshorylation of PHAS-I in heat-shocked rat hepatoma cells. *J. Biol. Chem.*, **272**, 26850–26856.

209. Sheikh, M. S. and Fornace, A. J. (1999) Regulation of translation initiation following stress. *Oncogene*, **18**, 6121–6128.

210. Uma, S., Hartson, S. D., Chen, J. J. and Matts, R. L. (1997) Hsp90 is obligatory for the heme-regulated eIF-2 alpha kinase to acquire and maintain an activable conformation. *J. Biol. Chem.*, **272**, 11648–11656.

211. Uma, S., Thulasiraman, V. and Matts, R. L. (1999) Dual role for Hsc70 in the biogenesis and regulation of the heme-regulated kinase of the a subunit of eukaryotic translation initiation factor 2. *Mol. Cell. Biol.*, **19**, 5861–5871.

212. Gallie, D. R., Caldwell, C. and Pitto, L. (1995) Heat shock disrupts cap and poly(A) tail function during translation and increases mRNA stability of introduced reporter mRNA. *Plant Physiol.*, **108**,1703–1713.

213. Gallie, D. R., Le, H., Caldwell, C., Tanguay, R. L., Hoang, N. X. and Browning, K. S. (1997) The phosphorylation state of translation initiation factors is regulated developmentally and following heat shock in wheat. *J. Biol. Chem.*, **272**, 1046–1053.

214. Nover, L., Scharf, K.-D. and Neumann, D. (1983) Formation of cytoplasmic heat shock granules in tomato cell cultures and leaves. *Mol. Cell. Biol.*, **3**, 1648–1655.

215. Nover, L., Scharf, K.-D. and Neumann, D. (1989) Cytoplasmic heat shock granules are formed from precursor particles and are associated with a specific set of mRNAs. *Mol. Cell. Biol.*, **9**, 1298–1308.

216. Neumann, D., zur Nieden, U., Manteuffel, R., Walter, G., Scharf, K.-D. and Nover, L. (1987) Intracellular localization of heat shock proteins in tomato cells cultures. *Eur. J. Cell. Biol.*, **43**, 71–81.

217. Neumann, D., Scharf, K.-D. and Nover, L. (1984) Heat shock induced changes of plant cell ultrastructure and autoradiographic localization of heat shock proteins. *Eur. J. Cell. Biol.*, **34**, 254–264.

218. Kirschner, M, Winkelhaus, S., Thierfelder, J. and Nover, L. (2000) Transient expression and heat stress induced aggregation of endogenous and heterologous small heat stress proteins in tobacco protoplasts. *Plant J.*, **24**, 397–412

219. Kedersha, N., Cho, M. R., Li, W., Yacono, P. W., Chen, S., Gilks, N., Golan, D. E. and Anderson, P. (2000) Dynamic shuttling of TIA-1 accompanies the recruitment of mRNA to mammalian stress granules. *J. Cell Biol.*, **151**, 1257–1268.

220. Kedersha, N. L., Gupta, M., Li, W., Miller, I. and Anderson, P. (1999) RNA-binding proteins TIA-1 and TIAR link the phosphorylation of eIF-2α to the assembly of mammalian stress granules. *J. Cell Biol.*, **147**, 1431–1442.

221. Cuesta, R., Laroia, G. and Schneider, R. J. (2000) Chaperone Hsp27 inhibits translation during heat shock by binding eIF4G and facilitating dissociation of cap-initiation complexes. *Genes Dev.*, **14**, 1460–1470.

222. Laroia, G., Cuesta, R., Brewer, G. and Schneider, R. J. (1999) Control of mRNA decay by heat shock-ubiquitin-proteasome pathway. *Science*, **284**, 499–502.

223. Gallouzi, I. E., Brennan, C. M. and Steitz, J. A. (2001) Protein ligands mediate the CRM1-dependent export of HuR in response to heat shock. *RNA-Publ. RNA Soc.*, **7**, 1348–1361.

224. Horton, L. E., James, P., Craig, E. A. and Hensold, J. O. (2001) The yeast hsp70 homologue Ssa is required for translation and interacts with Sis1 and Pab1 on translating ribosomes. *J. Biol. Chem.*, **276**, 14426–14433.

225. Hess, M. A. and Duncan, R. F. (1996) Sequence and structure determinants of *Drosophila* hsp70 mRNA translation – 5'-UTR secondary structure specifically inhibits heat shock protein mRNA translation. *Nucl. Acids Res.*, **24**, 2441–2449.

226. Yueh, A, and Schneider, R. J. (2000) Translation by ribosome shunting on adenovirus and hsp70 mRNAs facilitated by complementarity to 18S rRNA. *Genes Dev.*, **14**, 414–421.

227. Joshi, C. P. and Nguyen, H. T. (1995) 5' untranslated leader sequences of eukaryotic mRNAs encoding heat shock induced proteins. *Nucl. Acids Res.*, **23**, 541–549.

228. Pitto, L., Gallie, D. R. and Walbot, V. (1992) Role of the leader sequence during thermal repression of translation in maize, tobacco, and carrot protoplasts. *Plant Physiol.*, **100**, 1827–1833.

229. Ling, J., Wells, D. R., Tanguay, R. L., Dickey, L. F., Thompson, W. F. and Gallie, D. R. (2000) Heat shock protein HSP101 binds to the Fed-1 internal light regulatory element and mediates its high translational activity. *Plant Cell*, **12**, 1213–1227.

230. Brown, E. J. and Schreiber, S. L. (1996) A signaling pathway to translational control. *Cell*, **86**, 517–520.

231. Dufner, A. and Thomas, G. (1999) Ribosomal S6 kinase signaling and the control of translation. *Exp. Cell Res.*, **253**, 100–109.

232. Chou, M. M. and Blennis, J. (1995) The 70 kDa S6 kinase: regulation of a kinase with multiple roles in mitogenic signalling. *Curr. Opin. Cell. Biol.*, **7**, 806–814.

233. Pagano, M. (ed.) (1998) Cell cycle control. *Rearch Problems in Cell Differentiation*, Vol. 22. Springer, Berlin.

234. Kühl, N. M. and Rensing, L. (2000) Heat shock effects on cell cycle progression. *Cell. Mol. Life Sci.*, **57**, 450–463.

235. Joubes, J., Chevalier, C., Dudits, D., Heberle-Bors, E., Inze, D., Umeda, M. and Renaudin, J.-P. (2000) CDK-related protein kinases in plants. *Plant Mol. Biol.*, **43**, 607–621.

236. Meijer, M. and Murray, J. A. H. (2000) The role and regulation of D-type cyclins in the plant cell cycle. *Plant Mol. Biol.*, **43**, 623–635.

237. Stals, H., Casteels, P., van Montagu, M. and Inze, D. (2000) Regulation of cyclin-dependent kinases in *Arabidopsis thaliana*. *Plant Mol. Biol.*, **43**, 583–593.

238. Raboy, B., Marom, A., Dor, Y. and Kulka, R. G. (1999) Heat-induced cell cycle arrest of *Saccharomyces cerevisiae*: involvement of the RAD6/UBC2 and WSC2 genes in its reversal. *Mol. Microbiol.*, **32**, 729–739.

239. Trotter, E. W., Berenfeld, L., Krause, S. A., Petsko, G. A. and Gray, J. V. (2001) Protein misfolding and temperature up-shift cause G1 arrest via a common mechanism dependent on heat shock factor in *Saccharomyces cerevisiae*. *Proc. Natl Acad. Sci. USA*, **98**, 7313–7318.

240. Nakai, A. and Ishikawa, T. (2001) Cell cycle transition under stress conditions controlled by vertebrate heat shock factors. *EMBO J.*, **20**, 2885–2895.

241. Zarzov, P., Boucherie, H. and Mann, C. (1997) A yeast heat shock transcription factor (Hsf1) mutant is defective in both Hsc82/Hsp82 synthesis and spindle pole body duplication. *J. Cell Sci.*, **110**, 1879–1891.

242. Brown, C. R., Hong-Brown, L. Q., Doxsey, S. J. and Welch, W. J. (1996) Molecular chaperones and the centrosome. A role for HSP 73 in centrosomal repair following heat shock treatment. *J. Biol. Chem.*, **271**, 833–840.

243. Czar, M. J., Welsh, M. J. and Pratt, W. B. (1996) Immunofluorescence localization of the 90-kDa heat-shock protein to cytoskeleton. *Eur. J. Cell Biol.*, **70**, 322–330.

244. de Carcer, G., Avides, M. D., Lallena, M. J., Glover, D. M. and Gonzalez, C. (2001) Requirement of Hsp90 for centrosomal function reflects its regulation of Polo kinase stability. *EMBO J.*, **20**, 2878–2884.

245. Li, X. and Cai, M. (1999) Recovery of the yeast cell cycle from heat shock-induced G1 arrest involves a positive regulation of G1 cyclin expression by the S phase cyclin Clb5. *J. Biol. Chem.*, **274**, 24220–24231.

246. Rowley, A., Johnston, G. C., Butler, B., Werner-Washburne, M. and Singer, R. A. (1993) Heat shock-mediated cell cycle blockage and G1 cyclin expression in the yeast *Saccharomyces cerevisiae. Mol. Cell. Biol.*, **13**, 1034–1041.

247. Mai, B. and Breeden, L. (1997) Xbp1, a stress-induced transcriptional repressor of the *Saccharomyces cerevisiae* Swi/Mbp1 family. *Mol. Cell. Biol.*, **17**, 6491–6501.

248. Fuse, T., Yamada, K., Asai, K., Kato, T. and Nakanishi, M. (1996) Heat shock-mediated cell cycle arrest is accompanied by induction of p21 CKI. *Biochem. Biophys. Res. Commun.*, **225**, 759–763.

249. Miyata, Y., Ikawa, Y., Shibuya, M. and Nishida, E. (2001) Specific association of a set of molecular chaperones including HSP90 and Cdc37 with MOK, a member of the mitogen-activated protein kinase superfamily. *J. Biol. Chem.*, **276**, 21841–21848.

250. Morishima, Y., Kanelakis, K. C., Silverstein, A. M., Dittmar, K. D., Estrada, L. and Pratt, W. B. (2000) The Hsp organizer protein Hop enhances the rate of but is not essential for glucocorticoid receptor folding by the multiprotein Hsp90-based chaperone system. *J. Biol. Chem.*, **275**, 6894–6900.

251. Fisher, D. L., Mandart, E. and Doree, M. (2000) Hsp90 is required for c-Mos activation and biphasic MAP kinase activation in *Xenopus* oocytes. *EMBO J.*, **19**, 1516–1524.

252. Lewis, J., Devin, A., Miller, A., Lin, Y., Rodriguez, Y., Neckers, L. and Liu, Z. G. (2000) Disruption of Hsp90 function results in degradation of the death domain kinase, receptor-interacting protein (RIP), and blockage of tumor necrosis factor-induced nuclear factor-kappa B activation. *J. Biol. Chem.*, **275**, 10519–10526.

253. Helmbrecht, K., Zeise, E. and Rensing, L. (2000) Chaperones in cell cycle regulation and mitogenic signal transduction: a review. *Cell Prolif.*, **33**, 341–365.

254. Warner, J. R. (1999) The economics of ribosome biosynthesis in yeast. *Trends Biochem. Sci.*, **24**, 437–440.

255. Venema, J. and Tollervey, D. (1999) Ribosome synthesis in *Saccharomyces cerevisiae. Annu. Rev. Genet.*, **33**, 261–311.

256. Munsche, D., Nover, L., Ohme, K. and Scharf, K.-D. (1986) Ribosome biosynthesis in heat shocked tomato cell cultures I. Ribosomal RNA. *Eur J Biochem.*, **160**, 297–304.

257. Scharf, K.-D. and Nover, L. (1987) Control of ribosome biosynthesis in plant cell cultures under heat shock conditions II. Ribosomal proteins. *Biochem. Biophys. Acta*, **909**, 44–57.

258. Hadjiolov, A. A. (1985) The nucleolus and ribosome biogenesis. *Cell Biol Monographs*, Vol. 12. Springer, Wien.

259. Liu, Y., Liang, S. and Tartakoff, A. M. (1996) Heat shock disassembles the nucleolus and inhibits nuclear protein import and poly(a)(+) rna export. *EMBO J.*, **15**, 6750–6757.

260. Pederson, T. and Politz, J. C. (2000) The nucleolus and the four ribonucleoproteins of translation. *J. Cell Biol.*, **148**, 1091–1095.

261. Montane, M.-H., Tardy, F., Kloppstech, K., Havaux, M. (1998) Redox control of photo-synthetic acclimation to light irradiance: xanthophylls, light-harvesting chlorophyll a/b proteins and light-induced stress proteins in barley leaves grown in different light and gas environments. *Plant Physiol.*, **118**, 406–413.

262. Niyogi, K. K. (1999) Photoprotection revisited: genetic and molecular approaches *Annu. Rev. Plant Physiol. Plant Mol. Biol.*, **50**, 333–359.

263. Rintamäki, E., Kettunen, R. and Aro, E.-M. (1996) Differential D1 dephosphorylation in functional and photodamaged photosystem II centers. *J. Biol. Chem.*, **271**, 14870–14875.

264. Sharkey, T. D., Badger, M. R., von Caemmerer, S. and Andrews, T. J. (2001) Increased heat sensitivity of photosynthesis in tobacco plants with reduced Rubisco activase. *Photsynth. Res.*, **67**, 147–156.

265. Vani, B., Saradhi, P. P. and Mohanty, P. (2001) Alteration in chloroplast structure and thylakoid membrane composition due to *in vivo* heat treatment of rice seedlings: correlation with the functional changes. *J. Plant Physiol.*, **158**, 583–592.

266. Lu, C. M. and Zhang, J. H. (2000) Heat-induced multiple effects on PSII in wheat plants. *J. Plant Physiol.*, **156**, 259–265.

267. Rokka, A., Aro, E. M., Herrmann, R. G. andersson, B. and Vener, A. V. (2000) Dephosphorylation of photosystem II reaction center proteins in plant photosynthetic membranes as an immediate response to abrupt elevation of temperature. *Plant Physiol.*, **123**, 1525–1535.

268. Kumar, S. (ed.) (1999) Apoptosis: biology and mechanisms. *Research Problems in Cell Differentiation*, Vol. 23. Springer, Berlin.

269. Bratton, S. B. and Cohen, G. M. (2001) Apoptotic death sensor: an organelle's alter ego? *Trends Pharmacol. Sci.*, **22**, 306–315.

270. Adrain, C. and Martin, S. J. (2001) The mitochondrial apoptosome: a killer unleashed by the cytochrome seas. *Trends Biochem. Sci.*, **26**, 390–397.

271. Cryns, V. and Yuan, J. (1998) Proteases to die for. *Genes Dev.*, **12**, 1551–1570.

272. Gilchrist, D. G. (1998) Programmed cell death in plant disease: the purpose and promise of cellular suicide. *Annu. Rev. Phytopathol.*, **36**, 393–414.

273. Dangl. J. L., Dietrich, R. A. and Thomas, H. (2000) Senescence and programmed cell death. In Buchanan, B. B., Gruissem, W. and Jones, R. L. (eds.) Biochemistry & Molecular Biology of Plants. *Amer. Soc. Plant Physiologists*, Rockville, MD.

274. Lam, E. and del Pozo, O (2000) Caspase-like protease involvement in the control of plant cell death. *Plant Mol. Biol.*, **44**, 417–428.

275. Samali, A. and Orrenius, S. (1998) Heat shock proteins: regulators of stress response and apoptosis. *Cell Stress Chap.*, **3**, 228–236.

276. Bossy-Wetzel, E., Bakiri, L. and Yaniv, M. (1997) Induction of apoptosis by the transcription factor c-Jun. *EMBO J.*, **16**, 1695–1709.

277. Soldatenko, V. A. and Dritschilo, A. (1997) Apoptosis of Ewing's sarcoma cells is accompanied by accumulation of ubiquitinated proteins. *Cancer Res.*, **57**, 3881–3885.

278. Meriin, A. B., Gabai, V. L., Yaglom, J., Shifrin, V. I. and Sherman, M. Y. (1998) Proteasome inhibitors activate stress kinases and induce Hsp72 – diverse effects on apoptosis. *J. Biol. Chem.*, **273**, 6373–6379.

279. Kondo, T., Matsuda, T., Tashima, M., Umehara, H., Domae, N., Yokoyama, K., Uchiyama, T. and Okazaki, T. (2000) Suppression of heat shock protein-70 by ceramide in heat shock-induced HL-60 cell apoptosis. *J. Biol. Chem.*, **275**, 8872–8879.

280. Okazaki, T., Kondo, T., Kitano, T. and Tashima, M. (1998) Diversity and complexity of ceramide signalling in apoptosis. *Cell Signal.*, **10**, 685–692.

281. Kondo, T., Matsuda, T., Kitano,T., Takahashi,A., Tashima, M., Ishikura, H., Umehara, H., Domae, N., Uchiyama, T. and Okazaki, T. (2000) Role of c-jun expression increased by heat shock- and ceramide-activated caspase-3 in HL-60 cell apoptosis. *J. Biol. Chem.*, **275**, 7668–7676.

282. Gabai, V. L., Yaglom, J. A., Volloch, V., Meriin, A. B., Force, T., Koutroumanis, M., Massie, B., Mosser, D. D. and Sherman, M. Y. (2000) Hsp72-mediated suppression of c-Jun N-terminal kinase is implicated in development of tolerance to caspase-independent cell death. *Mol. Cell. Biol.*, **20**, 6826–6836.

283. Bijur, G. N., De Sarno, P. and Jope, R. S. (2000) Glycogen synthase kinase-3 beta facilitates staurosporine- and heat shock-induced apoptosis – protection by lithium. *J. Biol. Chem.*, **275**, 7583–7590.

284. Tournier, C., Hess, P., Yang, D. D., Xu, J., Turner, T. K., Nimnual, A., Bar-Sagi, D., Jones, S. N., Flavell, R. A. and Davis, R. J. (2000) Requirement of JNK for stress-induced activation of the cytochrome c-mediated death pathway. *Science*, **288**, 870–874.

285. Schett, G., Steiner, C. W., Groger, M., Winkler, S., Graninger, W., Smolen, J., Xu, Q. B. and Steiner, G. (1999) Activation of Fas inhibits heat-induced activation of HSF1 and up-regulation of hsp70. *FASEB J.*, **13**, 833–842.

286. Wang, G. H., Huang, H. G., Dai, R. J., Lee, K. Y., Lin, S. and Mivechi, N. F. (2001) Suppression of heat shock transcription factor HSF1 in zebrafish causes heat-induced apoptosis. *Genesis*, **30**, 195–197.

287. Beere, H. M. and Green, D. R. (2001) Stress management – heat shock protein-70 and the regulation of apoptosis. *Trends Cell Biol.*, **11**, 6–10.

288. Jolly, C. and Morimoto, R. I. (2000) Role of the heat shock response and molecular chaperones in oncogenesis and cell death. *J. Natl Cancer Inst.*, **92**, 1564–1572.

289. Garrido, C., Gurbaxani, S., Ravagnan, L. and Kroemer, G. (2001) Heat shock proteins: Endogenous modulators of apoptotic cell death. *Biochem. Biophys. Res. Commun.*, **286**, 433–422.

290. Sato, S., Fujita, N. and Tsuruo, T. (2000) Modulation of Akt kinase activity by binding to Hsp90. *Proc. Natl Acad. Sci. USA*, **97**, 10832–10837.

291. Bruey, J. M., Ducasse, C., Bonniaud, P., Ravagnan, L., Susin, S. A., Diaz-Latoud, C., Gurbaxani, S., Arrigo, A. P., Kroemer, G., Solary, E. and Garrido, C. (2000) Hsp27 negatively regulates cell death by interacting with cytochrome c. *Nat. Cell Biol.*, **2**, 645–652.

292. Pandey, P., Farber, R., Nakazawa, A., Kumar, S., Bharti, A., Nalin, C., Weichselbaum, R., Kufe, D. and Kharbanda, S. (2000) Hsp27 functions as a negative regulator of cytochrome c-dependent activation of procaspase-3. *Oncogene*, **19**, 1975–1981.

293. Samali, A., Robertson, J. D., Peterson, E., Manero, F., van Zeijl, L., Paul, C., Cotgreave, I. A., Arrigo, A.-P. and Orrenius, S. (2001) Hsp27 protects mitochondria of thermotolerant cells against apoptotic stimuli. *Cell Stress Chap.*, **6**, 49–58.

294. Beere, H. M., Wolf, B. B., Cain, K., Mosser, D. D., Mahboubi, A., Kuwana, T., Tailor, P., Morimoto, R. I., Cohen, G. M. and Green, D. R. (2000) Heat-shock protein 70 inhibits apoptosis by preventing recruitment of procaspase-9 to the Apaf-1 apoptosome. *Nat. Cell Biol.*, **2**, 469–475.

295. Saleh, A., Srinivasula, S. M., Balkir, L., Robbins, P. D. and Alnemri, E. S. (2000) Negative regulation of the Apaf-1 apoptosome by Hsp70. *Nat. Cell Biol.*, **2**, 476–483.

296. Li, C. Y., Lee, J. S., Ko, Y. G., Kim, J. I. and Seo, J. S. (2000) Heat shock protein 70 inhibits apoptosis downstream of cytochrome c release and upstream of caspase-3 activation. *J. Biol. Chem.*, **275**, 25665–25671.

297. McMillan, D. R., Xiao, X. Z., Shao, L., Graves, K. and Benjamin, I. J. (1998). Targeted disruption of heat shock transcription factor 1 abolishes thermotolerance and protection against heat-inducible apoptosis. *J. Biol. Chem.*, **273**, 7523–7528.

298. Nylandsted, J., Rohde, M., Brand, K., Bastholm, L., Elling, F. and Jäättelä, M. (2000) Selective depletion of heat shock protein 70 (Hsp70) activates a tumor-specific death program that is independent of caspases and bypasses Bcl-2. *Proc. Natl Acad. Sci. USA*, **97**, 7871–7876.

299. Nishimura, R. N., Santos, D., Esmaili, L., Fu, S. T. and Dwyer, B. E. (2000) Expression of antisense hsp70 is a major determining factor in heat-induced cell death of P-19 carcinoma cells. *Cell Stress Chap.*, **5**, 173–180.

300. Thress, K., Song, J. W., Morimoto, R. I. and Kornbluth, S. (2001) Reversible inhibition of Hsp70 chaperone function by Scythe and Reaper. *EMBO J.*, **20**, 1033–1041.
301. Ravagnan, L., Gurbaxani, S., Susin, S. A., Maisse, C., Daugas, E., Zamzami, N., Mak, T., Jaattela, M., Penninger, J. M., Garrido, C. and Kroemer, G. (2001) Heat-shock protein 70 antagonizes apoptosis-inducing factor. *Nature Cell Biol.*, **3**, 839–843.
302. Creagh, E. M., Carmody, R. J. and Cotter, T. G. (2000) Heat shock protein 70 inhibits caspase-dependent and -independent apoptosis in Jurkat T cells. *Exp. Cell Res.*, **257**, 58–66.
303. Xia, W. L., Voellmy, R. and Spector, N. L. (2000) Sensitization of tumor cells to Fas killing through overexpression of heat-shock transcription factor 1. *J. Cell Physiol.*, **183**, 425–431.
304. Mosser, D. D., Caron, A. W., Bourget, L., Meriin, A. B., Sherman, M. Y., Morimoto, R. I. and Massie, B. (2000) The chaperone function of hsp70 is required for protection against stress-induced apoptosis. *Mol. Cell. Biol.*, **20**, 7146–7159.
305. Daniel, X., Lacomme, C., Morel, J. B. and Roby, D. (1999) A novel myb oncogene homologue in *Arabidopsis thaliana* related to hypersensitive cell death. *Plant J.*, **20**, 57–66.
306. Che, F. S., Iwano, M., Tanaka, N., Takayama, S., Minami, E., Shibuya, N., Kadota, I. and Isogai, A. (1999) Biochemical and morphological features of rice cell death induced by *Pseudomonas avenae*. *Plant Cell. Physiol.*, **40**, 1036–1045.
307. Maccarrone, M., Van Zadelhoff, G., Veldink, G. A., Vliegenthart, J. F. G. and Finazzi-Agro, A. (2000) Early activation of lipoxygenase in lentil (*Lens culinaris*) root protoplasts by oxidative stress induces programmed cell death. *Eur. J. Biochem.*, **267**, 5078–5084.
308. Amor, Y., Babiychuk, E., Inze, D. and Levine, A. (1998) The involvement of poly(ADP-ribose) polymerase in the oxidative stress responses in plants. *FEBS Lett.*, **440**, 1–7.
309. Balk, J., Leaver, C. J. and McCabe, P. F. (1999) Translocation of cytochrome c from the mitochondria to the cytosol occurs during heat-induced programmed cell death in cucumber plants. *FEBS Lett.*, **463**, 151–154.
310. Chen, H. M., Zhou, J. and Dai, Y. R. (2000) Cleavage of lamin-like proteins in in vivo and in vitro apoptosis of tobacco protoplasts induced by heat shock. *FEBS Lett.*, **480**, 165–168.
311. Tian, R. H., Zhang, G. Y., Yan, C. H. and Dai, Y. R. (2000) Involvement of poly(ADP-ribose) polymerase and activation of caspase-3-like protease in heat shock-induced apoptosis in tobacco suspension cells. *FEBS Lett.*, **474**, 11–15.
312. Sun, Y. L., Zhao, Y., Hong, X. and Zhai, Z. H. (1999) Cytochrome c release and caspase activation during menadione-induced apoptosis in plants. *FEBS Lett.*, **462**, 317–321.
313. Korthout, HAAJ,, Berecki, G., Bruin, W., van Duijn, B. and Wang, M. (2000) The presence and subcellular localization of caspase 3-like proteinases in plant cells. *FEBS Lett.*, **475**, 139–144.
314. Delorme, V. G. R., McCabe, P. F., Kim, D. J. and Leaver, C. J. (2000) A matrix metallo-proteinase gene is expressed at the boundary of senescence and programmed cell death in cucumber. *Plant Physiol.*, **123**, 917–927.
315. Schmid, M., Simpson, D. and Gietl, C. (1999) Programmed cell death in castor bean endosperm is associated with the accumulation and release of a cysteine endopeptidase from ricinosomes. *Proc. Natl Acad. Sci. USA*, **96**, 14159–14164.
316. Sanchez, P., Zabala, M. D. and Grant, M. (2000) AtBI-1, a plant homologue of Bax Inhibitor-1, suppresses Bax-induced cell death in yeast and is rapidly upregulated during wounding and pathogen challenge. *Plant J.*, **21**, 393–399.

6 | Molecular mechanisms of signal transduction in cold acclimation

JULIO SALINAS

1. Introduction

Plants differ from other living organisms in their inability to move. As a consequence, plant development depends on a continuous adaptation to external, often unfavourable, environmental conditions. Low temperatures, including freezing, are one of the most common environmental conditions that have a major impact on plant survival (1, 2). Plant species exhibit extensive diversity in their response to low temperature. At one extreme are plants from tropical and subtropical regions which, in general, are injured when exposed to low non-freezing temperatures. By contrast, plants from temperate regions have developed natural adaptive mechanisms to tolerate low and freezing temperatures. Central to this adaptation is the process of cold acclimation, by which plants are able to adjust their metabolism to cold and increase their freezing tolerance in response to low non-freezing temperatures (1). Cold acclimation is a very complex process and involves a number of physiological and biochemical changes, including alterations in leaf cell structure (3), membrane composition (4, 5), protein composition (6), enzymatic activities, and the accumulation of sugars and polyamines (1, 7). It is widely accepted that most of these alterations are regulated by low temperatures through changes in gene expression and, in recent years, many cold-inducible genes have been isolated from a range of species. Although some of them have been shown to be involved in freezing tolerance (8–11), the relative importance of these genes in cold-acclimation is, in general, not well understood. Determining the molecular mechanisms that control the cold-acclimation response is of basic scientific interest and has, in addition, potential practical applications. Indeed, understanding the molecular basis of this response would not only increase our fundamental knowledge on how plants adapt to changes in the environment, but also contribute to develop new strategies to improve the tolerance of crop species to low and freezing temperatures, resulting in increased productivity and expanded areas of agricultural production (12).

In general, plant responses to endogenous physiological factors or environmental cues consist of three main events:

- perception of the stimulus;
- generation and transmission of signals; and
- subsequent changes in downstream biochemical processes.

In the case of cold acclimation, it can be assumed that a temperature transducer responds to low temperature by activating a signalling pathway which, in turn, institutes the biochemical and gene expression events needed for freezing tolerance. It is intriguing that, in spite of the high number of cold-induced genes that have been identified, how plants sense low temperatures and how the cold signal is transduced to activate the cold-aclimation response still remains far from complete comprehension. This chapter summarizes the current understanding on the signal transduction in cold acclimation and discusses recent developments regarding sensing and regulatory mechanisms involved in low-temperature signalling. Additional information about these and other aspects of cold acclimation can be found in recent revisions (13–16).

2. The complexity of the cold-acclimation response

Although signalling cascades are often viewed as linear chains of events, it is becoming increasingly apparent that plant responses to environmental factors involve extensive cross-talk between different pathways. In this regard, it is thus not appropriate to consider the cold-acclimation response as the result of a simple, linear signalling pathway activating the full set of processes required for increasing freezing tolerance. Instead, it should be contemplated as an event in which parallel and branched signalling pathways activate distinct sets of responses. This complexity is, perhaps, a consequence of the elevated number of deleterious effects that low and freezing temperatures produce in plant cells, which must be overcome during the process of cold acclimation to ensure plant survival. Among them, those compromising the integrity of cell membranes are the most noxious. Temperatures in a 0–10°C range provoke a decline in membrane fluidity, mainly due to a decrease in the level of fatty acid unsaturation in membrane lipids (17). Plants that have been exposed to low temperatures also show changes in lipid composition and in the ratio of lipid to protein in cell membranes (17). In the case of freezing temperatures, the freeze-induced membrane damage results primarily from the severe cellular dehydration associated with ice formation (18, 19). In fact, when temperatures drop below freezing, ice is initially formed extracellularly because the solute concentration is lower in the apoplast than in the intracellular fluid. A first consequence of ice accumulation in the intercellular spaces is the physical disruption of tissues and cells (20). Moreover, since the chemical potential of ice is lower than that of water, there is a movement of water from inside to outside the cells. Depending on the freezing temperatures and the corresponding degree of cellular dehydration, different forms of membrane damage can arise including expansion-induced-lysis, lamellar-to-hexagonal-II phase transition, and fracture jump lesions (21, 22).

Low non-freezing temperatures also cause dehydration, mainly due to a reduced water uptake in roots (23) and an impediment to close stomata (24) and protein denaturation (25). Similarly, freezing temperatures provoke protein denaturation and precipitation of solutes as a consequence of freeze-induced dehydration (1, 26). Moreover, low and freezing temperatures result in oxidative stress due to the generation of reactive oxygen species such as superoxide ($O_2^{\bullet-}$), hydrogen peroxide (H_2O_2), and hydroxyl radicals ($OH^{\bullet}$) (27–29). These reactive oxygen species emanate by disequilibrium in electron transfer reactions as light-harvesting reactions continue to function under these environmental conditions, while the accompanying biochemical reactions are severely restricted. The generation of reactive oxygen species leads to cellular injury and, ultimately, to death through damage to the photosystem II reaction centre and to membrane lipids (30).

In summary, when exposed to low and/or freezing temperatures, plants are confronted to a plethora of pernicious circumstances. Some plants are able to surmount this dangerous situation by triggering the cold-acclimation response, which is switched on by a sophisticated intracellular signalling network issued from the interaction among different, although related, pathways.

3. Sensing low temperatures

The first step in the signalling cascade leading to the cold-acclimation response is the recognition of low temperatures. Although it is clear that all organisms must be able to sense specific changes in temperatures, little is known about the sensors that detect cold in plant cells. Several possibilities have been suggested by different authors, probably reflecting the variety of effects that low temperatures have on plants.

Murata and Los (31) emphasized the role of membrane fluidity as the primary signal upon a change in surrounding temperature. They speculated that a physical phase transition occurs in microdomains of the plasma membrane upon a downward shift in temperature. The putative sensor for perception of such an alteration would be a membrane protein able to detect the conformational change in such microdomains, i.e. the transition of the physical phase from the liquid-crystalline to the gel state (31). These authors proposed that the hypothetical sensor might resemble a histidine kinase, as is the case of the osmolarity sensor identified in yeast (32). Several plant genes encoding plant histidine kinases have been cloned and shown to be involved in the perception of different environmental signals (33). Interestingly, one of these genes, *ATHK1*, is regulated by low temperature and high salinity in *Arabidopsis* (34). The ATHK1 protein, which is able to function as an osmosensor in yeast, contains both a transmitter and a receiver domain with conserved histidine and aspartate residues, respectively, as well as two hydrophobic transmembrane regions in the N-terminal half (34).

Monroy and Dhindsa (35) have proposed that in higher plants the membrane protein that senses the cold-induced shift in membrane fluidity could be a calcium-permeable channel. According to this hypothesis, the calcium channel would open at low temperatures upon a decrease in membrane fluidity and the entering calcium would activate the subsequent signal transduction pathways. Similarly, Minorsky (36) proposed that cold-induced increases in cytosolic free calcium are the primary event in sensing low temperatures. The hypothesis of a calcium channel as primary sensor of temperature decreases is supported by the existence of mechanosensitive calcium-selective cation channels in the plasmalemma that are activated in response to low temperatures (37). Although it has been proposed that either membrane or channel properties might be behind the mechanism by which the activity of these channels is cold induced (37), the exact nature of this mechanism remains to be determined.

In addition to its traditional role in energy transduction, the photosynthetic apparatus has also been proposed as an environmental sensor (38). In fact, changes in environmental conditions, such as temperature, results in an imbalance between the light energy absorbed through photochemistry versus the energy utilized through metabolism. This energy imbalance is sensed through alterations in the excitation pressure of photosystem II, which reflects the relative reduction state of the photosystem (38). Interestingly, the potential for an energy imbalance between photochemistry, electron transport and metabolism is exacerbated under cold conditions, which leads to an increased photosystem II excitation pressure (38). Thus, the redox state of the photosynthetic apparatus might act as a low temperature sensor by detecting imbalances between photochemical and biochemical reactions. Modulation of this chloroplastic redox-sensing mechanism would initiate a signalling cascade whereby the chloroplast would affect nuclear cold-regulated gene expression (39), influencing, thus, the cold-acclimation response.

It has already been mentioned that low temperatures result in an oxidative burst due to the generation of active oxygen species (27–29). A further possibility when considering mechanisms for sensing cold conditions may be the existence of sensors for the superoxides produced. In addition, changes in the physical tension of cytoskeleton produced as a result of the water stress imposed by low temperatures might be another trigger of the cold-acclimation response. The mechanisms by which plant cells would be able to sense the oxidative burst or the mechanical changes that take place in response to cold still remain unknown.

Although direct experimental evidence for low-temperature receptors are still lacking, the existence of different mechanisms that perceive cold simultaneously is a possibility that can not be ruled out. Indeed, when contemplating the enormous complexity of cold acclimation, it is difficult to conceive that low-temperature responses are triggered just through one sensing mechanism. Rather, it is reasonable to devise a scenario where low temperatures are sensed by a number of mechanisms that coordinately trigger different signal transduction pathways mediating cold acclimation.

4. Transducing the cold signal

Following cellular perception of low temperatures, signalling mechanisms must be activated to induce subsequent changes in downstream biochemical processes that contribute to freezing tolerance. As alluded to earlier, multiple signal transduction pathways are perhaps triggered in response to low temperatures to promote cold acclimation. In order to identify components that define these pathways, different experimental approaches have been pursued from which significant insights have begun to emerge.

4.1 Biochemical analysis of signal transduction in response to low temperature

Research carried out during the last years based on biochemical approaches has allowed the identifation of several intermediates in low-temperature signalling. Consistent with the suggestion of Monroy and Dhindsa (35) and Minorsky (36) that a calcium channel could be a primary cold sensor (see above), it has been shown that calcium is an important second messenger in a low-temperature signal transduction pathway involved in regulating the cold-acclimation response. By using transgenic plants of *Arabidopsis* and tobacco expressing the calcium-sensitive luminiscent protein aequorin, it was demonstrated that plants respond to low temperature by a large transient rise in cytosolic calcium concentration (40–42). Careful titration with different calcium chelators and calcium channel blockers led to the conclusion that this rise involves both an external influx through the plasma membrane and a release from internal stores (i.e. vacuoles), which seems to be regulated via inositol trisphosphate signalling (43). An increase in cytosolic calcium concentration in response to low temperature has also been reported in alfalfa protoplasts by monitoring $^{45}Ca^{2+}$ (35). Chelators and channel blockers were also used to show that this calcium influx, as in the case of tobacco and *Arabidopsis*, has two different sources: the cell wall, which is the main one, and intracellular (vacuolar) stores (35).

More significant evidence for the role of calcium as a second messenger in low-temperature signalling comes from the observation that the calcium chelator EGTA and the potent calcium-channel blockers lanthanum and verapamil affect the development of cold acclimation (44). Whereas EGTA inhibited the increased freezing tolerance of cold-acclimated alfalfa cultured cells by more than 70%, lanthanum and verapamil completely abolished that increase (44). Similar experiments carried out with *Arabidopsis* seedlings yielded nearly identical results (45). Interestingly, the inhibition of cold acclimation caused by calcium chelators and channel blockers is paralleled by an inhibition of cold-induced gene expression. Thus, the cold-dependent accumulation of transcripts corresponding to *CAS15* and *CAS18*, two low-temperature-regulated genes from alfalfa, was much reduced in the presence of the calcium chelator BAPTA, lanthanum or verapamil (35). In *Arabidopsis*, lanthanum and EGTA also caused a strong inhibition of cold-dependent expression of *KIN1* and *KIN2* (45), two genes that are coordinately regulated by low temperature (46). Furthermore, a positive correlation between cold-induced calcium influx and accumulation of cold-induced transcripts has been described (35). *CAS15* and *CAS18* transcripts accumulated at 25°C in protoplasts from alfalfa cell-suspension cultures if a calcium influx was experimentally provoked by treatment with the ionophore A23187 or with the calcium channel agonist Bay K8644 (35). All these results indicate that calcium plays an essential role in a low-temperature response leading to cold-regulated gene expression and cold acclimation. However, contrary to what is observed in low-temperature induction of *CAS* genes, *CAS* gene induction by A23187 and Bay K8644 treatment at 25°C declined with time, suggesting a role for as yet undetermined low-temperature-promoted processes in sustaining the calcium-induced *CAS* gene expression (35). Such processes might include other low-temperature, calcium-independent signalling pathways leading to either the inhibition of RNAses or the expression of mRNA-stabilizing proteins.

To elicit responses, calcium transients are, in general, efficiently linked, either directly via calcium-dependent protein kinases (CDPKs) (47) or indirectly via calmodulin (CaM) (48), to a protein phosphorylation signalling cascade. In order to investigate the role of CDPKs and protein phosphorylation in cold acclimation, Monroy and collaborators (44) studied the effect of W7, an antagonist of CDPKs and CaM, and H7, a protein kinase inhibitor, on alfalfa cell-suspension cultures. It was seen that W7 completely inhibited the capacity of the cultures to cold-acclimate, while H7 inhibited the development of cold acclimation by nearly 50%. Very similar results were also described in *Arabidopsis* (45). More insights on the role of protein phosphorylation in cold-acclimation response were obtained by directly analysing low-temperature-induced changes in the protein phosphorylation profiles of alfalfa cell-suspension cultures. The results obtained showed extensive changes in protein phosphorylation levels, increasing in several proteins whilst decreasing in others (44). Inhibitors of the cold-acclimation response, such as W7 and lanthanum, also prevented the low-temperature-induced changes in phosphoprotein profiles (44). Moreover, H7, W7, and lanthanum were also able to inhibit the cold-induced accumulation of low-temperature-regulated genes in both alfalfa cells and *Arabidopsis* plantlets (44, 45). All these data suggest that calcium-dependent protein phosphorylation may participate in a signal transduction pathway triggering the cold-acclimation response. In fact, calcium-binding proteins could couple the calcium response to downstream events, including cold-induced gene expression, leading to metabolic adaptation to cold and to the development of freezing tolerance. Interestingly, the phosphoprotein profile of alfalfa cells treated with abscisic acid (ABA), a plant hormone involved in stress response and cold acclimation (see below), was nearly identical to that of non-acclimated cells (44) which indicates that ABA should induced cold acclimation through separate signal transduction pathways.

In addition to CDPKs, a role for protein phosphatases as primary sensors of calcium signal in the cold-acclimation response is also emerging. By using the expression of *CAS15* gene as an end-point marker to study the role of protein phosphorylation in low-temperature signal transduction, Monroy *et al.* (49) first showed that cold induction of *CAS15* was prevented by the protein kinase inhibitor staurosporine. Conversely, okadaic acid, a specific inhibitor of protein phosphatases PP2A and PP1, induced *CAS15* expression at 25°C (49). This indicates that the activity of a staurosporine-sensitive protein kinase is required in a molecular switch for low-temperature signal transduction leading to *CAS15* expression, and that cold-mediated down-regulation of protein phosphatases may play an important role in low-temperature signalling. Cold-induced calcium influx was little affected by either staurosporine or okadaic acid (49), suggesting that both protein kinases and phosphatases act downstream of the calcium signal. Moreover, the same authors (49) showed that low temperatures provoked a rapid and important decrease in PP2A activity, and that this inhibition was mediated by cold-induced calcium influx and occurred at a post-transcriptional level. These observations indicate that PP2A is subject to cold inactivation. Additional evidence involving protein phosphatases in low-temperature signalling comes from the experiments carried out by Vazquez-Tello *et al.* (50). Aiming to understand the molecular mechanisms controlling the cold regulation of the wheat *WCS120* gene, these authors performed electrophoretic mobility shift assays with short promoter fragments and nuclear protein extracts from cold-acclimated and non-acclimated wheat plants. Surprisingly, they found several DNA-binding activities in the extracts from non-acclimated plants but none in the extracts from cold-acclimated plants. *In vitro* dephosphorylation with alkaline phosphatase restored the binding activity in the extracts derived from cold-acclimated plants, suggesting that the DNA-binding factors are present in the non-acclimated extracts in a non-/low-phosphorylated form, and interact with elements in the *WCS120* promoter repressing its transcription. On the other hand, okadaic acid markedly stimulated the *in vivo* accumulation of the WCS120 family of proteins under non-induced conditions (50), supporting the data obtained from the gel-shift assays, and suggesting that PP2A and/or PP1 might act as negative regulators of *WCS120* gene expression. Interestingly, cytoplasmic PP2A has been shown to inactivate MAP kinases, receptor protein kinases, and second messenger-dependent kinases including CDPKs (51–53). In the nucleus, PP2A has been shown to prevent transcription by inactivating transcription factors (54). It is tempting to speculate that inactivation of PP2A would promote low-temperature signal transduction indirectly by derepressing signal transducing protein kinases and/or directly by regulating the activity of *trans*-acting factors.

Taking all these biochemical results together, a working model can be proposed to describe the regulation of cold-acclimation response (Fig. 1). In this model, when a plant is exposed to low temperatures the signal is sensed and transduced into an increase in cytosolic calcium which, in turn, would inhibit protein phosphatase 2A (PP2A) activity and would activate a series of phosphorylation switches involving calcium-dependent protein kinases (CDPKs). These calcium-dependent phosphorylation/dephosphorylation events would lead to low-temperature-regulated gene expression that underlies the cold-acclimation response.

Although the involvement of ABA in plant responses to different environmental stresses has long been widely accepted, it was Chen *et al.* (55) who found that in *Solanum commersonii*, a species that can cold-acclimate, exogenous application of ABA to plants grown at 20°C resulted in increased freezing tolerance. Furthermore, they also reported that the levels of ABA significantly increased in *S. commersonii* plants exposed to low temperatures, but not in *Solanum tuberosum*, a plant species

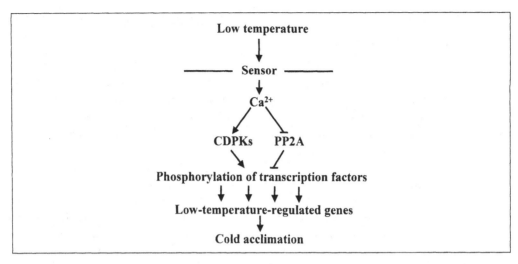

Fig. 1 Suggested model from biochemical analysis describing how low temperature may regulate gene expression leading to cold acclimation. CDPDs, calcium-dependent protein kinases; PP2A, protein phosphatase 2A. Arrows and bars indicate positive and negative regulation, respectively.

that is unable to cold acclimate (55). Subsequent studies have extended these observations (56–61), leading to the hypothesis that an ABA-dependent pathway may play a role in triggering the cold-acclimation response and the mechanisms of freezing tolerance. Nevertheless, the significance of such a role still remains unclear. Although many low-temperature-inducible genes are also positively regulated by exogenous ABA treatment, most of them are also induced by cold in ABA-deficient (*aba*) or -insensitive (*abi*) *Arabidopsis* mutants (16). This indicates that these genes do not require ABA for their expression under low-temperature conditions but that they respond to ABA. In addition, the endogenous ABA that accumulates in response to low temperature seems not to be sufficient to induce genes whose expression is regulated by ABA but not by cold stress (16). All these data would suggest that the ABA-signalling pathway is not very important in low-temperature responses. By using single-cell microinjection experiments, the ABA signalling pathway leading to the induction of *RD29A* (also termed *LTI78* and *COR78*) and *KIN2*, two genes of *Arabidopsis* regulated by low temperature, dehydration, and ABA, has been uncovered to some extent (62). In the proposed model, the pathway would be repressed by a hypothetical protein phosphatase that is sensitive to okadaic acid. Recognition of ABA by a putative receptor would initiate downstream events leading to inactivation of the phosphatase. Derepression would result in the generation of cyclic ADP-ribose (cADPR) which, in turn, would release calcium, most likely from vacuolar stores. The increased levels of cytoplasmic calcium would promote a cascade of phosphorylation/dephosphorylation episodes that would conclude with the induction of *RD29A* and *KIN2* (62). In order to elucidate the implication of ABA in low-temperature responses, it would be of special interest to determine how the abolishment of this pathway affects cold acclimation.

4.2 Molecular analysis of signal transduction in response to low temperature

Early work at molecular level on low-temperature signalling focused primarily on the identification of plant genes showing both homology to factors involved in signal transduction in other organisms and up-regulation in response to cold. As a result, many genes fulfilling these conditions have been identified that may be involved in the cold-acclimation response by amplifying the transduction efficiency of the signalling cascades.

Two-component systems play important roles in plant signal transduction in response to environmental cues and growth regulators (33). Typically, a two-component system consists of a sensory histidine kinase and a response regulator. In some cases, however, two-component systems include additional signalling modules or motifs containing an active histidine residue that mediates phospho-transfer (HPt) constituting more complicated phosphorelay circuits (33). Genes encoding response regulators (63) and intermediates containing an HPt domain (64) have been identified, whose corresponding transcripts accumulate in response to low temperature. Urao *et al.* (63) isolated two cDNAs from *Arabidopsis*, *ATRR1* and *ATRR2*, that encoded two-component response regulator-like proteins. *ATRR1* and *ATRR2* are preferentially expressed in roots and transiently induced by cold treatment. Miyata *et al.* (64) isolated three *Arabidopsis* cDNAs, *ATHP1-3*, that encoded two-component phosphorelay-mediator-like proteins. These proteins contain an HPt-like domain and function as phosphorelay mediators in yeast cells. Similar to *ATRR1* and *ATRR2*, the expression of *ATHP* genes is also higher in roots than in other tissues, and is significantly induced in cold conditions (64). This similarity suggests that the corresponding factors may function together in a phosphorelay cascade(s) involved in low-temperature signalling and cold acclimation.

In the previous section, the crucial role that calcium plays, as second messenger, in a low-temperature signal transduction pathway that regulates cold-acclimation was substantiated. Cytoplasmic calcium levels increase rapidly in response to low temperature due to an influx of calcium from both extracellular and, to a lesser extent, intracellular stores (see above). In mammalian cells, the release of calcium from the intracellular stores is mainly controlled by the phosphatidylinositol (PI) turnover system in which the PI-specific phospholipase C (PI-PLC) plays a key function. PI-PLC hydrolyses phosphatidylinositol 4,5-bisphosphate (PIP_2) to produce 1,2-diacylglycerol (DG) and inositol 1,4,5-trisphosphate (IP_3) that binds to an IP_3 receptor and opens a calcium channel localized in the membrane of the endoplasmic reticulum to release calcium to the cytoplasm (65). Alexandre *et al.* (66) proposed that a calcium channel coupled with an IP_3 receptor could exist in the vacuolar membrane of plant cells. Interestingly, an *Arabidopsis* cDNA corresponding to a PI-PLC, *AtPLC1*, has been cloned (67), confirming that the PI turnover system may also work in plant cells. The AtPLC1 protein contains the X and Y domains of phospholipase C, an EF hand structure, and shows calcium-dependent activity when expressed in bacteria (67). The *AtPLC1* gene is expressed at very low levels under standard conditions but is markedly induced by cold exposition (67). These observations suggest that AtPLC1 can act as an intermediate in the signal transduction pathway mediated by low temperature, leading to the increase in cytoplasmic calcium from the vacuole and triggering the cold-acclimation response.

Molecular analyses have provided further data indicating that calcium-binding proteins have a role in cold acclimation. There is evidence that the transcript levels of *NpCaM-1*, a tobacco-CaM gene, accumulate in response to low temperature (68). Consistent with a function for *NpCaM-1* in cold acclimation, its expression is regulated by a calcium-signalling pathway predominantly operational in the cytoplasm (68). Low-temperature exposure also results in a rapid, strong, and transient increase in expression of *Arabidopsis TCH2* and *TCH3* genes, which encode proteins closely related to CaM (42). Again, as might be anticipated if these genes participate in cold acclimation, their inducibility under low-temperature conditions requieres an intracellular increase of calcium (42). Furthermore, CDPKs whose transcript levels accumulate significantly at 4°C have been isolated (35, 45). Thus, in alfalfa, Monroy *et al.* (35) identified a *CDPK* cDNA whose corresponding mRNA transiently increased by more than eightfold in response to cold. In *Arabidopsis*, several *CDPK* genes were also identified which exhibited up-regulation by low temperature (45).

Molecular studies have also shown that, besides CDPKs, other protein kinases participate as intermediates in low-temperature signalling, corroborating the important role of phosphorylation in cold acclimation. Among them, proteins related to mitogen-activated protein kinases (MAPKs) seem to have a significant function. MAPKs perform their function as part of protein kinase modules, which are mainly composed of MAPKs, MAPK kinases (MAPKKs), and MAPKK kinases (MAPKKKs). It has been reported that *ATMPK3* and *ATMEKK1*, two *Arabidopsis* genes encoding a MAPK and a MAPKKK, respectively, are transcriptionally induced by cold (69, 70). Two other genes from *Arabidopsis*, *ATPK6* and *ATPK19*, are also positively regulated in response to low temperature (71). These genes encode proteins related to ribosomal S6 kinases that are phosphorylated and activated by MAPKs in mammalian systems (71). Interestingly, *ATPK19* displays a similar pattern of cold induction to *ATMPK3* and *ATMEKK1*, suggesting that these genes could be part of a MAPK cascade implicated in the adaptive response of plant cells to low-temperature conditions. In accordance with a role of *ATMPK3* in cold acclimation, a closely related alfalfa homologue, *MMK4*, is also transcriptionally up-regulated upon cold stress (72). Besides the transcriptional regulation of the gene, MMK4 is activated at the post-translational level (72), indicating that MAPK cascades involved in low-temperature signalling can be regulated at both transcriptional and post-translational levels. The *Arabidopsis ATMAP3Kβ3* gene, that encodes a MAPKKK, also shows cold-induced transcript accumulation (73). In addition to MAPK-encoding genes, genes encoding other protein kinases are also up-regulated at 4°C. This is the case of the *PKABA1* gene from wheat, which is also regulated by ABA (74). The same occurred with *RPK1*, an *Arabidopsis* gene that encodes a receptor-like protein kinase and is induced by ABA (75). The *Arabidopsis ATDBF2* gene is able to complement the yeast *dbf2* mutant (76), evidencing its homology to the *DBF2* gene from yeast that encodes a protein kinase involved in the CCR4 general transcriptional complex (76). When *ATDBF2* is expressed constitutively, several cold-responsive genes from *Arabidopsis* are expressed under control conditions (76), suggesting that it could control cold-regulated gene expression.

Other genes encoding factors that are implicated in signal transduction cascades are also induced in response to low temperature, indicating that they may have a role in cold acclimation. Two of these genes are *RCI1A* and *RCI1B* from *Arabidopsis*, which encode proteins belonging to the 14-3-3 family (77, 78). Interestingly, it has been described that members from this family of highly conserved regulatory proteins can bind and activate *Arabidopsis* CDPK isoforms (79). Moreover, sugar beet 14-3-3 proteins have been shown to preferentially associate with the plasma membrane under low-temperature conditions promoting the formation of H^+-ATPase/14-3-3 complexes and, consequently, an increase in H^+-ATPase activity (80). These data suggest that some 14-3-3 proteins can participate in cold acclimation through the interaction with other signalling factors. *RCI2A* and *RCI2B* are also two homologous genes from *Arabidopsis* whose corresponding transcripts accumulate in response to low temperature and encode proteins that may be involved in cold acclimation (81). These genes encode small, highly hydrophobic proteins containing two transmembrane domains each (81). Recently, a yeast plasma membrane protein, PMP3, has been isolated which shows high similarity to RCI2 proteins (82). Deletion of *PMP3* induces a hyperpolarization of the membrane potential that can be abolished by expressing the *RCI2A* gene (82). These observations demonstrate that both genes share a conserved function, enabling one to hypothesize that RCI2 proteins can be involved in the transient depolarization that takes place in the plasma membrane of plant cells as a consequence of cold exposure (83).

Work at a molecular level on low-temperature signalling has also focused on the identification of DNA elements that mediate cold-induced gene expression. Yamaguchi-Shinozaki and Shinozaki (84) first identified a 9 bp conserved sequence, TACCGACAT, in the promoter of the *Arabidopsis RD29A* gene that activated gene expression in response to low temperature and drought when fused to a reporter gene. This sequence, termed dehydration-responsive element (DRE), contains a 5 bp core motif, CCGAC, named C-repeat (CRT) (85) and low-temperature-reponse element (LTRE) (86), that is essential for the low-temperature responsiveness of several cold-induced genes, including the *Arabidopsis COR15A* gene (85), the *Brassica napus BN115* gene (86), and the wheat *WCS120* gene (87). The CRT/DRE/LTRE sequence is not responsive to ABA (84), indicating that it imparts cold-regulated gene expression through an ABA-independent pathway. A step forward in understanding how this sequence regulates cold-induced gene expression was the isolation of a cDNA from *Arabidopsis*, *CBF1*, encoding a CRT/DRE/LTRE-binding protein (88). The CBF1 protein has a putative bipartite nuclear localization sequence, an acidic region that potentially serves as an activation domain, and an AP2 DNA-binding domain (88). Expression analysis in yeast confirmed that CBF1 functions as a transcriptional activator up-regulating CRT/DRE/LTRE-dependent transcription (88). In fact, over-expression of *CBF1* in *Arabidopsis* transgenic plants induced the expression of some cold-regulated genes and increased the freezing tolerance of non-acclimated transformed plants (9). These results have been extended independently by different groups (10, 89, 90), establishing that *CBF1* is a member of a small gene family encoding at least three closely related genes, *CBF1–3*, which are organized in tandem on chromosome 4 of *Arabidopsis*. CBF2 and CBF3 are also transcription factors, and their over-expression in transgenic *Arabidopsis* provoked constitutive expression of *RC29A* and enhanced freezing tolerance (10, 11). In addition, over-expression of *CBF3* results in several biochemical changes associated with cold acclimation, including elevated levels of compatible osmolytes, proline, and soluble sugars (91). These observations suggest that CBF3 not only activates low-temperature-induced gene expression but also different components of the cold-acclimation response. The three *CBF* genes show identical expression patterns, being induced rapidly in response to low temperature but not in response to other related treatments (i.e. dehydration, ABA, or high salt) (10, 89, 90). The induction of *CBF* genes seems to be regulated at the transcriptional level, as reporter genes fused to *CBF* promoters are induced by cold (J. Medina and J. Salinas, unpublished results). However, the available evidences suggest that they are not self-regulated since the CRT/DRE/LTRE motif, CCGAC, is not found in the *CBF* promoters (89, 90), and over-expression of *CBF1* does not result in accumulation of *CBF3* transcripts (89). Gilmour *et al.* (89) proposed that, under low temperatures, a signal transduction pathway should be turned on, resulting in modifying either a CBF activator that would be inactive at warm temperatures, or an associated protein that would allow the activator to induce *CBF* expression. Interestingly, the cold-inducible genes whose expression was described to be controlled through a series of phosphorylation/dephosphorylation events mediated by a cytosolic calcium increase (45, see above), have also been reported to be regulated by CBF proteins (9, 11). This occurrence allows the speculation that the signal transduction pathway leading to the induction of *CBF* expression and, finally, to cold acclimation may be mediated by calcium-binding proteins (i.e. CDPKs) that would couple the calcium increase originated in response to low temperature (Fig. 2).

Although the CBF regulatory pathway seems to be rather conserved in plants, as shown by the fact that the CRT/DRE/LTRE motif accounts for the cold responsiveness of different genes from several species (84–87), it is intriguing that some other cold-inducible genes do not contain this sequence in their promoter regions. Thus, the core sequence CCGAC is not only absent from the *CBF* promoters (89, 90) but also from the 5′ regions of the *Arabidopsis TCH2*, *TCH3*, and *TCH4* cold-inducible genes (42). It is neither found in the promoter region that is able to confer low-temperature responsiveness to the *Arabidopsis LHCB1*3* gene (92), nor in the promoter of the barley cold-inducible gene *BLT101.1* (93). These data indicate that other element(s) in adittion to the CRT/DRE/LTRE motif and, consequently, other transcription factors distinct to the CBFs should be involved in mediating low-temperature-regulated gene expression. Electrophoretic mobility shift assays allowed the identification of a hexanucleotide, CCGAAA, within the promoter region of the cold-regulated *BLT4.9* gene from barley as the binding site of a low-mobility nuclear protein complex (94). Mutation of the hexanucleotide reduced the low-temperature responsiveness of the *BLT4.9* promoter to basal levels (94). In spite of its similarity to the CRT/DRE/LTRE

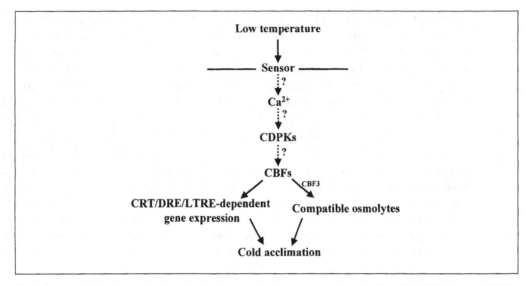

Fig. 2 Proposed signal transduction pathway in response to low temperature mediated by CBF proteins. CDPKs, calcium-dependent protein kinases; CRT/DRE/LTRE, low-temperature-responsive motif (CCGAC) bound by CBF transcriptional activators. Solid arrows indicate experimentally established relationships. Broken arrows indicate components that have not yet been identified.

sequence, the CCGAAA motif is not bound by CBF1 (93). A CCGAC element also exists in the *BLT4.9* promoter, upstream of the CCGAAA motif. However, electrophoretic mobility shift assays indicated that this sequence is not a binding site for nuclear factors (93). Although it is clear that CBF proteins are positive regulators of cold acclimation through a CRT/DRE/LTRE-dependent pathway, from the results described above the existence of a single molecular mechanism for regulating cold-induced gene expression is unlikely. Rather, an intricate scenario emerges where the regulation of gene expression in response to low temperature is mediated by different signal transduction pathways through distinct *cis*- and *trans*-regulatory elements. In this way, several transcription factors have been identified whose transcripts are induced by cold. Thus, LIP19 and MLIP15, two bZip transcription factors from rice and maize, respectively (95, 96), and several zinc fingers from *Arabidopsis* and petunia (97) are encoded by low-temperature-responsive genes. Although the regulatory motifs bound by these factors are at present unknown, they are good candidates to be implicated in regulating cold-induced gene expression.

4.3 Genetic analysis of signal transduction in response to low temperature

An engaging approach to uncover signal components required for cold-acclimation response is to isolate and characterize mutants affected in their ability to tolerate freezing temperatures. Mutants of *Arabidopsis* that are either sensitive or tolerant to freezing have been isolated, and their phenotypes analysed in detail. *Arabidopsis* plants carrying mutations in ABA biosynthesis (*aba1*) or the ability to respond to ABA (*abi1*) are less tolerant to freezing than wild-type plants (98–100). These observations constitute additional evidence to that obtained from biochemical analysis (see above) that ABA has a role in cold acclimation. Recently, Llorente *et al.* (101) identified a freezing sensitive (*frs1*) mutant of *Arabidopsis* that showed both a reduced constitutive freezing tolerance as well as a reduced freezing tolerance after cold acclimation. The *frs1* was an ABA-deficient mutant and the *frs1* mutation a new allele of the *ABA3* locus, strengthing the conclusion that ABA is required for full development of cold acclimation. Some authors (99), however, cautioned that the decrease in freezing tolerance shown by ABA mutants could be a side-effect caused by the lack of active ABA signal transduction required for important functions in plant growth and development. To gain information on how ABA can mediate cold acclimation, the expression patterns of different genes that are regulated by low temperature, dehydration, and ABA were analysed in *frs1/aba3* mutants exposed to both cold and dehydration (101). The results obtained indicate that gene expression is affected under dehydration but not cold conditions, consistent with the fact that, in most cases, genes regulated by water stress are also responsive and mediated by ABA (102). Similarly, *aba1* and *abi1* mutations have little or no effect on low-temperature-induced gene expression (103). These data suggest that cold acclimation in *Arabidopsis* depends, in part, on ABA-regulated proteins that would allow the plant to survive the cellular dehydration imposed by low, including freezing, temperatures. Although ABA has an evident role in activating the cold-acclimation response, its implication in regulating specific low-temperature-induced gene expression is still uncertain.

By screening M3 families derived from 1804 EMS mutagenized M2 plants, Warren *et al.* (104) identified seven non-allelic sensitive-to-freezing (*sfr*) mutants which only acquired partial freezing tolerance after cold acclimation. The *sfr* mutations were recessive or co-dominant (in one case), according to the expectation that they should be loss-of-function mutants. The identity of the *SFR* genes as well as the molecular basis of how mutations in these genes affect cold acclimation and freezing tolerance remains unknown. A preliminary characterization (105) revealed that four mutations, *sfr3*, *-4*, *-6*, and *-7*, reduce or block anthocyanin accumulation during cold acclimation. Moreover, *sfr4* prevents the low-temperature-induced increase in sucrose and glucose levels, and both *sfr4* and *sfr7* alter fatty acid composition after cold acclimation. The *sfr1*, *-2*, and *-5* mutations do not provoke any obvious alteration in fatty acid composition, or sucrose and anthocyanin levels (105), indicating that they have other, perhaps highly specific, effects on low-temperature responses. When mutants *sfr2* to *sfr7* were tested for cold-induced gene expression, only *sfr6* plants were deficient in induction of *KIN1*, *COR15A*, and *LTI78* genes (106), all of which contain the CRT/DRE/LTRE motif in their promoters. Low-temperature-induced expression of *CBF* and *ATPSC1* genes, which lack the motif, was not affected in *sfr6* mutants (106). Calcium measurements demonstrated that the failure in cold induction of CRT/DRE/LTRE-containing genes in *sfr6* plants can not be ascribed to altered low-temperature calcium signalling (106). It is significant that the CRT/DRE/LTRE-dependent gene expression is unpaired in *sfr6* but not the expression of the *CBF* genes whose corresponding transcripts accumulate normally in the mutant in response to low temperature. Likely, the SFR6 protein acts downstream of CBFs in the signal transduction pathway mediated by low temperature through these factors. The isolation and molecular characterization of the products encoded by *SFR* genes should provide significant new insights to the understanding of the cold-acclimation response.

By screening 800 000 EMS mutagenized M2 seedlings of *Arabidopsis* for mutants showing constitutive freezing tolerance, Xin and Browse (107) isolated a series of mutants that, in the absence of cold acclimation, showed enhanced freezing tolerance compared to wild-type plants. Allelism tests identified mutations in at least six different loci. One of these mutants, named *eskimo1* (*esk1*), was produced by a single recessive mutation and showed very high freezing tolerance compared with wild-type non-acclimated plants (low temperature that resulted in 50% of plant survival (LT_{50}) of −10.6°C and −5.5°C, respectively). Cold-acclimated *esk1* mutants also had higher freezing tolerance than wild-types, although the differences were less pronounced (LT_{50} of −14.8°C and −12.6°C, respectively) (107). Plants with mutations at the *esk1* locus acumulate around 30-fold more proline and twofold more soluble sugars than wild-type plants (107). These alterations may contribute to the increased freezing tolerance of *esk1* mutants since, as mentioned before, both proline and soluble sugars are compatible osmolytes that typically accumulate during cold acclimation and seem to be effective cryoprotectants (108–110). The expression of low-temperature-inducible genes *KIN2*, *COR15A*, *COR47*, and *LTI78* from *Arabidopsis*, which is controlled by the CRT/DRE/LTRE motif, is not affected by the *esk1* mutation (107). These results indicate that ESK1 functions in a different pathway from that mediated by the CBF proteins. The fact that *esk1* is a recessive mutation suggests that ESK1 may act as a negative regulator in a pathway leading to the synthesis of proline and sugars, which would result in a significant increase in freezing tolerance.

An alternative genetic approach to identify genes involved in low-temperature signalling and cold acclimation was recently carried out by Zhu and collaborators (111). They used EMS-mutagenized M2 *Arabidopsis* transgenic plants containing the *RD29A promoter::luciferase* chimeric gene to screen for mutants affected in cold-regulated *RD29A* expression. A large number of constitutive (*cos*), high- (*hos*), and low-expressor (*los*) mutants were identified which should define positive or negative regulators in low-temperature signalling. Some of the identified mutants were also affected in *RD29A* expression in response to ABA and/or NaCl, suggesting that signal transduction pathways mediated by cold, ABA, and osmotic stress converge to activate or repress gene expression. Very similar mutations have been isolated when screening for mutants altered in *RCI2A* gene expression (Medina, Rodriguez, Neuhaus, and Salinas, unpublished results), which demonstrates that cross-talk among low-temperature-, ABA-, and osmotic-stress-mediated pathways is not specific of *RD29A* regulation, but rather a general circumstance. One of the mutants isolated by Zhu and colleagues, *hos1*, shows enhanced expression of *RD29A*, *COR47*, *COR15A*, *KIN1* and *ADH* in response to low temperature (112). Non-acclimated *hos1* plants are less cold hardy than the wild type ones (112). Nonetheless, after cold acclimation they acquire the same degree of freezing tolerance as do the wild-type plants (112). The recessive character of *hos1* suggests that *HOS1* is an important negative regulator of low-temperature signalling in plant cells and that it plays a critical role in controlling gene expression during cold acclimation. Another mutant identified, *hos2*, also exhibits enhanced expression of *RD29A* and other stress genes under low-temperature conditions (113). However, this enhanced gene induction in the cold does not bring about an increase in freezing tolerance. In fact, compared with wild-type plants, *hos2* mutants are defective in developing freezing tolerance during cold acclimation (113). The authors suggest that, since *hos2* is a recessive mutation, besides being a negative regulator for cold-induced gene expression, *HOS2* may play a positive role in the regulation of some other cellular changes (i.e. membrane composition and sugar metabolism) that contribute to freezing tolerance (113). Determining the molecular nature of the products encoded by the mutated genes will not only shed light on the signal transduction pathways mediated by low temperature, but also help to better understand the interactions between these pathways and those mediated by ABA and other related environmental stresses.

5. Concluding remarks and future perspectives

This chapter focuses on recent advances on how plants respond to low temperature, triggering the cold-acclimation response. The ability to endure low temperatures and freezing is a major determinant of the geographical distribution and productivity of agricultural crops. Only modest increases (1–2°C) in the freezing tolerance of crop species would have a dramatic impact on agricultural productivity and profitability, minimizing crop losses and diminishing the use of energy-costly practices to modify the microclimate. Determining how cold acclimation is initiated and coordinated at the molecular level is the first step in developing molecular approaches towards improving the freezing tolerance of agricultural crop species.

Complementary biochemical, molecular, and genetic analyses are starting to unravel the complexity of the low-temperature responses that lead to cold acclimation. The results obtained have revealed the existence of multiple signal transduction pathways linking the perception of the cold stress signal to the activation of the full set of processes required for increasing freezing tolerance (Fig. 3). Many signalling molecules have been implicated in low-temperature responses and several genes encoding factors involved in signalling have been shown to be up-regulated by low temperatures. These factors may have a role in the amplification of the stress signals and the adaptation of plant cells to cold stress conditions. *Cis-* and *trans-*acting elements involved in low-temperature-regulated gene expression have been studied and a major transcription system uncovered. Genetic analyses have allowed the

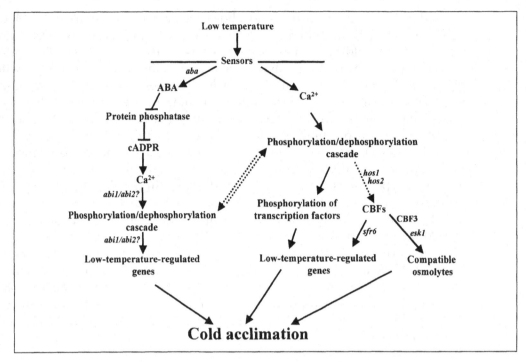

Fig. 3 Outline of the major signalling network that mediates the cold-acclimation response. *aba*, *abi1*, and *abi2* are mutations involved in ABA signalling. *sfr6*, *hos1*, and *hos2* are mutations related to CBF signalling. *esk1* is a mutation involved in compatible osmolyte signalling. Solid arrows indicate experimentally established pathways. Broken arrows indicate pathways that have not yet been experimentally established. ABA, abscisic acid; cADPR, cyclic ADP-ribose.

isolation of freezing-tolerant and -sensitive mutants, and the confirmation that low-temperature signalling pathways interact with other pathways, increasing or attenuating the responses to different environmental stresses. It is evident that cross-talk between pathways plays a critical role in coordinating the execution of the appropriated cold-acclimation response.

To gain a better understanding on how the identified signalling components function and communicate, it will be important to develop novel methods of analysis. In this regard, the *Arabidopsis* genome has just been completely sequenced, which means that all the low-temperature-regulated genes can be identified by systematic analysis of gene expression by DNA chip/microarray technology. Mutant lines obtained by insertion of T-DNA and transposons will become of the utmost importance to analyse the function of the identified genes. Complementary classical and reverse genetic approaches will be crucial to uncover novel components of cold signalling, including those affecting pathway regulation, and for the precise comprehension of the multiple facets of cold-acclimation response. In summary, new routes to attain a complete understanding on how cold-acclimation response is activated by low temperature are being traced, which ensures a very exciting time ahead in this field of research.

Acknowledgements

I thank J. J. Sánchez-Serrano and G. Salcedo for helpful discussions and comments on the manuscript. Work in my laboratory is supported by grants from CICYT (BIO98-0189) and EU (QLK3-CT2000-00328).

References

1. Levitt, J. (1980) Responses of plant to environmental stresses. Vol. 1. *Chilling, Freezing, and High Temperature Stresses*, 2nd edn. Academic Press, New York, pp 497.
2. Boyer, J. S. (1982) Plant productivity and environment. *Science*, **218**, 443.
3. Ristic, Z. and Ashworth, E. N. (1993) Changes in leaf ultrastructure and carbohydrates in *Arabidopsis thaliana* (L.) *Heyn* cv. Columbia during rapid cold-acclimation. *Protoplasma*, **172**, 111.
4. Lynch, D. V. and Steponkus, P. L. (1987) Plasma membrane lipid alterations associated with cold acclimation of winter rye seedlings (*Secale cererale* L. cv. Puma). *Plant Physiol.*, **83**, 761.
5. Miquel, M., James Jr, D., Dooner, H. and Browse, J. (1993) *Arabidopsis* requires poly-unsaturated lipids for low-temperature survival. *Proc. Natl Acad. Sci. USA*, **90**, 6208.
6. Raison, J. K. (1973) The influence of temperature-induced phase changes on kinetics of respiratory and other membrane-associated enzymes. *J. Bioenerg.*, **4**, 258.
7. Strand, A., Hurry, V., Gustafsson, P. and Gardeström, P. (1997) Development of *Arabidopsis thaliana* leaves at low temperatures releases the supression of photosynthesis and photosynthetic gene expression despite the accumulation of soluble carbohydrates. *Plant J.*, **12**, 605.
8. Artus, N. N., Uemura, M., Steponkus, P. L., Gilmour, S. J., Lin, C. and Thomashow, M. F. (1996) Constitutive expression of the cold-regulated *Arabidopsis thaliana COR15a* gene affects both chloroplast and protoplast freezing tolerance. *Proc. Natl Acad. Sci. USA*, **93**, 13404.
9. Jaglo-Ottosen, K. R., Gilmour, S. J., Zarka, D. G., Schabenberger, O. and Thomashow, M. F. (1998) *Arabidopsis CBF1* overexpression induces *COR* genes and enhances freezing tolerance. *Science*, **280**, 104.
10. Liu, Q., Kasuga, M., Sakuma, Y., Abe, H., Miura, S., Yamaguchi-Shinozaki, K. and Shinozaki, K. (1998) Two transcription factors, DREB1 and DREB2, with an EREBP/AP2 DNA binding domain separate two cellular signal transduction pathways in drought- and low-temperature responsive gene expression, respectively, in *Arabidopsis*. *Plant Cell*, **10**, 1391.
11. Kasuga, M., Liu, Q., Miura, S., Yamaguchi-Shinozaki, K. and Shinozaki, K. (1999) Improving plant drought, salt, and freezing tolerance by gene transfer of a single stress-inducible transcription factor. *Nat. Biotech.*, **17**, 287.
12. Steponkus, P. L., Uemura, M., Joseph, R. A., Gilmour, S. J. and Thomashow, M. F. (1998) Mode of action of the *COR15a* gene on the freezing tolerance of *Arabidopsis thaliana*. *Proc. Natl Acad. Sci. USA*, **95**, 14570.
13. Hughes, M. A. and Dunn, M. A. (1996) The molecular biology of plant acclimation to low temperature. *J. Exp. Bot.*, **47**, 291.
14. Thomashow, M. F. (1998) Role of cold-responsive genes in plant freezing tolerance. *Plant Physiol.*, **118**, 1.

15. Thomashow, M. F. (1999) Plant cold acclimation: freezing tolerance genes and regulatory mechanisms. *Annu. Rev. Plant Physiol. Mol. Biol.*, **50**, 571.

16. Shinozaki, K. and Yamaguchi-Shinozaki, K. (2000) Molecular responses to dehydration and low temperature: differences and cross-talk between two stress signaling pathways. *Curr. Opin. Biotechnol.*, **3**, 217.

17. Harwood, J. L., Jones, A. L., Perry, H. J., Rutter, A. J., Smith, K. L. and Williams, M. (1994) Changes in plant lipids during temperature adaptation. In Cossins, A. R. (ed.), *Temperature Adaptation of Biological Membranes*. Portland Press Proceedings, London, p.107.

18. Steponkus, P. L. (1984) Role of the plasma membrane in freezing injury and cold acclimation. *Annu. Rev. Plant Physiol.*, **35**, 543.

19. Steponkus, P. L., Uemura, M. and Webb, M. S. (1993) Membrane destabilizatión during freeze-induced dehydration. *Curr. Topics Plant Physiol.*, **10**, 37.

20. Olien, C. R. (1981) Analysis of midwinter freezing stress. In Olien, C. R. and Smith, M. N. (eds.), *Analysis and Improvement of Plant Cold Hardiness*. CRC Press Inc., Boca Raton, FL, p. 61.

21. Steponkus, P. L., Uemura, M. and Webb, M. S. (1993) A contrast of the cryostability of the plasma membrane of winter rye and spring oat. Two species that widely differ in their freezing tolerance and plasma membrane lipid composition. In Steponkus, P. L. (ed.), *Advances in Low-temperature Biology*. JAI Press Ltd, London, Vol. 2, p. 211.

22. Uemura, M. and Steponkus, P. L. (1997) Effect of cold acclimation on membrane lipid composition and freezing-induced membrane destabilization. In Li, P. H. and Chen, T. H. H. (eds.), *Plant Cold Hardiness, Molecular Biology, Biochemistry and Physiology*. Plenum Press, New York, p. 171.

23. Pardossi, A., Vernieri, P., Mori, B. and Toguoni, F. (1994) Water relations and ABA in chilled plants. In Dörffling, K., Brettschneider, B., Tantau, H. and Pithan, K. (eds.), *Crop Adaptation to Cool Climates*. ECSP-EEC-AEEC, Brussels, p.215.

24. Patterson, B. D. and Graham, D. (1987) Temperature and Metabolism. In Davis, D. D. (ed.), *The Biochemistry of Plants*. Academic Press, New York, Vol. 12, p. 152.

25. Guy, C. L., Haskell, D. and Li, Q. B. (1998) Association of proteins with the stress 70 molecular chaperones at low temperature: evidence for the existence of cold labile proteins in spinach. *Cryobiology*, **36**, 301.

26. Guy, C. L. (1990) Cold acclimation and freezing stress tolerance: role of protein metabolism. *Annu. Rev. Plant Physiol. Plant. Mol. Biol.*, **41**, 187.

27. Kendall, E. J. and McKersie, B. D. (1989) Free radical and freezing injury to cell membranes of winter wheat. *Physiol. Plant.*, **86**, 740.

28. Foyer, C. H., Lopez-Delgado, H., Dat, J. F. and Scott, I. M. (1997) Hydrogen peroxide- and glutathione-associated mechanisms of acclimatory stress tolerance and signalling. *Physiol. Plant.*, **100**, 241.

29. Prasad, T.K,. Anderson, M. D., Martin, B. A. and Stewart, C. R. (1994) Evidence for chilling-induced oxidative stress in maize seedlings and a regulatory role for hydrogen peroxide. *Plant Cell*, **6**, 65.

30. McKersie, B. D. and Bowley, S. R. (1997) Active oxygen and freezing tolerance in transgenic plants. In Li, P. H. and Chen, T. H. H. (eds.), *Plant Cold Hardiness, Molecular Biology, Biochemistry and Physiology*. Plenum Press, New York, p. 203.

31. Murata, N. and Los, D. A. (1997) Membrane fluidity and temperature perception. *Plant Physiol.*, **115**, 875.

32. Maeda, M., Wurgler-Murphy, S. M. and Saito, H. (1994) A two-component system that regulates san osmosensing MAP kinase cascade in yest. *Nature*, **309**, 242.

33. Urao, T., Yamaguchi-Shinozaki, K. and Shinozaki, K. (2000) Two component systems in plant signal transduction. *Trends Plant Sci.*, **5**, 67.

34. Urao, T., Yakubov, B., Satoh, R., Yamaguchi-Shinozaki, K., Seki, M., Hirayama, T. and Shinozaki, K. (1999) A transmembrane hybrid-type histidine kinase in *Arabidopsis* functions as an osmosensor. *Plant Cell*, **11**, 1473.

35. Monroy, A. F. and Dhindsa, R. S. (1995) Low-temperature signal transduction: induction of cold acclimation-specific genes of alfalfa by calcium at 25°C. *Plant Cell*, **7**, 321.

36. Minorsky, P. V. (1989) Temperature sensing by plants: a review and hypothesis. *Plant Cell Environ.*, **12**, 119.

37. Ping Ding, J. and Pickard, B. G. (1993) Modulation of mechanosensitive calcium-selective cation channels by temperature. *Plant J.*, **3**, 713.

38. Huner, N. P. A., Öquist, G. and Sarhan, F. (1998) Energy balance and acclimation to light and cold. *Trends Plant Sci.*, **3**, 224.

39. Gray, G. R., Chauvin, L. P., Sarhan, F. and Huner, N. P. A. (1997) Cold acclimation and freezing tolerance. A complex interaction of light and temperature. *Plant Physiol.*, **114**, 467.

40. Knight, M. R., Campbell, A. K., Smith, S. M. and Trewavas, A. J. (1991) Transgenic plant aequorin reports the effect of touch and cold-shock and elicitors on cytoplasmic calcium. *Nature*, **352**, 524.

41. Knight, H., Trewavas, A. J. and Knight, M. R. (1996) Cold calcium signaling in *Arabidopsis* involves two cellular pools and a change in calcium sigature after acclimation. *Plant Cell*, **8**, 489.

42. Polisensky, D. H. and Braam, J. (1996) Cold-shock regulation of the *Arabidopsis TCH* genes and the effects of modulating intracellular calcium levels. *Plant Physiol.*, **111**, 1271.

43. Minorsky, P. V. and Spanswick, R. M. (1989) Electrophysiological evidence for a role for calcium in temperature sensing by roots of cucumber seedlings. *Plant Cell Environ.*, **12**, 137.

44. Monroy, A. F., Sarhan, F. and Dhindsa, R. S. (1993) Cold-induced changes in freezing tolerance, protein phosphorylation, and gene expression: evidence for a role of calcium. *Plant Physiol.* **102**, 1227.

45. Tähtiharju, S., Sangwan, V., Monroy, A. F., Dhindsa, R. S. and Borg, M. (1997) The induction of *kin* genes in cold-acclimating *Arabidopsis thaliana*. Evidence of a role for calcium. *Planta*, **203**, 442.

46. Kurkela, S. and Borg-Franck, M. (1992) Structure and expression of *kin2*, one of two cold- and ABA-induced genes of *Arabidopsis thaliana*. *Plant Mol. Biol.*, **19**, 689.

47. Roberts, D. M. and Harmon, A. C. (1992) Calcium-modulated proteins: targets of intracellular calcium signals in higher plants. *Annu. Rev. Plant Physiol. Plant Mol. Biol.*, **43**, 375.

48. Schulman, H., Hanson, P. I. and Meyer, T. (1992) Decoding calcium signals by multifuncional CaM kinase. *Cell. Calcium*, **13**, 401.

49. Monroy, A. F., Sangwan, V. and Dhindsa, R. S. (1998) Low temperature signal transduction during cold acclimation: protein phosphatase 2A as an early target for cold-acclimation. *Plant J.*, **13**, 653.

50. Vazquez-Tello, A., Ouellet, F. and Sarhan, F. (1998) Low temperature-stimulated phosphorylation regulated the binding of nuclear factors to the promoter of *Wcs120*, a cold-specific gene wheat. *Mol. Gen. Genet.*, **257**, 157.

51. Barnes, G. N., Slevin, J. T. and Vanaman, T. C. (1995) Rat brain protein phosphatase 2A: an enzyme that may regulate autophosphorylated protein kinases. *J. Neurochem.*, **64**, 340.

52. Chen, J., Martin, B. L. and Brautigan, D. L. (1992) Regulation of protein serine-threonine phosphatase type-2A by tyrosine phosphorylation. *Science*, **257**, 1261.

53. Cohen, P., Holmes, C. F. B. and Tsukitani, Y. (1990) Okadaic acid: a new probe for the study of cellular regulation. *Trends Biochem. Sci.*, **15**, 98.

54. Wheat, W. H., Roesler, W. J. and Klemm, D. J. (1994) Simian virus 40 small tumor antigen inhibits dephosphorylation of protein kinase A-phosphorylated CREB and regulates CREB transcriptional stimulation. *Mol. Cell. Biol.*, **14**, 5881.

55. Chen, H. H., Li, P. H. and Brenner, M. L. (1983) Involvement of abscisic acid in potato cold acclimation. *Plant Physiol.*, **71**, 362.

56. Orr, W., Keller, W. A. and Singh, J. (1986) Induction of freezing tolerance in an embryogenic cell suspension culture of *Brassica napus* by abscisic acid at room temperature. *J. Plant Physiol.*, **126**, 23.

57. Lang, V., Heino, P. and Palva, E. T. (1989) Low temperature acclimation and treatment with exogenous abscisic acid induce common polypeptides in *Arabidopsis thaliana* (L.) Heynh. *Theor. Appl. Genet.*, **79**, 801.

58. Ishikawa, M., Robertson, A. J. and Gusta, L. V. (1990) Effect of temperature, light, nutrients and dehardening on abscisic acid-induced cold hardiness in *Bromus inermis* Leyss suspension cultured cells. *Plant Cell Physiol.*, **31**, 51.

59. Lalk, I. and Dörffling, K. (1985) Hardening, abscisic acid, proline, and freezing resistance in two winter wheat varieties. *Physiol. Plant.* **63**, 287.

60. Luo, M., Liu, J., Mohapatra, S., Hill, R. D. and Mohapatra, S. S. (1992) Characterization of a gene family encoding abscisic acid and environmental stress-inducible proteins of alfalfa. *J. Biol. Chem.*, **267**, 15367.

61. Lang, V., Mäntylä, E., Welin, B., Sundberg, B. and Palva, E. T. (1994) Alterations in water stress status, endogenous abscisic acid content, and expression of *rab18* gene during the development of freezing tolerance in *Arabidopsis thaliana*. *Plant Physiol.*, **104**, 1341.

62. Wu, Y., Kuzma, J., Marechal, E., Graeff, R., Lee, H. C., Foster, R. and Chua, N-H. (1997) Abscisic acid signaling through cyclic ADP-ribose in plants. *Science*, **278**, 2126.

63. Urao, T., Yakubov, B., Yamaguchi-Shinozaki, K. and Shinozaki, K. (1998) Stress-responsive expression of genes for two-component response regulator-like proteins in *Arabidopsis thaliana*. *FEBS Lett.*, **427**, 175.

64. Miyata, S., Urao, T., Yamaguchi-Shinozaki, K. and Shinozaki, K. (1998) Characterization of genes for two-component phosphorelay mediators with a single Hpt domain in *Arabidopsis thaliana*. *FEBS Lett.*, **437**, 11.

65. Berridge, M. J. (1993) Inositol trisphosphate and calcium signalling. *Nature*, **361**, 315.

66. Alexandre, J., Lassalles, J. P. and Kado, R. T. (1990) Opening of Ca^{2+} channels in isolated red beet root vacuole membrane by inositol 1,4,5-trisphosphate. *Nature*, **343**, 567.

67. Hirayama, T., Ohto, C., Mizoguchi, T. and Shinozaki, K. (1995) A gene encoding a phosphatidylinositol-specific phospholipase C is induced by dehydration and salt stress in *Arabidopsis thaliana*. *Proc. Natl Acad. Sci. USA*, **92**, 3903.

68. Van der Luit, A. H., Olivari, C., Haley, A., Knight, M. R. and Trewavas, A. J. (1999) Distinct calcium signaling pathways regulate calmodulin gene expression in tobacco. *Plant Physiol.*, **121**, 705.

69. Mizoguchi, T., Gotoh, Y., Nishida, E., Yamaguchi-Shinozaki, K., Hayashida, N., Iwasaki, T., Kamada, H. and Shinozaki, K. (1994) Characterization of two cDNAs that encode MAP kinase homologues in *Arabidopsis thaliana* and analysis of the possible role of auxin in activating such kinase activities in cultured cells. *Plant J.*, **5**, 111.

70. Mizoguchi, T., Irie, K., Hirayama, T., Hayashida, N., Yamaguchi-Shinozaki, K., Matsumoto, K. and Shinozaki, K. (1996) A gene encoding a mitogen-activated protein kinase kinase kinase is induced simultaneously with genes for a mitogen-activated

protein kinase and an S6 ribosomal protein kinase by touch, cold, and water stress in *Arabidopsis thaliana*. *Proc. Natl Acad. Sci. USA*, **93**, 765.

71. Mizoguchi, T., Hayashida, N., Yamaguchi-Shinozaki, K., Kamada, H. and Shinozaki, K. (1995) Two genes that encode ribosomal-protein S6 kinase homologs are induced by cold or salinity stress in *Arabidopsis thaliana*. *FEBS Lett.*, **358**, 199.

72. Jonak, C., Kiegerl, S., Ligterink, W., Barker, P. J., Huskisson, N. S. and Hirt, H. (1996) Stress signaling in plants: a mitogen-activated protein kinase pathway is activated by cold and drought. *Proc. Natl Acad. Sci. USA*, **93**, 11274.

73. Jouannic, S., Hamal, A., Leprince, A. S., Tregear, J. W., Kreiss, M. and Henry, Y. (1999) Characterization of novel plant genes encoding MEKK/STE11 and RAF-related protein kinases. *Gene*, **229**, 171.

74. Holappa, L. D. and Walker-Simmons, M. K. (1995) The wheat abscisic acid-responsive protein kinase mRNA, PKABA1, is up-regulated by dehydration, cold temperature, and osmotic stress. *Plant Physiol.*, **108**, 1203.

75. Hong, S. W., Jon, J. H., Kwak, J. M. and Nam, H. G. (1997) Identification of a receptor-like protein kinase gene rapidly induced by abscisic acid, dehydration, high salt, and cold treatments in *Arabidospsis thaliana*. *Plant Physiol.*, **113**, 1203.

76. Lee, J. H., Van Montagu, M. and Verbruggen, N. (1999) A highly conserved kinase is an essential component for stress tolerance in yeast and plant cells. *Proc. Natl Acad. Sci. USA*, **96**, 5873.

77. Jarillo, J. A., Capel, J., Leyva, A., Martínez-Zapater, J. M. and Salinas, J. (1994) Two related low-temperature-inducible genes of *Arabidopsis* encode proteins showing high homology to 14-3-3 proteins, a family of putative kinase regulators. *Plant Mol. Biol.*, **25**, 693.

78. Abarca, M. D., Madueño, F., Martínez-Zapater, J. M. and Salinas, J. (1999) Dimerization of *Arabidopsis* 14-3-3 proteins: structural requirements within the N-terminal domain and effect of calcium. *FEBS Lett.*, **462**, 377.

79. Camoni, L., Harper, J. F. and Palmgren, M. G. (1998) 14-3-3 proteins activate a plant calcium-dependent protein kinase (CDPK). *FEBS Lett.*, **430**, 381.

80. Chelysheva, V. V., Smolenskaya, I. N., Trofimova, M. C., Babakov, A. V. and Muromtsev, G. S. (1999) Role of the 14-3-3 proteins in the regulation of H^+-ATPase activity in the plasma membrane of suspension-cultured sugar beet cells under cold stress. *FEBS Lett.*, **456**, 22.

81. Capel, J., Jarillo, J. A., Salinas, J. and Martínez-Zapater, J. M. (1997) Two homologous low-temperature-inducible genes from Arabidopsis encode highly hydrophobic proteins. *Plant Physiol.*, **115**, 569.

82. Navarre, C. and Goffeau, A. (2000) Membrane hyperpolarization and salt sensitivity induced by deletion of PMP3, a highly conserved small protein of yeast plasma membrane. *EMBO J.*, **19**, 2515.

83. Lewis, B. D., Karlin-Neumann, G., Davis, R. W. and Spalding, E. P. (1997) Ca^{2+}-activated anion channels and membrane depolarizations induced by blue light and cold in Arabidopsis seedlings. *Plant Physiol.*, **114**, 1327.

84. Yamaguchi-Shinozaki, K. and Shinozaki, K. (1994) A novel *cis*-acting element in an *Arabidopsis* gene is involved in responsiveness to drought, low-temperature or high-salt stress. *Plant Cell*, **6**, 251.

85. Baker, S. S., Wilhelm, K. S. and Thomashow, M. F. (1994) The 5'-region of *Arabidopsis thaliana cor15a* has cis-acting elements that confer cold-, drought- and ABA-regulated gene expression. *Plant Mol. Biol.*, **24**, 701.

86. Jiang, C., Iu, B. and Singh, J. (1996) Requirement of a CCGAC *cis*-acting element for cold induction of the *BIN115* gene from winter *Brassica napus*. *Plant Mol. Biol.*, **30**, 679.

87. Ouellet, F., Vazquez-Tello, A. and Sarhan, F. (1998) The wheat *wcs120* promoter is cold-inducible in both monocotyledonous and dicotyledonous species. *FEBS Lett.*, **423**, 324.

88. Stockinger, E. J., Gilmour, S. J., Thomashow, M. F. (1997) *Arabidopsis thaliana CBF1* encodes an AP2 domain-containing transcriptional activator that binds to the C-repeat/DRE, a *cis*-acting DNA regulatory element that stimulates transcription in response to low temperature and water deficit. *Proc. Natl Acad. Sci. USA*, **94**, 1035.

89. Gilmour, S. J., Zarka, D. G., Stockinger, E. J., Salazar, M. P., Houghton, J. M. and Thomashow, M. F. (1998) Low temperature regulation of the *Arabidopsis* CBF family of AP2 transcriptional activators as an early step in cold-induced *COR* gene expression. *Plant J.*, **16**, 433.

90. Medina, J., Bargues, M., Terol, J., Pérez-Alonso, M. and Salinas, J. (1999) The *Arabidopsis CBF* family is composed of three genes encoding AP2 domain-containing proteins whose expression is regulated by low temperature but not by abscisic acid or dehydration. *Plant Physiol.*, **119**, 463.

91. Gilmour, S. J., Sebolt, A. M., Salazar, M. P., Everard, J. D. and Thomashow, M. F. (2000) Overexpression of the *Arabidopsis CBF3* transciptional activator mimics multiple biochemical changes associated with cold acclimation. *Plant Physiol.*, **124**, 1854.

92. Capel, J., Jarillo, J. A., Madueño, F., Jorquera, M. J., Martínez-Zapater, J. M. and Salinas, J. (1998) Low temperature regulates Arabidopsis *Lhcb* gene expression in a light-independent manner. *Plant J.*, **13**, 411.

93. Hughes, M. A., Brown, A. P. C., Vural, S. and Dunn, M. A. (1999) Low temperature regulation of gene expression in cereals. In Smallwood, M. F., Calvert, C. M. and Bowles, D. J. (eds.), *Plant Responses to Environmental Stress*. BIOS Scientific Publishers, Oxford, p. 89.

94. Dunn, M. A., White, A. J., Vural, S. and Hughes, M. A. (1998) Identification of promoter elements in a low temperature-responsive gene (*blt4.9*) from barley. *Plant Mol. Biol.*, **38**, 551.

95. Aguan, K., Sugawara, K., Suzuki, N. and Kusano, T. (1993) Low-temperature-dependent expression of a rice gene encoding a protein with a leucine-zipper motif. *Mol. Gen. Genet.*, **240**, 1.

96. Kusano, T., Berberich, T., Harada, M., Suzuki, N. and Sugawara, K. (1995) A maize DNA-binding factor with a bZIP motif is induced by low temperature. *Mol. Gen. Genet.*, **248**, 507.

97. Takatsuji, H. (1999) Zinc-finger proteins: the classical zinc finger emerges in contemporary plant science. *Plant Mol. Biol.*, **39**, 1073.

98. Heino, P., Sandman, G., Lang, V., Nordin, K. and Palva, E. T. (1990) Abscisic acid deficiency prevents development of freezing tolerance in *Arabidopsis thaliana* (L.) Heynh. *Theor. Appl. Genet.*, **79**, 801.

99. Gilmour, S. J. and Thomashow, M. F. (1991) Cold acclimation and cold-regulated gene expression in ABA mutants of *Arabidopsis thaliana*. *Plant. Mol. Biol.*, **17**, 1233.

100. Mäntylä, E., Lang, V. and Palva, E. T. (1995) Role of abscisic acid in drought-induced freezing tolerance, cold acclimation, and accumulation of LTI78 and RAB18 proteins in *Arabidopsis thaliana*. *Plant Physiol.*, **107**, 141.

101. Llorente, F., Oliveros, J. C., Martínez-Zapater, J. M. and Salinas, J. (2000) A freezing-sensitive mutant of *Arabidopsis*, *frs1*, is a new *aba3* allele. *Planta*, **211**, 648.

102. Shinozaki, K. and Yamaguchi-Shinozaki, K. (1997) Gene expression and signal transduction in water-stress response. *Plant Physiol.*, **115**, 327.

103. Nordin, K., Heino, P. and Palva, E. T. (1991) Separate signal pathways regulate the expression of a low-temperature-induced gene in *Arabidopsis thaliana* (L.) Heynh. *Plant Mol. Biol.*, **16**, 1061.

104. Warren, G., Mckown, R., Marin, A. L. and Teutonico, R. (1996) Isolation of mutations affecting the development of freezing tolerance in *Arabidopsis thaliana* (L.) Heynh. *Plant Physiol.*, **111**, 1011.

105. Mckown, R., Kuroki, G. and Warren, G. (1996) Cold responses of *Arabidopsis* mutants impaired in freezing tolerance. *J. Exp. Bot.*, **47**, 1919.

106. Knight, H., Veale, E. L., Warren, G. J. and Knight, M. R. (1999) The *sfr6* mutation in *Arabidopsis* suppresses low-temperature induction of genes dependent on the CRT/DRE sequence motif. *Plant Cell*, **11**, 1.

107. Xin, Z. and Browse, J. (1998) *Eskimo1* mutants of *Arabidopsis* are constitutively freezing-tolerant. *Proc. Natl Acad. Sci. USA*, **95**, 7799.

108. Rudolph, A. S. and Crowe, J. H. (1985) Membrane stabilization during freezing: the role of two natural cryoprotectants, trehalose and proline. *Cryobiology*, **22**, 367.

109. Anchordoguy, T. J., Rudolph, A. S., Carpenter, J. F. and Crowe, J. H. (1987) Modes of interaction of cryoprotectants with membrane phospholipids during freezing. *Cryobiology*, **24**, 324.

110. Nanjo, T., Kobayashi, M., Yoshiba, Y., Kakubari, Y., Yamaguchi-Shinozaki, K. and Shinozaki, K. (1999) Antisense suppression of proline degradation improves tolerance to freezing and salinity in *Arabidopsis thaliana*. *FEBS Lett.*, **461**, 205.

111. Ishitani, M., Xiong, L., Stevenson, B. and Zhu, J-K. (1997) Genetic analysis of osmotic and cold stress signal transduction in Arabidopsis: interactions and convergence of abscisic acid-dependent and abscisic acid-independent pathways. *Plant Cell*, **9**, 1935.

112. Ishitani, M., Xiong, L., Lee, H., Stevenson, B. and Zhu, J-K. (1998) *HOS1*, a genetic locus involved in cold-responsive gene expression in Arabidopsis. *Plant Cell*, **10**, 1151.

113. Lee, H., Xiong, L., Ishitani, M., Stevenson, B. and Zhu, J-K. (1999) Cold-regulated gene expression and freezing tolerance in an *Arabidopsis thaliana* mutant. *Plant J.*, **17**, 301.

7 | Dehydration-stress signal transduction

HANS-HUBERT KIRCH, JONATHAN PHILLIPS AND DOROTHEA
BARTELS

1. Introduction: dehydration-stress studies in plants

Water deficit is an important factor limiting plant productivity and geographical distribution. Water deficit leads to dehydration of plant tissues, but not only water supply also low temperature (in particular freezing stress) and salt conditions cause cellular water loss. Through evolution a broad range of adaptations of plants has occurred in order to cope with dehydration and maintain plant water balance. The response to water deficit and the adaptation process require *de novo* gene transcription. It is the process from sensing water deficit to the transcription of genes that allow adaptation, which will be described here at the molecular level.

Water deficit arises when the rate of transpiration is higher than water uptake. Stomatal transpiration is the main cause of water loss and varies widely between plant species: values range between 1 litre and 10 ml of water loss per m^2 leaf area per hour. Most flowering plants do not survive a loss greater than 5–15 % of their relative water content according to Gaff (1). Responses to dehydration vary depending on the rate of water loss, the degree of water deficit, the duration of the stress, the cell type, and the developmental stage of the plant considered. The severest form of water loss is desiccation when most of the protoplasmic water is lost from the cell and only a very low amount of tightly bound water is left. Desiccation is part of the developmental programme of most higher plants when they form desiccation tolerant seeds. In this chapter two types of adaptive responses are considered:

- reactions that allow plant cells to continue growth; and
- reactions that allow plants to survive but involve some type of metabolic arrest.

The adaptation mechanisms are genetically determined, and one research objective is to discover the relevant genes in order to modify non-adapted plants. The response of plants to water deficit is complex but the accumulation of small osmoprotectant molecules and hydrophilic proteins have been conserved in many plants during evolution. These molecules are supposed to be involved in protection of the cells during dehydration. The synthesis of stress-protective molecules is under transcriptional control.

Following the isolation of many genes with up-regulated expression in response to dehydration, a major research goal has been the discovery of proteins that regulate gene transcription that leads to dehydration tolerance. The current knowledge of signalling pathways in plants during dehydration has mainly been derived from studies with mutants from the genetic model plant *Arabidopsis thaliana* and from examining systems tolerant to dehydration, such as seeds and desiccation-tolerant angiosperm species, in particular the resurrection plant *Craterostigma plantagineum* and lower organisms (e.g. the moss *Tortula ruralis*).

2. Signal perception

In contrast to bacteria or yeast, very little is known about dehydration perception in plant cells. Changes in membrane properties and elevations of Ca^{2+} levels have been postulated as possible primary sensing mechanisms, but there is no direct proof for these mechanisms (2). In bacteria and yeast, mutants have facilitated the identification of osmosensor molecules. Osmosensors are two-component systems that contain a histidine kinase as sensor and a response regulator which relays the phosphorylation signal and leads to transcriptional activation of downstream genes (3). Recently, the gene *ATHK1* from *Arabidopsis* has been identified, and the gene product shows significant homologies to the yeast osmosensor SLN1 (4). The molecule contains a histidine kinase which can complement *sln1* yeast mutants. This implies that ATHK1 functions as an osmosensor and transmits the stress signal to a downstream kinase cascade. However, its functional role in osmosensing in plants has yet to be determined.

Another candidate for mediating primary events could be a member of the phospholipase D (PLD) family that was cloned from *C. plantagineum* (5). PLD cleaves phospholipids within membranes, resulting in the formation of an apolar head group and a phosphorylated lipid moiety, termed phosphatidic acid. Phosphatidic acid has been shown to be a second messenger in signalling pathways in animal cells. PLD activity is strikingly induced within minutes by dehydration in *C. plantagineum* and does not require *de novo* transcription or protein synthesis.

3. The role of abscisic acid: ABA-dependent and ABA-independent signalling pathways

Following cellular perception of water deficit, signalling processes have to be initiated, which finally are translated into molecular and cellular responses, leading to biochemical reactions, metabolic and physiological adjustments, allowing a plant to reprogramme its developmental state and thus to cope with stress (for reviews see references 6–10) (Fig. 1). The phytohormone abscisic acid plays an important role in regulating many physiological plant processes, particularly during seed maturation and germination, and it is also a major signal in response to environmental stresses such as dehydration, high salinity, and low temperature (7, 11, 12). Dehydration

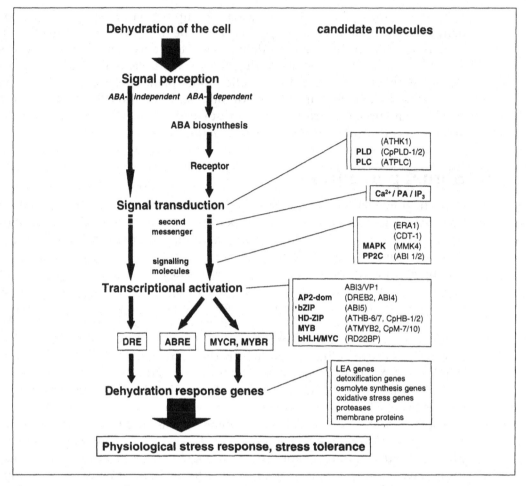

Fig. 1 General overview of elements and molecules involved in dehydration-stress signalling.

probably serves as a trigger for the accumulation of ABA, which in turn activates various stress-associated genes. Increased endogenous ABA levels in response to water stress have been reported in many physiological studies (13, 14). The role of ABA is further supported by the fact that many stress-activated genes can also be induced by exogenous application of ABA (9, 11, 15). Particularly interesting in this respect are callus cultures from the resurrection plant *Craterostigma plantagineum*, which are *per se* not desiccation tolerant but require exogenous ABA to survive drying (16).

All studies to date regarding the molecular events taking place in response to water deficit suggest complex regulatory systems mediating stress-induced gene expression, which involve both ABA-dependent as well as ABA-independent signal transduction pathways (Fig. 1).

3.1 ABA-independent regulation of dehydration-stress response

The expression of the dehydration-inducible gene *RD29A* (*LTI78/COR78*) from *Arabidopsis thaliana* involves two separate signalling pathways in response to water deficit:

- a rapid response and
- a slow response involving ABA.

Yamaguchi-Shinozaki and Shinozaki (17) identified a *cis*-acting promoter element, termed dehydration-responsive element/C-repeat (DRE/CRT) that responds to osmotic stress but not ABA and is essential for the initial rapid induction of *RD29A* (see also Section 5.2). DRE-related motifs have been reported in several other osmotic stress-responsive genes (18–20). The promoter of the maize *RAB17* gene contains two DRE-like elements which participate in ABA-induction, in contrast to the ones identified in *Arabidopsis*. This points to different roles of the DRE element in maize and *Arabidopsis* (21). Protein factors that specifically interact with the DRE/CRT sequence element have been isolated independently and are classified into two groups, CBF1/DREB1 and DREB2 (see Section 5.2).

The three *Arabidopsis* genes *RD19*, *RD21*, and *ERD1*, which encode thiol proteases and a Clp protease regulatory subunit, respond to dehydration but not to ABA (22, 23). The promoter analysis of the *ERD1* gene indicates that *cis*-acting elements other than DRE/CRT are involved in this ABA-independent stress response (24), implying another ABA-independent signalling cascade in dehydration-stress response. None of the components in this pathway has yet been identified.

3.2 ABA-dependent regulation of dehydration-stress response

Gene expression during water stress is regulated by at least two ABA-dependent systems. Expression of *RD29A*, in addition to the fast response (see Section 3.1), is also regulated by a slower ABA-response complex, mediated through a bZIP/ABRE (ABA-responsive element) system (see Section 5.1) and requires the accumulation of endogenous ABA in a dehydration-stress situation (17, 25). Many drought stress- and ABA-responsive gene promoters contain potential ABRE operating as *cis*-acting DNA elements. ABREs were first identified in the wheat *Em1a* and rice *RAB* genes and multiple copies of elements fused to a minimal promoter conferred ABA response to a reporter gene, supporting the model that ABREs are critical for ABA induction (26, 27). Evidence from the barley *HVA22* gene suggests that a coupling element is probably required to specify the function of ABREs and constitute an ABA-response complex (28). ABRE DNA-binding proteins EMBP-1 from wheat were shown to encode bZIP proteins (26). In addition, several other bZIP transcription factors from different species that respond to osmotic stress and ABA have been analysed recently (see also Section 5.1).

A second ABA-dependent signalling pathway requires *de novo* protein synthesis for the expression of dehydration-stress-inducible genes. The *Arabidopsis RD22* gene is dehydration- and ABA-inducible, but its ABA-responsive expression is dependent on protein synthesis (29). A specific region of the *RD22* promoter is necessary for the ABA-dependent gene expression and contains *cis*-acting elements that are homologous to MYC and MYB recognition sequences (30). Two cDNAs coding for MYC/ MYB transcriptional activators have been isolated and may function cooperatively in the ABA-dependent expression of *RD22* (97, see also Section 5.4). Both potential *RD22* transcriptional regulators are also induced by dehydration and ABA, which is consistent with the requirement of protein synthesis for the *RD22* gene expression.

For the desiccation-associated gene *CDeT27-45* from *C. plantagineum* ABREs are probably not the main determinants for ABA- or drought response (15). A novel promoter element of the *CDeT27-45* gene was identified that functions in ABA-inducible gene expression and is necessary but not sufficient for ABA-response. Nuclear proteins from ABA-treated tissue specifically bind to this region (32).

The existence of several pathways confers flexibility to the plant response and allows immediate reactions to stress as well as slow adaptive responses. The DREB/ CBF system responds to dehydration and low temperature and acts in a rapid ABA-independent way early in the stress response. By contrast, ABA-dependent pathways are probably important for slow adaptive processes during dehydration stress. ABA biosynthesis is induced by water deficit and activates the two ABA-dependent signalling systems ABRE/bZIP and MYC/MYB, which require *de novo* protein synthesis in response to ABA.

4. Mutants as tool to dissect the signalling pathways

Genetic analysis is one main approach to examine the molecular mechanisms of dehydration tolerance in plants. The isolation and characterization of mutants affected in environmental stress responses has been proven to be an important tool to dissect osmotic stress signal transduction. Several experimental approaches have been used to isolate mutants relevant to dehydration-stress signal transduction processes:

- the selection and analysis of hormonal mutants affected in either hormone biosynthesis or showing an altered hormone response;

- the search for mutants that show an altered response to increasing osmotic stress conditions, i.e. altered sensitivity to drought, high salinity or cold;

- a third approach based on the use of chimeric gene constructs containing a reporter gene fused to an osmotic stress responsive promoter (33, 34).

The following section gives a short description about the main features and possible functions of mutants, mainly from *Arabidopsis thaliana*, that affect dehydration-stress signal transduction. The features of the mutants are summarized in Table 1.

Table 1 Characteristics of mutations affecting ABA and dehydration stress signal transduction

Species	Mutation	Dominance[a]	Phenotype	Gene product[b]	Responses to stress	Reference
A. thaliana	aba1	R	ABA deficiency	Zeaxanthin epoxidase	yes	41, 43, 44
	aba2	R	ABA deficiency	Xanthoxin oxidase	yes	41, 42, 45
	aba3	R	ABA deficiency	AB aldehyde oxidase[c]	yes	41, 42, 45
Z. mays	vp14	R	Viviparous, ABA deficiency	9-cis-epoxycarotenoid dioxygenase	yes	46, 47
A. thaliana	abi1	SD	ABA insensitivity	Protein phosphatase 2C	yes	49, 52, 53
	abi1-1R[d]	R	ABA supersensitivity	Protein phosphatase 2C	yes	58
	abi2	SD	ABA insensitivity	Protein phosphatase 2C	yes	49,54,56
	abi3	R	ABA insensitivity	VP1 homologous seed-specific transcription factor	no	49, 54, 56
	abi4	R	ABA insensitivity in seeds	APETALA2 domain protein	unknown	49, 51, 61
	abi5	R	ABA insensitivity in seeds	bZIP class transcription factor	unknown	50, 66
	ade1	R	Deregulated ABA response	Unknown	yes	34
	era1	R	ABA hypersensitivity[e]	β-Subunit of farnesyl transferase	yes[f]	71, 72
	hos5	R	Altered osmotic stress response	Unknown	yes	33, 76
C. plantagineum	cdt-1	D	ABA-independent callus desiccation tolerance	Regulatory RNA or short polypeptide	yes	77

[a] Dominance of mutant alleles over wild-type: D, dominant; SD, semi-dominant; R, recessive.
[b] Molecular function of the protein of the wild-type gene.
[c] Defect in the molybdenum cofactor required for aldehyde oxidase activity (35).
[d] Intragenic suppressor mutation of abi1-1.
[e] Seed germination sensitivity to exogenous ABA treatment.
[f] era1 plants exhibit a significant reduction in transpirational water loss during drought treatment.

4.1   ABA-deficient mutants

ABA is thought to be synthesized in higher plants by an indirect (C_{40}) pathway from carotenoid precursors, via violaxanthin, xanthoxin, and ABA-aldehyde (35). The conversion from violaxanthin to xanthoxin probably is the rate-limiting step during water stress (36). ABA-deficient mutants have been isolated from a number of species, like the *flacca* mutant from tomato (37) or *droopy* from potato (38). ABA-biosynthesis mutants have demonstrated that endogenous ABA is required for the regulation of many genes in abiotic stress signalling cascades (39). ABA-deficient mutants also revealed that dehydration-stress-associated genes can be responsive to exogenous ABA, but still be induced by water stress independent of ABA (15, 40).

Here we would like to focus on *aba* mutants from *A. thaliana* and on one mutant from maize (Table1). ABA-deficient (*aba1–aba3*) mutants from *Arabidopsis* were identified in seed germination screens by selection of revertants in non-germinating gibberellin-sensitive *Arabidopsis* lines (41, 42). A characteristic phenotype of most *aba* mutants is an increased transpirational water loss and an enhanced tendency to wilt under water-stress conditions (41, 42), indicating that endogenous ABA levels regulate stomatal closure (12). The *Arabidopsis aba1* mutant is blocked in the epoxidation of zeaxanthin to violaxanthin, the first step of the ABA biosynthesis pathway (43). The *ABA1* gene was cloned through homology with the *ABA2* gene from *Nicotiana plumbaginifolia*, which was isolated by transposon-tagging and encodes the chloroplast-imported zeaxanthin epoxidase (44). The last two steps in the biosynthesis pathway are defined by *Arabidopsis aba2* and *aba3* mutants, respectively, and involve the conversion of xanthoxin to ABA-aldehyde and the oxidation of ABA-aldehyde to ABA, catalysed by constitutively expressed enzymes (35, 42, 45).

Particularly interesting is the isolation and cloning of the viviparous and ABA-deficient mutant from maize, *vp14*, by transposon-tagging (46). VP14 encodes a 9-*cis*-epoxycarotenoid dioxygenase and catalyses the cleavage of violaxanthin or neo-xanthin to xanthoxin, supposed to be the key regulatory step in ABA biosynthesis (46, 47). Transcript analysis revealed that wild-type *VP14* was constitutively expressed in embryos and roots, but heavily induced by dehydration in seedling leaves and thus supports a role for ABA in water-stress gene activation (46). A similar 9-*cis*-epoxycarotenoid dioxygenase was recently characterized in the tomato mutant *notabilis* (48). It is tempting to suggest that the dioxygenases are important modulators in dehydration-stress responses whenever increased ABA levels are involved.

4.2   ABA-insensitive and hypersensitive mutants

Mutants that have an altered response to ABA revealed the involvement of ABA in dehydration-stress signal transduction. ABA-response mutants that either decrease or increase sensitivity to ABA, do have normal endogenous ABA levels and their phenotype cannot be reverted to wild-type by exogenous ABA application. Although such mutants have been described in various plant species, many mutations have essentially been characterized in *Arabidopsis* (Table 1). The ABA-insensitive mutants *abi1–abi5* were identified by the ability of seeds to germinate on normally inhibitory ABA concentrations (49, 50), whereas the *era1–era3* (enhanced response to ABA) mutants were isolated, because seeds did not germinate in the presence of low, non-inhibitory ABA concentrations (35).

4.2.1 *abi1*, *abi2*, and *abi1-1R*

The dominant mutants *abi1* and *abi2* show a significant reduction in seed dormancy, and also display pleiotropic effects in vegetative tissue, as abnormal stomatal regulation and defects in ABA-dependent signal transduction during water stress (39, 49, 51). The *ABI1* and *ABI2* genes have been cloned and encode homologous type 2C Ser/Thr protein phosphatases, indicating that protein phosphorylation and dephosphorylation are involved in ABA signal transduction (52–54). Both mutant proteins have identical amino acid substitutions at equivalent positions, which significantly reduces their PP2C activity (55, 56). Although the similarity between *ABI1* and *ABI2* suggests that their gene products have overlapping functions, both mutants exhibit different effects on some responses to exogenous ABA and water stress (39). Transient expression experiments in maize protoplasts with the wild-type ABI1 PP2C domain suggest a negative regulation of gene expression in response to ABA (57). The role of ABI1 as a negative regulator of ABA signalling has been verified by Gosti *et al.* (58), who isolated seven individual loss-of-function alleles of the *ABI1* gene as intragenic revertants of the *abi1-1* mutant. In contrast to the ABA-insensitive *abi1-1*, *abi1-1R* mutants displayed a hypersensitive phenotype regarding inhibition of seed germination and seedling root growth after exogenous ABA application. The mutant plants also exhibited an increased seed dormancy and enhanced dehydration-stress tolerance. All *abi1-1R* loss-of-function alleles resulted from missense mutations in conserved regions of the ABI1 PP2C domain. The corresponding ABI1-1R proteins lack any detectable PP2C activity *in vitro*, indicating that the loss of ABI1 PP2C function in the mutants is responsible for the enhanced ABA responsiveness.

4.2.2 *abi3* (see also Section 5.7)

A. thaliana abi3 mutants are defective in various aspects of seed development and desiccation tolerance (49, 59–61). The *ABI3* gene has been isolated by positional cloning (62) and encodes a transcriptional activator with homology to the seed-specific ABA-sensitive *VP1* gene from maize (63). *ABI3* expression is seed specific (62). However, ectopic expression of *ABI3* in vegetative *Arabidopsis* tissues activates several LEA genes and can be suppressed by the *abi1* mutation (64), indicating that ABI3 directly participates in ABA signalling. Furthermore, the *ABI3* gene product leads to ABA responsiveness of the *CDeT27-45* promoter from the desiccation tolerant *C. plantagineum* in transgenic *Arabidopsis* (65). Although the biochemical role of ABI3 has not been revealed, evidence from different experimental approaches suggests that ABI3 activates downstream genes and thus is a major player in the ABA-signalling pathways.

4.2.3 *abi4* and *abi5*

abi4 and *abi5* were selected on the basis of ABA-resistant germination, and mutations resulted in pleiotropic effects in seed development but did not alter vegetative growth (50). Both genes have recently been isolated by map-based cloning. *ABI4* encodes a transcription factor with homology to APETALA2 (AP2) domain proteins (66) and *ABI5* is a member of the bZIP transcription factor family (67) with highest homologies to DPBF-1 from sunflower (68) and TRAB1, an ABA regulated rice protein that interacts with VP1 (69). Expression analysis revealed that despite the seed-specific nature of the *abi4* and *abi5* phenotypes, expression of both genes is not seed specific. Low levels of *ABI4* transcript were discovered in vegetative shoots and it is possible that *ABI4* plays a previously undetected role in vegetative growth (66). *ABI4* may not only be involved in stress responses but also in sugar sensing, as a mutation in *ABI4* leads to a reduced sensitivity to sugars (70).

Like *ABI4*, *ABI5* is also expressed at low levels in vegetative tissue, and its function is necessary for the induction of some LEA genes by ABA. Furthermore, ectopic expression of *ABI3* confers a strong ABA-inducible vegetative expression, indicating an interaction of *ABI5* with other ABA response loci (67). Whether *ABI4* and *ABI5* play any role in dehydration stress has still to be determined.

4.2.4 *era1*

The *Arabidopsis* T-DNA insertion mutant *era1-1* confers an exaggerated response to exogenous ABA at the level of germination and stomatal closure and the corresponding *ERA1* gene encodes the β-subunit of farnesyl transferase, an enzyme that catalyses the attachment of a 15-carbon farnesyl lipid to specific proteins (71). *era1* mutants exhibit a significant reduction in transpirational water loss during drought stress and in both *abi1/era1* and *abi2/era1* double mutants, *era1* suppresses the ABA-insensitive phenotype (72). Therefore *ERA1* may function as a negative regulator of ABA signal transduction by modifying signalling proteins through farnesylation for membrane localization. The target of the ERA1 farnesyl transferase may be at or downstream of the ABI1 and ABI2 phosphatases (72, 73).

4.3 Mutants with an altered response to ABA and osmotic stress

Ishitani *et al.* (33) recently presented an efficient approach towards the generation of mutants in dehydration-stress signalling. Their strategy is based on the expression of a luciferase reporter gene driven by the dehydration-stress-responsive *RD29A* promoter, containing the DRE/CRT-repeat and ABRE *cis*-acting element (17), in transgenic *Arabidopsis* plants. A mutagenized seed population was screened for plants showing an altered luciferase expression in response to ABA, low temperature, dehydration or high salinity, resulting in the detection of a large number of either *cos*, *los*, or *hos* (constitutive, low, or high expression of osmotically responsive genes) mutant plants. The identified mutants were categorized into different classes according to the response defects to one or a combination of stress and ABA signals, challenging the current models of parallel ABA-dependent and ABA-independent stress signal transduction pathways (40). Up to now alleles of three different recessive *hos* mutations have been analysed in detail (74–76). One of these mutants, *hos5-1*, showed an increased expression of the *RD29A* gene only in response to osmotic stress and ABA, whereas low-temperature activation was not affected (76). Genetic analysis of *hos5-1 / aba1-1* and *hos5-1 / abi1-1* double mutants revealed that the osmotic stress hypersensitivity was ABA-independent and combined treatments of *hos5-1* with ABA and NaCl or polyethylene glycol exhibited an additive effect on the reporter gene expression. The *HOS5* locus therefore seems to function as a negative regulator of osmotic stress-responsive gene expression shared by ABA-dependent and ABA-independent signalling cascades.

In a similar mutant screen Foster and Chua (34) isolated an ABA-deregulated expression mutant (*ade1*). The recessive *ade1* mutant is ABA-response specific and exhibits a hypersensitive ABA response for both the reporter and endogenous stress-induced genes, thus indicating a negative regulatory function in ABA signalling for *ADE1*. It is likely that the identification and further genetic analysis of other mutants based on the screens reported by Ishitani *et al.* (33) and Foster and Chua (34) will provide a much better understanding about cross-talk and convergence of different stress-signalling pathways.

Furini *et al.* (77) developed another experimental system in the desiccation tolerant resurrection plant *Craterostigma plantagineum*, which allows the identification of dominant mutants in the ABA-mediated dehydration signal transduction pathway. Desiccation tolerance of *C. plantagineum* callus is dependent on the exogenous application of ABA (16). Based on T-DNA activation tagging, a dominant mutation *CDT-1* was isolated. The ectopic activation in transgenic callus cultures conferred desiccation tolerance in the absence of ABA and resulted in the constitutive expression of several ABA- and dehydration-inducible genes. Structural features of the *CDT-1* gene indicate that its function in the ABA signal transduction pathway could be transmitted via a regulatory RNA or a short polypeptide.

5. Identification of transcription factors important for dehydration signalling pathways

Putative transcription factors that may be involved in dehydration signalling pathways have been obtained by way of a variety of experimental approaches. Some genes coding for these factors were isolated based on their DNA binding properties (via Southwestern screening with oligonucleotides as a probe or the yeast one-hybrid system), their expression characteristics (via differential cDNA screening) or based on their capability to interact with other proteins (via the yeast two-hybrid system). Three genes encoding transcription factors that when mutated result in pleiotropic defects in ABA responses have also been identified via a positional cloning approach (62, 66, 67).

Table 2 Transcription factors involved in dehydration signalling

Transcription factor	Recognition DNA consensus sequence	References
bZIP = basic region/leucine zipper protein	ACGT (G-box ABRE)	13
AP2 (APETELA2) EREBP domain	TACCGACAT	60, 84
Homeodomain-leucine zipper (HD-Zip) protein	HDE consensus CAAT (X) ATTG	89, 90
MYB-protein	TGAACT	99
Basic helix-loop helix protein	CANNTG	102
ABI3/VP1		

To date, transcription factors that have been linked to dehydration tolerance belong to six families of structurally related proteins. The corresponding genes encoding these molecules are responsive to dehydration and/or the gene products specifically interact with functional *cis*-acting elements present within dehydration responsive promoter regions, suggesting an association with the dehydration process. However, it should be emphasized that the exact role of most factors in gene regulation is unknown. Here, aspects of each transcription factor family that is thought to be important in dehydration signalling pathways is reviewed. We also draw attention to data concerning the mechanistic properties of the proteins, such as the influence of possible protein–protein interactions, to illustrate how dehydration-responsive gene expression might be regulated. A summary of the transcription factors described in this review is given in Table 2.

5.1 Basic region/leucine zipper (bZIP) proteins

The bZIP domain is composed of DNA binding and dimerization motifs and has been found in transcription factors from a wide range of eukaryotes. This motif comprises of a stretch of about 25 amino acids rich in basic residues, that is the basic region and the DNA-binding motif. Immediately adjacent is the leucine zipper (ZIP) where leucine residues appear every seventh residue over three to six repeats, which is responsible for dimer formation and is folded into an α-helical structure that forms a coiled coil when two ZIPs are arranged in parallel.

To date, most plant bZIPs bind to DNA elements with the ACGT core sequence, otherwise known as the G-box (78). The most characterized G-box *cis*-element in the context of dehydration tolerance is the *ABA-r*esponsive element (ABRE), which contains the palindromic motif CACGTG (13). Electromobility shift assays and DNA footprinting techniques were instrumental in demonstrating that a bZIP protein is capable of specifically interacting with an ABRE (26). Subsequently, dehydration and/or ABA-responsive cDNAs encoding bZIP proteins that bind to ABREs *in vitro* have been cloned from a wide variety of plant species including sunflower (68, 79), *Arabidopsis* (80) and tobacco (81).

That bZIPs have a role in the ABA response was confirmed when the *ABI5* gene from *Arabidopsis*, isolated using a positional cloning approach, was found to encode a member of the bZIP family (67). The previously characterized *abi5-1* allele encodes a protein that lacks the DNA binding and dimerization domains required for ABI5 function. Analysis of *ABI5* expression provides evidence for ABA regulation, cross-regulation by other ABI genes, and possibly autoregulation. Comparison of seed and ABA-inducible vegetative gene expression in wild-type and *abi5-1* plants indicates that *ABI5* regulates a subset of late embryogenesis-abundant (*LEA*) genes.

The next task is to assign individual bZIP transcription factors to specific roles within promoters. This can be problematic, especially in systems such as the G-box, where multiple factors with overlapping DNA-binding specificity exist (82).

5.2 AP2/EREBP domain proteins

The *Arabidopsis* floral homeotic gene *APETELA2* (*AP2*) encodes a protein that contains two copies of a 68 amino acid repeat unit (83), now termed the AP2 domain. In a second report, a family of tobacco proteins named EREBPs that bind to an ethylene-responsive DNA element of pathogenesis-related protein genes, via a 59 amino acid domain, were identified (84). The AP2 and EREBP DNA binding motifs share a high degree of homology and consequently it is now known as the AP2/EREBP domain.

As previously discussed, one ABA-independent pathway overlaps with the cold-acclimation response and regulates the expression of dehydration inducible genes (85). A *cis*-acting element was defined in the promoter of a *LEA*-like gene (*rd29A*) and is responsible for both dehydration and cold induced expression (17). This sequence (TACCGACAT), termed the *dehydration-responsive element* (DRE) or C-repeat, stimulates transcription in response to water deficit and is found in the promoter regions of other dehydration and cold responsive genes (20). Using the yeast one-hybrid system, Liu *et al.* (86) isolated two *Arabidopsis* cDNAs that encode DRE-binding proteins, DREB1A and DREB2A. Analysis of the deduced amino acid sequences indicates that both proteins have an AP2/EREBP domain, a potential nuclear localization sequence, and a possible acidic activation domain. Several lines of evidence led the authors to suggest that both proteins can function as transcriptional activators that bind to the DRE DNA regulatory element and, thus, is likely to have a role in cold- and dehydration-regulated gene expression in *Arabidopsis*. Earlier a cDNA clone that encodes a C-repeat/DRE binding factor, CBF1, was isolated, again via the yeast one-hybrid screening technique (87). Expression of *CBF1* did not, however, change appreciably in plants exposed to low temperature or in detached leaves subjected to water deficit. In contrast to *CBF1*, expression of *DREB1A* and *DREB2A* was induced by low-temperature and dehydration, respectively. These results indicate that two independent families of C-repeat/DRE binding factors function as transcription factors in two separate signal transduction pathways (25).

Simultaneously, direct evidence that an AP2/EREBP domain protein plays a role in the ABA-signalling network came from a positional cloning approach. The *ABI4* gene (see Section 4.2.3) was isolated and the predicted protein product was found to contain an AP2/EREBP DNA binding domain (66). The single mutant allele identified has a single base pair deletion, resulting in a frameshift that disrupts the C-terminal half of the protein but leaves the DNA binding domain intact.

5.3 Homeodomain-leucine zipper (HD-Zip) proteins

The homeodomain (HD) is a specific DNA-binding domain that is highly conserved throughout eukaryotes (88). One class of HD proteins, homeodomain-leucine zipper (HD-Zip) proteins, are putative transcription factors encoded by a class of homeobox genes as yet found only in plants (89, 90). The HD-Zip proteins have a ZIP motif immediately following the HD, similar to that found in the bZIP motif. Based on *in vitro* and *in vivo* DNA-binding assays, HD-Zips specifically bind to pseudo-palindromic sequences, termed *homeodomain recognition elements* (HDEs) (91, 92); however, the identification of a functional HDE in a gene promoter has not been published.

To date five HD-Zip transcripts that are responsive to dehydration have been reported. The first report, made by Söderman *et al.* (93), concerned the HD-Zip gene *ATHB7* from *Arabidopsis*. A closely related gene *ATHB12* was later isolated, which shares many of the characteristics of *ATHB7* (94). *ATHB7* transcripts are present in all organs of the plant at low levels but expression is up-regulated by water deficit as well as by exogenous treatment with ABA. The induction of *ATHB7* is mediated strictly via ABA, since no induction of *ATHB7* was detectable in the ABA-deficient mutant *aba3* subjected to drought treatment. Induction levels in two ABA-insensitive mutants *abi2* and *abi3* were similar to the wild-type response. In the *abi1* mutant, however, induction was impaired as 100-fold higher concentrations of ABA were required for a maximum induction as compared with wild-type. In this mutant the *ATHB7* response was reduced also after dehydration treatment. These results indicate that *ATHB7* is transcriptionally regulated in an ABA-dependent manner and may act in a signal transduction pathway, which mediates a drought response and also includes the ABI1 protein.

A second ABA and dehydration responsive HD-Zip gene, *ATHB6*, from *Arabidopsis* has also been reported by Söderman *et al.* (95). In contrast to *ATHB7*, *ATHB6* induction is impaired in the two ABA-insensitive mutants, *abi1* and *abi2*, but unaffected in the *abi3* mutation. Again, the induction is ABA-dependent since no increase in *ATHB6* transcript is detectable in the ABA-deficient mutant *aba3* subjected to drought treatment. These results suggest that ATHB6 may act downstream of both ABI1 and ABI2 in a signal transduction pathway that mediates a drought stress response. A fusion of the *ATHB6* promoter with the reporter gene GUS in transgenic *Arabidopsis* showed high-level expression in leaf primordia. Expression in developing cotyledons, leaves, roots, and carpels was restricted to regions of cell division and/or differentiation. The expression in the cotyledons was detectable in the epidermis and high in the stomatal cells. In mature cotyledons and leaves, the marker gene was expressed only in the vascular tissue. In addition to a role in dehydration signalling, ATHB6 may therefore also play a role in plant development.

The molecular dissection of desiccation tolerance in *C. plantagineum* led to the isolation of two dehydration-stress-inducible HD-Zip genes (*CpHB-1* and 2) (96). When the coding region of *CpHB-1* was used as bait in the yeast two-hybrid system, the ability of CpHB-1 to form homodimers was demonstrated. The two-hybrid system was also used to isolate *CpHB-2*, the gene product of which heterodimerizes with CpHB-1. Both transcripts are inducible by dehydration in leaves and roots, but steady-state levels vary in response to exogenously applied ABA. Although expression of *CpHB-1* is not inducible by ABA, the transcript level of *CpHB-2* increases during ABA treatment. Both genes are expressed at very early stages of dehydration and thus may be involved in the regulation of gene expression during dehydration. *CpHB-1* and 2 differential expression in response to ABA suggests that they act in different branches of the dehydration-induced signalling network. *In vitro* binding studies revealed that CpHB-1 specifically binds to HDEs. Using an HDE for *in vitro* binding studies with nuclear proteins from dehydrated leaves, an inducible DNA-protein complex was identified.

What are the implications of the possible interaction between HD-Zips that are thought to be regulated by two independent branches of the dehydration signalling network? In general, factors capable of homo- and/or heterodimerization are thought to increase the length and variability of DNA target sequences. This hypothesis is supported by the apparent change in DNA binding properties of CpHB-1 and CpHB-2 in electrophoretic mobility shift assays. CpHB-2 apparently has a weak affinity to one variant of the consensus homeodomain element (HDE-1) relative to that observed for CpHB-1. Intermediate binding affinities are observed when both HD-Zips are combined in different ratios. The titration of DNA-binding activity suggests that cross-talk between pathways may result in a 'fine control mechanism' that in turn may govern the transcriptional regulation of target genes.

5.4 MYB proteins

The MYB domain, defined as the DNA-binding domain of the proto-oncogene Myb, includes three copies of a repeat composed of approximately 50 amino acids with three conserved tryptophan residues, which are spaced at 18–19 amino acid intervals and build up a hydrophobic core (97, 98). Plants have a large number of *MYB* genes with various proposed functions (99).

The *Arabidopsis* gene *Atmyb2* encodes a transcription factor, which is homologous to Myb, and was cloned from a cDNA library prepared from dehydrated *Arabidopsis* rosette plants (100). *Atmyb2* mRNA is induced by dehydration and disappears upon rehydration. The *Atmyb2* mRNA also accumulates upon salt stress and after application of ABA. A GUS reporter gene driven by the *Atmyb2* promoter is induced by dehydration and salt stress in transgenic *Arabidopsis* plants, indicating that *Atmyb2* is responsive to dehydration at the transcriptional level. Atmyb2 is capable of binding specifically to oligonucleotides that contain a consensus Myb recognition sequence (TAACTG), such as that found in the *simian virus 40* enhancer and the maize *bronze-1* promoter. In a second study (see Section 5.5), the same research group showed that the Atmyb2 protein activates transcription of the ABA and dehydration responsive *rd22* promoter when fused to the GUS reporter gene in a transient transactivation experiment using *Arabidopsis* leaf protoplasts (31). This discovery leads to the hypothesis that MYB transcription factors are involved in the regulation of genes that are responsive to water stress, at least in *Arabidopsis*.

Three *Myb* genes isolated from the resurrection plant *Craterostigma plantagineum* (*cpm 5/7/10*) and *Atmyb2* have closely related MYB domains (101). The remainder of the deduced protein sequences have no similarity to other reported sequences. The *Myb* genes in the *C. plantagineum* genome comprise a small gene family of 6–8 members. RNA blot experiments showed that *cpm10* is specifically expressed in undifferentiated callus tissue and up-regulated by treatment with exogenous ABA that renders the callus desiccation tolerant. The expression of *cpm7*, on the other hand, is localized in roots and up-regulated by dehydration.

5.5   Basic helix–loop–helix (bHLH) proteins

The bHLH domain is composed of two subdomains, the basic region (similar to that found in bZIPs) responsible of DNA binding and the HLH region for dimerization. bHLH proteins recognize DNA elements with a CANNTG sequence; however, regions flanking the core element are also important for binding (102).

As already mentioned (see Section 5.4), promoter analysis of the dehydration-responsive gene, *rd22*, established that a 67 bp DNA fragment is sufficient for dehydration- and ABA-induced gene expression and that this DNA fragment contains two closely located putative recognition sites for the bHLH protein Myc and one putative recognition site for Myb. Analysis of the 67 bp region of the *rd22* promoter in transgenic tobacco plants revealed that both the first Myc site and the Myb recognition site function as *cis*-acting elements in the dehydration-induced expression of the *rd22* gene. A cDNA encoding a Myc-related DNA-binding protein was isolated by DNA-ligand-binding screening, using the 67 bp region as a probe, and designated *rd22BP1* (31). The *rd22BP1* gene encodes a 68 kDa protein that has a typical DNA-binding domain of a basic region helix–loop–helix leucine zipper motif in Myc-related transcription factors. The rd22BP1 protein binds specifically to the first Myc recognition site in the 67 bp fragment. RNA gel-blot analysis revealed that transcription of the *rd22BP1* gene is induced by dehydration stress and ABA treatment, and its induction precedes that of *rd22*. Thus the current hypothesis is that the rd22BP1 (Myc) and Atmyb2 (Myb) proteins function as transcriptional activators in the dehydration- and ABA-inducible expression of the *rd22* gene.

5.6   ABI3/VP1 proteins

Genetic and molecular evidence supports the hypothesis that the maize VP1 and *Arabidopsis* ABI3 proteins are homologous transcription factors essential for ABA action in seeds (see Section 4.2.2; 12). VP1 has been shown to transactivate the promoter of *Em*, an ABA-responsive LEA gene, in transient assay systems in maize and rice (63, 103). The transcription factor VP1 regulates maturation and dormancy in plant seeds by activating genes responsive to the stress hormone ABA. Although activation involves ABA-responsive elements (ABREs), VP1 itself does not specifically bind ABREs. This led to the hypothesis that VP1/ABI3 proteins are transcriptional co-activators that physically interact with G-box binding proteins (104, 105). Further evidence was provided when a basic region leucine zipper (bZIP) factor, TRAB1, which interacts with both VP1 and ABREs, was cloned (69). Transcription from a chimeric promoter with GAL4-binding sites was ABA-inducible if cells expressed a GAL4 DNA-binding domain/TRAB1 fusion protein. Results indicate that TRAB1 is a *trans*-acting factor involved in ABA-regulated transcription and reveal a molecular mechanism for the VP1-dependent, ABA-inducible transcription that controls maturation and dormancy in plant embryos.

6. Signalling components involved in water stress

There are a number of factors that are likely to act downstream of signal perception and upstream of promoter-binding proteins. Protein kinases activated by ABA and/or dehydration have been isolated from wheat (106), *Arabidopsis thaliana* (107) and the resurrection plant *Craterostigma plantagineum* (108). Furthermore, calcium-dependent protein kinases were identified in dehydrated *A. thaliana* plants (109). Mitogen-activated kinases (MAPK) are thought to be involved not only in environmental stress-response signal transduction pathways in yeast, but also in plants (110). MAPK genes activated by drought stress and/or ABA have been identified in animals and yeast *Arabidopsis* (111) and *C. plantagineum* (H. H. Kirch, R. Hübner, and D. Bartels, unpublished data). An alfalfa MAPK, MMK4, has recently been shown to be activated transcriptionally and post-translationally by drought and low temperature (112). Phospholipid signalling involving phospholipases has been suggested to play a role in several developmental and physiological processes in plants (113). Evidence that it may also be involved in dehydration response was supported by the isolation of a rapidly drought-induced phospholipase C from *Arabidopsis* (*AtPLC*) (114) and a dehydration-responsive phoshatidylinositol 4-phosphate kinase (PIP5K) (115). Ritchie and Gilroy (116) recently reported that ABA treatment results in a quick increase of phospholipase D (PLD) activity in barley aleurone, and desiccation triggers induction and activity of two PLDs in *C. plantagineum* (5).

7. Cross-talk of the dehydration-stress signalling pathway

Mutant analysis and promoter studies of dehydration-regulated genes provide evidence that dehydration-activated signalling pathways interact with other stress pathways, in particular with low temperature and salt stress.

The concept of communicating signalling pathways inducing osmotic stress responses is strongly supported by results obtained from the analysis of mutants that influence the effect of ABA, dehydration, high salinity, and low temperature on a stress-activated promoter-reporter gene fusion. Ishitani *et al.* (33) propose that dehydration, salt and cold signal transduction pathways might interact and converge to activate stress-associated genes based on the observation that many mutants concomitantly show reduced or enhanced responses to low temperature, osmotic stress, and ABA. This view is further corroborated by investigations of connective effects of abiotic stress treatments and ABA on stress-associated gene expression in *Arabidopsis*. Xiong *et al.* (76) can show that combined stress treatments have both synergistic as well as antagonistic effects on the regulation of stress-responsive gene expression, depending on duration of stress treatments and the nature of the stresses involved. Recent analysis of the sugar-sensing mutant *SUN6* revealed that this mutant is identical to the *ABI4* mutant (70). This observation even points to a connection between osmotic stress and sugar sensing.

8. Exploiting the knowledge of dehydration-stress signal transduction to engineer stress tolerance

Improvement in yield in relation to limiting water supply is agronomically desirable. In environments where water deficits can occur at any stage of growth, knowledge of the mechanisms of desiccation tolerance should play a role in survival of the crop until soil moisture levels improve. Transgenic approaches offer a powerful means to gain information towards a better understanding of the mechanisms that govern stress tolerance. They also open up new opportunities to improve stress tolerance by incorporating genes involved in stress protection from any source into agriculturally important crop plants. To date, the 'transgenic approach' has been to transfer a single gene into plants and then observe the phenotypic and biochemical changes before and after a specific stress treatment. A major limitation of this strategy is that the functions of very few genes involved in desiccation tolerance have been established, or not enough is known about the regulatory mechanisms. Molecular marker analysis has been used to study dehydration tolerance principally in cereals (117). Results from this type of approach and analysis of gene expression patterns suggest that tolerance to water deficit is a complex quantitative trait, since no single diagnostic marker for tolerance has yet been found. The transformation of plants using regulatory genes is therefore an attractive approach for producing dehydration-tolerant plants. Since the products of these genes regulate gene expression and signal transduction under stress conditions, the over-expression of these genes can activate the expression of many stress tolerance genes simultaneously.

As discussed earlier (see Section 5.2), expression of a subset of dehydration- and cold-regulated genes is mediated by transcription factors belonging to the AP2/ EREBP domain family of proteins via a DNA regulatory element termed the DRE (dehydration-responsive element)/CRT (C-repeat) *Arabidopsis* (86, 118). A major scientific breakthrough in drought tolerance research was then made when Kasuga *et al.* (119) transformed *Arabidopsis thaliana* with a cDNA encoding the transcriptional regulator DREB1a driven by either the constitutive CaMV 35S promoter or an abiotic stress-inducible promoter. The over-expression of this gene activated the expression of many stress tolerance genes such as *LEA* genes and *P5CS*. In all cases, the transgenic plants were more tolerant to drought, salt, and freezing stresses. However, the constitutive over-expression of DREB1a also resulted in severe growth retardation under normal growth conditions. By contrast, the stress-inducible expression of this gene had minimal effects on plant growth and provided greater tolerance to stress conditions than genes driven by a strong constitutive promoter. These data were in agreement with an earlier report that demonstrated that over-expression of another C-repeat/DRE-binding factor, *CBF1*, in transgenic *Arabidopsis* plants at non-acclimating temperatures induces *COR* gene expression and increases plant freezing tolerance (120).

Acknowledgements

The work in the laboratory of D. Bartels was supported by the DFG Schwerpunktprogramm 'Molekulare Analyse der Phytohormonwirkung'.

References

1. Gaff, D. F. (1971) Desiccation tolerant flowering plants in southern Africa. *Science* **174**, 1033.
2. Knight, H. (2000) Calcium signalling during abiotic stress in plants *Int. Rev. Cytol.* **195**, 269.
3. Wurgler-Murphy S. M. and Saito, S. (1997) Two-component signal transducers and MAPK cascades. *Trends Biochem. Sci.*, **22**, 172.
4. Urao, T., Yakubov, B., Yamaguchi-Shinozaki, K., Seki, M., Hirayama, T. and Shinozaki, K. (1999) A transmembrane hybrid-type histidine kinase in Arabidopsis functions as an osmosensor. *Plant Cell*, **11**, 1743.
5. Frank, W., Munnik, T., Kerkmann, K., Salamini, F. and Bartels, D. (2000) Water deficit triggers phospholipase D activity in the resurrection plant *Craterostigma plantagineum*. *Plant Cell*, **12**, 123.
6. Bohnert, H. J., Nelson, D. E. and Jensen, R. G. (1995) Adaptations to environmental stresses. *Plant Cell*, **7**, 1099.
7. Bray, E. A. (1997) Plant responses to water deficit. *Trends Plant Sci.*, **2**, 48.
8. Bohnert, H. J. and Shevleva, E. (1998) Plant stress adaptations – making metabolisms move. *Curr. Opin. Plant Biol.*, **1**, 267.
9. Tabaeizadeh, Z. (1998) Drought-induced responses in plant cells. *Int. Rev. Cytol.*, **182**, 193.
10. Chandler, J. and Bartels, D. (1999) Plant Desiccation. In Lerner, H. R. (ed.), *Plant Responses to Environmental Stresses: from Phytohormones to Genome Reorganization*. Marcel Dekker Inc., New York, p. 575.
11. Chandler, P. M. and Robertson, M. (1994) Gene expression regulated by abscisic acid and its relation to stress tolerance. *Ann. Rev. Plant Physiol. Plant Mol. Biol.*, **45**, 113.
12. Leung, J. and Giraudat, J. (1998) Abscisic acid signal transduction. *Ann. Rev. Plant Physiol. Plant Mol. Biol.*, **49**, 199.

13. Busk, P. K. and Pages, M. (1998) Regulation of abscisic acid-induced transcription. *Plant Mol. Biol.*, **37**, 425.

14. Swamy, P. M. and Smith, B. (1999) Role of abscisic acid in plant stress tolerance. *Curr. Sci.*, **76**, 1220.

15. Ingram, J. and Bartels, D. (1996) The molecular basis of dehydration tolerance in plants. *Annu. Rev. Plant Physiol. Plant Mol. Biol.*, **47**, 377.

16. Bartels, D., Schneider, K., Terstappen, G., Piatkowski, D. and Salamini, F. (1990) Molecular cloning of abscisic acid-modulated genes which are induced during desiccation of the resurrection plant *Craterostigma plantagineum*. *Planta*, **175**, 485.

17. Yamaguchi-Shinozaki, K. and Shinozaki, K. (1994) A novel *cis*-acting element in an *Arabidopsis* gene is involved in responsiveness to drought, low-temperature, or high-salt stress. *Plant Cell*, **6**, 251.

18. Nordin, K., Heino, P. and Palva, E. T. (1991) Separate signaling pathways regulate the expression of a low-temperature-induced gene in *Arabidopsis thaliana* (L.) Heynh. *Plant Mol. Biol.*, **16**, 1061.

19. Baker, S. S., Wilhelm, K. S. and Thomashow, M. F. (1994) The 5'-region of *Arabidopsis thaliana cor15a* has *cis*-acting elements that confer cold-, drought-, and ABA-regulated gene expression. *Plant Mol. Biol.*, **24**, 701.

20. Wang, H., Datla, R., Georges, F., Loewen, M. and Cutler, A. J. (1995) Promoters from *kin1* and *cor6.6*, two homologous *Arabidopsis thaliana* genes: transcriptional regulation and gene expression induced by low temperature, ABA, osmoticum and dehydration. *Plant Mol. Biol.*, **28**, 605.

21. Busk, P. K., Jensen, A. B. and Pagès, M. (1997) Regulatory elements *in vivo* in the promoter of the abscisic acid responsive gene *rab17* from maize. *Plant J.*, **11**, 1285.

22. Koizumi, M., Yamaguchi-Shinozaki, K., Tsuyi, H. and Shinozaki, K. (1993) Structure and expression of two genes that encode distinct drought-inducible cysteine proteases in *Arabidopsis thaliana*. *Gene*, **129**, 175.

23. Kiyosue, T., Yamaguchi-Shinozaki, K. and Shinozaki, K. (1993) Characterization of a cDNA for a dehydration-inducible gene that encodes a Clp A, B like protein in *Arabidopsis thaliana*. *Biochem. Biophys. Res. Commun.*, **196**, 1214.

24. Nakashima, K., Kiyosue, T., Yamaguchi-Shinozaki, K. and Shinozaki, K. (1997) A nuclear gene, *erd1*, encoding a chloroplast-targeted Clp protease regulatory subunit homolog is not only induced by water stress but also developmentally upregulated during senescence in *Arabidopsis thaliana*. *Plant J.*, **12**, 851.

25. Shinozaki, K. and Yamaguchi-Shinozaki, K. (2000) Molecular responses to dehydration and low temperature: differences and cross-talk between two stress-signaling pathways. *Curr. Opin. Plant Biol.*, **3**, 217.

26. Guiltinan, M. J., Marcotte Jr, W. R. and Quatrano, R. S. (1990) A plant leucine zipper protein that recognizes an abscisic acid response element. *Science*, **250**, 267.

27. Skriver, K., Olsen, P. L., Rogers, J. C. and Mundy, J. (1991) *Cis*-acting DNA elements responsive to gibberellin and its antagonist abscisic acid. *Proc. Natl Acad. Sci. USA*, **88**, 7266.

28. Shen, Q. and Ho, T. H. D. (1995). Functional dissection of an abscisic acid (ABA)-inducible gene reveals two independent ABA-responsive complexes each containing a G-box and a novel *cis*-acting element. *Plant Cell*, **7**, 295.

29. Yamaguchi-Shinozaki, K. and Shinozaki, K. (1993) The plant hormone abscisic acid mediates drought-induced expression but not seed-specific expression of *rd22*, a gene responsive to dehydration stress in *Arabidopsis thaliana*. *Mol. Gen. Genet.*, **238**, 17.

30. Iwasaki, T., Yamaguchi-Shinozaki, K. and Shinozaki, K. (1995) Identification of a *cis*-regulatory region of a gene in *Arabidopsis thaliana* whose induction by dehydration is mediated by abscisic acid and requires protein synthesis. *Mol. Gen. Genet.*, **247**, 391.

31. Abe, H., Yamaguchi-Shinozaki, K., Urao, T., Iwasaki, T., Hosokawa, D. and Shinozaki, K. (1997) Role of Arabidopsis Myc and Myb homologs in drought-and abscisic acid-regulated gene expression. *Plant Cell*, **9**, 1859.

32. Nelson, D., Salamini, F. and Bartels, D. (1994) Abscisic acid promotes novel DNA-binding activity to a desiccation-related promoter of *Craterostigma plantagineum*. *Plant J.*, **5**, 451.

33. Ishitani, M., Xiong, L., Stevenson, B. and Zhu, J.-K. (1997) Genetic analysis of osmotic and cold stress signal transduction in *Arabidopsis*: Interactions and convergence of abscisic acid-dependent and abscisic acid-independent pathways. *Plant Cell*, **9**, 1935.

34. Foster, R. and Chua, N.-H. (1999) An *Arabidopsis* mutant with deregulated ABA gene expression: implications for negative regulator function. *Plant J.*, **17**, 363.

35. Cutler, A. J. and Krochko, J. E. (1999) Formation and breakdown of ABA. *Trends Plant Sci.*, **4**, 472.

36. Kende, H. and Zeevaart, J. A. D. (1997) The five 'classical' plant hormones. *Plant Cell*, **9**, 1197.

37. Taylor, I. B., Linforth, R. S. T., Al Naieb, R. J., Bowman, W. R. and Marples, B. A. (1988) The wilty tomato mutants *flacca* and *sitiens* are impaired in the oxidation of ABA-aldehyde to ABA. *Plant Cell Environ.*, **11**, 739.

38. Quarrie, S. A. (1982) *Droopy*: a wilty mutant of potato deficient in abscisic acid. *Plant Cell Environ.*, **5**, 23.

39. Merlot, S. and Giraudat, J. (1997) Genetic analysis of abscisic acid signal transduction. *Plant Physiol.*, **114**, 751.

40. Shinozaki, K. and Yamaguchi-Shinozaki, K. (1996) Molecular responses to drought and cold stress. *Curr. Opin. Biotechnol.*, **7**, 161.

41. Koornneef, M., Jorna, M. L., Brinkhorst-van der Swan, D. C. L. and Karssen, C. M. (1982) The isolation of abscisic acid (ABA) deficient mutants by selection of induced revertants in non-germinating gibberellin sensitive lines of *Arabidopsis thaliana* (L.) Heynh. *Theor. Appl. Genet.*, **61**, 385.

42. Léon-Kloosterziel, K. M., Gil, M. A., Ruijs, G. J., Jacobsen, S. E., Olszewski, N. E., Schwartz, S. H., Zeevaart, J. A. and Koornneef, M. (1996) Isolation and characterization of abscisic acid-deficient Arabidopsis mutants at two new loci. *Plant J.*, **10**, 655.

43. Rock, C. D. and Zeevaart, J. A. D. (1991) The *aba* mutant of *Arabidopsis thaliana* is impaired in epoxy-carotenoid biosynthesis. *Proc. Natl Acad. Sci. USA*, **88**, 7496.

44. Marin, E., Nussaume, L., Quesada, A., Gonneau, M., Sotta, B., Hugueney, P., Frey, A. and Marion-Poll, A. (1996) Molecular identification of zeaxanthin epoxidase of *Nicotiana plumbaginifolia*, a gene involved in abscisic acid biosynthesis and corresponding to *ABA* locus of *Arabidopsis thaliana*. *EMBO J.*, **15**, 2331.

45. Schwartz, S. H., Leon-Kloosterziel, K. M., Koornneef, M. and Zeevaart, J. A. D. (1997a) Biochemical characterization of the *aba2* and *aba3* mutants in *Arabidopsis thaliana*. *Plant Physiol.*, **114**, 161.

46. Tan, B. C., Schwartz, S. H., Zeevaart, J. A. D. and McCarty, D. R. (1997) Genetic control of abscisic acid biosynthesis in maize. *Proc. Natl Acad. Sci. USA*, **94**, 12235.

47. Schwartz, S. H., Tan, B. C., Gage, D. A., Zeevaart, J. A. D. and McCarty, D. R. (1997) Specific oxidative cleavage of carotenoids by *VP14* of maize. *Science*, **276**, 1872.

48. Burbidge, A., Grieve, T. M., Jackson, A., Thompson, A., McCarty, D. R. and Taylor, I. B. (1999) Characterization of the ABA-deficient tomato mutant *notabilis* and its relationship with maize *Vp14*. *Plant J.*, **17**, 427.

49. Koornneef, M., Reuling, G. and Karssen, C. M. (1984) The isolation and characterization of abscisic acid-insensitive mutants of *Arabidopsis thaliana*. *Physiol. Plant.*, **61**, 377.

50. Finkelstein, R. R. (1994) Mutations at two new *Arabidopsis* ABA response loci are similar to the *abi3* mutations. *Plant J.*, **5**, 765.

51. Giraudat, J., Parcy, F., Bertauche, N., Gosti, F., Leung, J., Morris, P.-C., Bouvie-Durand, M. and Vartanian, N. (1994) Current advances in abscisic acid action and signaling. *Plant Mol. Biol.*, **26**, 1557.

52. Leung, J., Bouvier-Durand, M., Morris, P.-C., Guerrier, D., Chefdor, F. and Giraudat, J. (1994) *Arabidopsis* ABA-response gene ABI1: Features of a calcium-modulated protein phosphatase. *Science*, **264**, 1448.

53. Meyer, K., Leube, M. P. and Grill, E. (1994) A protein phosphatase 2C involved in ABA signal transduction in *Arabidopsis thaliana*. *Science*, **264**, 1452.

54. Rodriguez, P. L., Benning, G. and Grill, E. (1998) ABI2, a second protein phosphatase 2C involved in abscisic acid signal transduction in *Arabidopsis*. *FEBS Lett.*, **421**, 185.

55. Bertauche, N., Leung, J. and Giraudat, J. (1996) Protein phosphatase activity of abscisic acid insensitive 1 (ABI1) protein from *Arabidopsis thaliana*. *Eur. J. Biochem.*, **241**, 193.

56. Leung, J., Merlot, S. and Giraudat, J. (1997) The *Arabidopsis* ABSCISIC ACID–INSENSITIVE2 (ABI2) and ABI1 genes encode redundant protein phosphatases 2C involved in abscisic acid signal transduction. *Plant Cell*, **9**, 759.

57. Sheen, J. (1996) Ca^{2+}-dependent protein kinases and stress signal transduction. *Science*, **274**, 1900.

58. Gosti, F., Beaudoin, N., Serizet, C., Webb, A. A. R., Vartanian, N. and Giraudat, J. (1999) ABI1 protein phosphatase 2C is a negative regulator of abscisic acid signaling. *Plant Cell*, **11**, 1897.

59. Koornneef, M., Hanhart, C. J., Hilhorst, H. W. M. and Karssen, C. M. (1989) *In vivo* inhibition of seed development and reserve protein accumulation in recombinants of abscisic acid biosynthesis and responsiveness mutants in *Arabidopsis thaliana*. *Plant Physiol.*, **90**, 463.

60. Finkelstein, R. R. and Somerville, C. R. (1990) Three classes of abscisic acid (ABA)-insensitive mutations of *Arabidopsis* define genes that control overlapping subsets of ABA responses. *Plant Physiol.*, **94**, 1172.

61. Nambara, E., Naito, S. and McCourt, P. (1992) A mutant of *Arabidopsis* which is defective in seed development and storage protein accumulation is a new *abi3* allele. *Plant J.*, **2**, 435.

62. Giraudat, J., Hauge, B. M., Valon, C., Smalle, J., Parcy, F. and Goodman, H. M. (1992) Isolation of the *Arabidopsis ABI3* gene by positional cloning. *Plant Cell*, **4**, 251.

63. McCarty, D. R., Hattori, T., Carson, C. B., Vasil, V., Lazar, M. and Vasil, I. K. (1991) The *Viviparous-1* developmental gene of maize encodes a novel transcriptional activator. *Cell*, **66**, 895.

64. Parcy, F., Valon, C., Raynal, M., Gaubie-Comella, P., Delseny, M. and Giraudat, J. (1994) Regulation of gene expression programs during *Arabidopsis* seed development: role of the *ABI3* locus and of endogenous abscisic acid. *Plant Cell*, **6**, 1567.

65. Furini, A., Parcy, F., Salamini. F. and Bartels, D. (1996) Differential regulation of two ABA-inducible genes from *Craterostigma plantagineum* in transgenic *Arabidopsis* plants. *Plant Mol. Biol.*, **30**, 343.

66. Finkelstein, R. R., Wang, M. L., Lynch, T. J., Rao, S. and Goodman, H. M. (1998) The *Arabidopsis* abscisic acid response locus *ABI4* encodes an APETALA2 domain protein. *Plant Cell*, **10**, 1043.

67. Finkelstein, R. R. and Lynch, T. J. (2000) The Arabidopsis abscisic acid response gene *ABI5* encodes a basic leucine zipper transcription factor. *Plant Cell*, **12**, 599.

68. Kim, S. Y., Chung, H.-J. and Thomas, T. L. (1997) Isolation of a novel class of bZIP transcription factors that interact with ABA-responsive and embryo-specification elements in the *Dc3* promoter using a modified yeast one-hybrid system. *Plant J.*, **11**, 1237.

69. Hobo, T., Kowyama, Y. and Hattori, T. (1999) A bZIP factor, TRAB1, interacts with VP1 and mediates abscisic acid-induced transcription. *Proc. Natl Acad. Sci.*, **96**, 15348.

70. Huijser, C., Kortstee, A., Pego, J., Weisbeek, P., Wismann, E. and Smeekens, S. (2000) The Arabidopsis *Sucrose uncoupled-6* gene is identical to *Abscisic acid insensitive-4*: involvement of abscisic acid in sugar responses. *Plant J.*, **23**, 577.

71. Cutler, S., Ghassemian, M., Bonetta, D., Cooney, S. and McCourt, P. (1996) A protein farnesyl transferase involved in abscisic acid signal transduction in *Arabidopsis. Science*, **273**, 1239.

72. Pei, Z.-M., Ghassemian, M., Kwak, C. M., McCourt, P. and Schroeder, J. I. (1998) Role of farnesyl transferase in ABA regulation of guard cell anion channels and plant water loss. *Science*, **282**, 287.

73. Nambara, E. and McCourt, P. (1999) Protein farnelysation in plants: a greasy tale. *Curr. Opin. Plant Biol.*, **2**, 388.

74. Ishitani, M., Xiong, L., Lee, H., Stevenson, B. and Zhu, J.-K. (1998) *HOS1*, a genetic locus involved in cold-responsive gene expression in *Arabidopsis. Plant Cell*, **10**, 1151.

75. Lee, H., Xiong, L., Ishitani, M., Stevenson, B. and Zhu, J.-K. (1999) Cold-regulated gene expression and freezing tolerance in an *Arabidopsis thaliana* mutant. *Plant J.*, **17**, 301.

76. Xiong, L., Ishitani, M., Lee, H. and Zhu, J.-K. (1999) *HOS5* – a negative regulator of osmotic stress-induced gene expression in *Arabidopsis thaliana. Plant J.*, **19**, 569.

77. Furini, A., Csaba, K., Salamini, F. and Bartels, D. (1997) High level transcription of a member of a repeated gene confers dehydration tolerance to callus tissue of *Craterostigma plantagineum. EMBO J.*, **16**, 3599.

78. Williams, M. E., Foster, R. and Chua, N. H. (1992) Sequences flanking the hexameric G-box core CACGTG affect the specificity of protein binding. *Plant Cell*, **4**, 485.

79. Kim, S. Y. and Thomas, T. L. (1998) A family of novel basic leucine zipper proteins binds to seed-specification elements in the carrot *Dc3* gene promoter. *J. Plant Physiol.*, **152**, 607.

80. Choi, H. I., Hong, J. H., Ha, J. O., Kang, J. Y. and Kim, S. Y. (2000) ABFs, a family of ABA-responsive element binding factors. *J. Biol. Chem.*, **275**, 1723.

81. Oeda, K., Salinas, J. and Chua, N. H. (1991) A tobacco bZip transcription activator (TAF-1) binds to a G-box-like motif conserved in plant genes. *EMBO J.*, 10, 1793.

82. Lu, G., Paul, A-L., McCarty, D. R. and Ferl, R. J. (1996) Transcription factor veracity: Is GBF3 responsible for ABA-regulated expression of Arabidopsis *Adh. Plant Cell*, **8**, 847.

83. Jofuku, K. D., Den Boer, B. G. W., Van Montagu, M. and Okamura, J. K. (1994) Control of *Arabidopsis* flower and seed development by the homeotic gene *APETALA2. Plant Cell*, **6**, 1211.

84. Ohme-Takaghi, M. and Shinshi, H. (1995) Ethylene-inducible DNA binding proteins that interact with an ethylene-responsive element. *Plant Cell*, **7**, 173.

85. Yamaguchi-Shinozaki, K. and Shinozaki, K. (1993) Characterization of the expression of a desiccation-responsive *rd29* gene of *Arabidopsis thaliana* and analysis of its promoter in transgenic plants. *Mol. Gen. Genet.*, **236**, 331.

86. Liu, Q., Kasuga, M., Sakuma, Y., Abe, H., Miura, S., Yamaguchi-Shinozaki, K. and Shinozaki K. (1998) Two transcription factors, DREB1 and DREB 2, with an EREBP/AP2 DNA binding domain separate two cellular signal transduction pathways in drought-

and low-temperature-responsive gene expression, respectively, in *Arabidopsis*. *Plant Cell*, **10**, 1391.

87. Stockinger, E. J., Gilmour, S. J. and Thomashow, M. F. (1997) *Arabidopsis thaliana* CBF1 encodes an AP2 domain-containing transcriptional activator that binds to the C-repeat/ DRE, a *cis*-acting DNA regulatory element that stimulates transcription in response to low temperature and water deficit *Proc. Natl Acad. Sci.*, **94**, 1035.

88. Gehring, W. J., Affolter, M. and Burglin, T. (1994) Homeodomain proteins. *Ann. Rev. Biochem.*, **63**, 487.

89. Ruberti, I., Sessa, G., Lucchetti, S. and Morelli, G. (1991) A novel class of plant proteins containing a homeodomain with a closely linked leucine zipper motif. *EMBO J.*, **10**, 1787.

90. Schena, M. and Davis, R. W. (1992) HD-Zip proteins members of an *Arabidopsis* homeo-domain protein superfamily. *Proc. Natl Acad. Sci.*, **89**, 3894.

91. Sessa, G., Morelli, G. and Ruberti, I. (1993) The Athb-1 and -2 HD-Zip domains homo-dimerize forming complexes of different DNA binding specificities. *EMBO J.*, **12**, 3507.

92. Meijer, A. H., de Kam, R. J., d'Erfurth, I., Shen, W. and Hoge, J. H. (2000) HD-Zip proteins of families I and II from rice: interactions and functional properties. *Mol. Gen. Genet.*, **263**, 12.

93. Söderman, E., Mattsson, J. and Engstrom, P. (1996) The *Arabidopsis* homeobox gene *ATHB-7* is induced by water deficit and by abscisic acid. *Plant J.*, **10**, 375.

94. Lee, Y-H. and Chun, J-Y. (1998) A new homeodomain-leucine zipper gene from *Arabidopsis thaliana* induced by water stress and abscisic acid treatment. *Plant Mol. Biol.*, **37**, 377.

95. Söderman, E., Hjellstrom, M., Fahleson, J. and Engstrom, P. (1999) The HD-Zip gene *ATHB6* in *Arabidopsis* is expressed in developing leaves, roots and carpels and up-regulated by water deficit conditions. *Plant Mol. Biol.*, **40**, 1073.

96. Frank, W., Phillips, J., Salamini, F. and Bartels, D. (1998) Two dehydration inducible transcripts from the resurrection plant *Craterostigma plantagineum* encode interacting homeodomain-leucine zipper proteins. *Plant J.*, **15**, 413.

97. Ogata, K., Hojo, H., Aimoto, S., Nakai, T., Nakamura, H., Sarai, A., Ishii, S. and Nishimura, Y. (1992) Solution structure of a DNA-binding unit of Myb a helix-turn-helix-related motif with conserved tryptophans forming a hydrophobic core. *Proc. Natl Acad. Sci.*, **89**, 6428.

98. Ogata, K., Morikawa, S., Nakamura, H., Sekikawa, A., Inoue, T., Kanai, H., Sarai, A., Ishii, S. and Nishimura, Y. (1994) Solution structure of a specific DNA complex of the Myb DNA-binding domain with cooperative recognition helices. *Cell*, **79**, 639.

99. Meissner, R. C., Jin, H., Cominelli, E., Denekamp, M., Fuertes, A., Greco, R., Kranz, H. D., Penfield, S., Petroni, K., Urzainqui, A., Martin, C., Paz-Ares, J., Smeekens, S., Tonelli, C., Weisshaar, B., Baumann, E., Klimyuk, V., Marillonnet, S., Patel, K., Speulman, E., Tissier, A. F., Bouchez, D., Jones, J. J. D., Pereira, A., Wisman, E. and Bevan, M. (1999) Function search in a large transcription factor gene family in Arabidopsis: Assessing the potential of reverse genetics to identify insertional mutations in R2R3 MYB genes. *Plant Cell*, **11**, 1827.

100. Urao, T., Yamaguchi-Shinozaki, K., Urao, S. and Shinozaki, K. (1993) An *Arabidopsis myb* homolog is induced by dehydration stress and its gene product binds to the conserved MYB recognition sequence. *Plant Cell*, **5**, 1529.

101. Iturriaga, G., Leyns, L., Villegas, A., Gharaibeh, R., Salamini, F. and Bartels, D. (1996) A family of novel *Myb*-related genes from the resurrection plant *Craterostigma plantagineum*

are specifically expressed in callus and roots in response to ABA or desiccation. *Plant Mol. Biol.*, **32**, 707.

102. Meshi, T. and Iwabuchi, M. (1995) Plant transcription factors. *Plant Cell Physiol.*, **36**, 1405.
103. Hattori, T., Terada, T. and Hamasuna, S. T. (1994) Sequence and functional analyses of the rice gene homologous to the maize *Vp1*. *Plant Mol. Biol.*, **24**, 805.
104. Vasil, V., Marcotte, W. R.Jr., Rosenkrans, L., Cocciolone, S. M., Vasil, I. K., Quatrano, R. S. and McCarty, D. R. (1995) Overlap of Viviparous1 (VP1) and abscisic acid response elements in the Em promoter: G-box elements are sufficient but not necessary for VP1 transactivation. *Plant Cell*, **7**, 1511.
105. Hill, A., Nantel, A., Rock, C. D. and Quatrano, R. S. (1996) A conserved domain of the viviparous-1 gene product enhances the DNA binding activity of the bZIP protein EmBP-1 and other transcription factors. *J. Biol. Chem.*, **271**, 3366.
106. Holappa, L. D. and Walker-Simmons, M. K. (1995) The wheat abscisic acid-responsive protein kinase mRNA, PKABA1, is up-regulated by dehydration, cold temperature, and osmotic stress. *Plant Physiol.*, **108**, 1203.
107. Hwang, I. and Goodman, H. M. (1995) An *Arabidopsis thaliana* root-specific kinase homolog is induced by dehydration, ABA, and NaCl. *Plant J.*, **8**, 37.
108. Heino, P., Nylander, M., Palva, T. and Bartels, D. (1998) Isolation of a cDNA clone corresponding to a protein kinase differentially expressed in the resurrection plant *Craterostigma plantagineum*. *J. Exp. Bot.*, **49**, 1773.
109. Urao, T., Katagiri, T., Mizoguchi, T., Yamaguchi-Shinozaki, K., Hayashida, N. and Shinozaki, K. (1994) Two genes that encode Ca^{2+}-dependent protein kinases are induced by drought and high-salt stresses in *Arabidopsis thaliana*. *Mol. Gen. Genet.*, **224**, 331.
110. Meskiene, I. and Hirt, H. (2000) MAP kinase pathways: molecular plug-and-play chips for the cell. *Plant Mol. Biol.*, **42**, 791.
111. Mizoguchi, T., Irie, K., Hirayama, T., Hayashida, N., Yamaguchi-Shinozaki, K., Matsumoto, K. and Shinozaki, K. (1996) A gene encoding a MAP kinase kinase is induced simultaneously with genes for a MAP kinase and an S6 kinase by touch, cold and water stress in *Arabidopsis thaliana*. *Proc. Natl Acad. Sci. USA*, **93**, 765.
112. Jonak, C., Kiegerl, M., Ligterink, W., Barker, P. J., Huskisson, N. S. and Hirt, H. (1996) Stress signaling in plants: a MAP kinase pathway activated by cold and drought. *Proc. Natl Acad. Sci. USA*, **93**, 11274.
113. Munnik, T., Irvine, R. F. and Musgrave, A. (1998) Phospholipid signaling in plants. *Biochim. Biophys. Acta*, **1389**, 222.
114. Hirayama, T. Ohto, C., Mizoguchi, T. and Shinozaki, K. (1995) A gene encoding a phosphatidyl-inositol-specific phospholipase C is induced by dehydration and salt stress in *Arabidopsis thaliana*. *Proc. Natl Acad. Sci. USA*, **92**, 3903.
115. Mikami, K., Katagiri, T., Iuchi, S., Yamaguchi-Shinozaki, K. and Shinozaki, K. (1998) A gene encoding phosphatidylinositol-4-phosphate 5-kinase is induced by water stress and abscisic acid in *Arabidopsis thaliana*. *Plant J.*, **15**, 563.
116. Ritchie, S. and Gilroy, S. (1998) Abscisic acid signal transduction in the barley aleurone is mediated by phospholipase D activity. *Proc. Natl Acad. Sci. USA*, **95**, 2697.
117. Quarrie, S. A. (1996) New molecular tools to improve the efficiency of breeding for increased drought resistance. *Plant Growth Reg.*, **20**, 167.
118. Stockinger, E. J., Gilmour, S. J. and Thomashow, M. F. (1997) *Arabidopsis thaliana CBF1* encodes an AP2 domain-containing transcription activator that binds to the C-repeat/DRE, a *cis*-acting DNA regulatory element that stimulates transcription in response to low temperature and water deficit. *Proc. Natl Acad. Sci. USA*, **94**, 1035.

119. Kasuga, M., Liu, Q., Miura, S., Yamaguchi-Shinozaki, K. and Shinozaki, K. (1999) Improving plant drought, salt, and freezing tolerance by gene transfer of a single stress-inducible transcription factor. *Nature Biotech.*, **17**, 287.

120. Jaglo-Ottosen, K. R., Gilmour, S. J., Zarka, D. G., Schabenberger, O. and Thomashow, M. F. (1998) *Arabidopsis* CBF1 overexpression induces COR genes and enhances freezing tolerance. *Science*, **280**, 104.

8 | Salt-stress signal transduction in plants

LIMING XIONG AND JIAN-KANG ZHU

Plants need essential mineral nutrients to grow and develop. However, excessive mineral salts in the soil can have detrimental effects on plants due to their interference on plant metabolism and perturbation on plant water relations. Besides primary salinilization derived from natural soil genesis under arid or semiarid climate or in coastal areas, many cultivated soils in temperate regions also accumulate increasing amounts of salts as a result of inappropriate irrigation and nutrient management practices. This secondary salinilization may constitute a greater threat to agricultural production than the primary salinilization. It is estimated that around one-quarter of arable lands throughout the world are salt-affected (1). To be able to continue cultivating these lands, and increase the productivity of crops on salt-affected soils, one solution that has been tempting to scientists is to increase the tolerance of crop plants to salinity. Towards this ultimate goal, considerable effort has been made to understand salt-tolerance mechanisms in plants. These research works include the following aspects:

- characterizing plant responses to salt stress, either at molecular, cellular or whole-plant levels (for recent review, see references 1–6);
- comparing the physiology and biochemistry of salt-tolerant and sensitive lines of the same or similar plant species (for review, see reference 7);
- understanding the mechanisms of salt uptake, translocation, and distribution in plants. Major research in this aspect is focused on the physiological, electro-physiological, and molecular characterization of various ion transporters and assessment of their relevance to salt accumulation and salt tolerance in plants. This represents the most active research area in the field and much progress has been made (for review, see references 8–11);
- understanding the signal transduction pathways that lead to plant adaptation to salt stress. Relatively, research on the signal transduction processes of salt stress is limited.

This chapter mainly deals with the signal transduction aspect of salt stress. Some background on plant salt uptake and tolerance mechanisms will be briefly introduced. Information on salt-stress signal transduction in plants, albeit limited, has suggested that salt-stress signalling pathways between plants and yeast as well as other organisms do show some extent of conservation. Therefore, related studies in non-plant systems, particularly in yeast, are also covered briefly.

1. Salt uptake and determinants of salt tolerance in plants

1.1 Salt-uptake mechanisms

Excessive salts in the soil solution manifest their damaging effects mainly after entering into plant cells. Salts, most notably, sodium and chloride ions, are not required for the growth of most plants. Different from animals that use Na^+ for generating electrochemical gradient, higher plants have adopted H^+ to do the job. Land plants are also not equipped with the necessary machinery to actively absorb these ions. However, Na^+ can still enter plant cells via several routes.

Usually, Na^+ concentration in the external soil solution is higher than that in the cytosol of root cells, a passive movement of Na^+ is thus thermostatically feasible. It is even suspected that Na^+ ions may leak into the cell via non-ion channels such as various permeases or ABC transporters, yet direct evidence for such entry has not been reported in plants. Current information supports that Na^+ enters plant root cells via various cation channels. These could be voltage-dependent cation channels or voltage-independent cation channels (VIC). Due to the similarity between Na^+ and K^+, voltage-dependent K^+ inward rectifiers or outward rectifiers become a reasonable path for Na^+ entry into root cells (for review see reference 9). For example, in barley roots, the K^+ transporter HKT1 may function as a low-affinity Na^+ transporter at high external Na^+ [12].

Whereas the molecular identities of voltage-dependent ion channels that facilitate Na^+ entry have been understood to some extent, those of the VIC-type non-selective cation channels are not yet identified. It is thought that VIC channels may well represent the major path for Na^+ entry into plant cells (8, 10, 11, 13)

1.2 Salt extrusion and compartmentation

Besides the restriction of Na^+ entry into the root cells, there are several other routes that could lead to a reduced accumulation of salts in plants (or in the cytosol) once the ions enter the root cells. Although glycophytic plants do not have mechanisms to exude salts via salt glands, salts can still be pumped out of the cell, be prevented from loading into the xylem vessel, or be compartmentalized into the vacuole. The importance of these various strategies in the overall salt tolerance may vary with plant varieties and the severity of salt stress.

1.2.1 Salt compartmentation

In higher plant cells, Na^+ ions are extruded from the cells or compartmentalized in their vacuoles mainly by Na^+/H^+ antiporters. These antiporters make use of the pH gradient generated by P-type H^+-ATPase (for plasma membrane-localized antiporters) or V-type H^+-ATPases or H^+-pyrophosphatases (PPase) (for tonoplast antiporters). In the *Arabidopsis* genome, there are a number of genes encoding Na^+/H^+ antiporters. The first plant Na^+/H^+ antiporter gene functionally characterized was *AtNHX1*, which encodes a tonoplast antiporter homologous to the yeast Na^+/H^+ antiporter Nhx1 (14). Over-expression of *AtNHX1* complemented some of the salt-sensitive phenotypes of the yeast *nhx1* mutant (15). The role of AtNHX1 in plant salt tolerance was demonstrated by work showing that over-expression of this Na^+/H^+ antiporter confers salt tolerance to the transgenic *Arabidopsis* plants (16). This indicates that the activity of tonoplast Na^+/H^+ antiporter may limit the ability of this glycophytic plant to compartmentalize Na^+.

As the V-type ATPase functions in the generation of proton gradient that drives tonoplast Na^+/H^+ antiporters, defects in V-ATPase may compromise salt tolerance. In yeast mutants defective in VMA3 subunit of the V-ATPase, introducing the Na^+/H^+ antiporter AtNHX1 did not increase its salt tolerance, indicating that V-ATPase is required for full function of Na^+/H^+ antiporter in salt sequestration (17). *Arabidopsis DET3* gene was found to encode a B subunit of V-ATPase. *det3* mutants exhibited a de-etiolated phenotype in the dark. This B subunit may be needed for the assembly of the V-ATPase holoenzyme (18). Phenotypic studies with *det3* suggested that V-ATPase is required for cell expansion. Apart from its short statue, it is not known if this mutation has any impact on salt tolerance of the mutant plants.

1.2.2 Restriction of long-distance transport of salt

Not surprisingly, the above ground parts of plants are more sensitive to excessive salts than are the roots. Therefore, one strategy for glycophytic plants to increase salt tolerance is to restrict salt transport to aerial parts. This long-distance transport of salts consists of several processes, and hence may be regulated at each of these steps (19): the ions need first to enter root epidermal or cortical cells, and then must continue travelling through layers of cortical cells through intercellular connections and leave the symplast to be loaded into xylem vessels. Ions inside the xylem fluid are carried upward through transpiration flow, and once they reach the leaves, need to cross the plasma membrane of leaf cells to re-enter the cytoplasm. Channels that load K^+ into the root xylem have been identified (e.g. reference 20). Yet those responsible for Na^+ loading have not been isolated. A likely candidate responsible for this Na^+ loading/unloading is the SOS1 (salt overly sensitive 1) protein identified recently (21). *SOS1* encodes a putative plasma membrane Na^+/H^+ antiporter. Loss-of-function mutations in *SOS1* confer salt hypersensitivity and *sos1* mutants also cannot grow well under low K^+ conditions. Indirect evidence suggesting that SOS1 might function in Na^+ unloading is twofold. First, *SOS1* is mainly expressed in the pericycle cells surrounding the xylem vessels and also in the leaf veins. This expression pattern in the root is reminiscent of that of SKOC1 (20), which functions in leading K^+ into the xylem. Second, it was found that when plants were supplied with NaCl, $[Na^+]$ in the xylem sap of *sos1* mutant plants is higher than that in the wild-type plants (H. Shi and J.-K. Zhu, unpublished observation). This suggests that SOS1 may actually prevent Na$^+$ from entering the xylem vessel. Interestingly, *SOS1* gene was also expressed at root tips. It is thus likely that SOS1 may have additional functions other than regulating long-distance transport of salts. As the root tip cells are not well developed and are deficient in prototype vacuoles, this expression pattern of *SOS1* is consistent with the idea that SOS1 is localized on the plasma membrane, as suggested by its sequence characteristics (21). It is interesting that as the root cells differentiate, the expression of *SOS1* becomes restricted to specific cell files, suggesting that positional information is involved in the regulation of *SOS1* expression.

1.3 Production of osmolytes and detoxification of free radicals

Increasing the production of osmolytes is a general phenomenon found in all plants in response to salt stress. While it is not known whether this is simply a result of stress damage or an adaptive strategy, increased osmolyte production does have measurable benefits for plants under salt or other stresses. When plants encounter hyperosmolarity, accumulation of ions such as Na$^+$ in the vacuoles can serve as a means to lower osmotic potential of the cells, yet, as discussed above, glycophytes are not so equipped to accomplish this job to an extent that significantly improves salt tolerance without harming themselves. On the other hand, various compatible solutes such as proline, glycine betaine, and polyols could greatly reduce the damage caused by stress. Besides reducing the osmotic potential of the cytosol to facilitate water uptake, many osmolytes have additional functions such as protecting proteins from misfolding and alleviating the toxic effect of reactive oxygen species generated by the primary salt stress (22, 23). In fact, the direct protective effect of osmolytes on cellular processes against oxidative stress may outweigh their role in osmotic adjustment for increasing plant tolerance to salt stress.

The beneficial roles of osmolytes in plant salt tolerance were previously demonstrated by some biochemical studies where it was shown that the activities of certain enzymes were less inhibited by excessive salts in the presence of osmolytes (7). In recent years, the role of osmolytes was further fortified with experimental evidence that salt tolerance was increased to various extents in transgenic plants that were engineered to synthesize new osmolytes absent in the parental lines or in plants that over-expressed the genes whose products limit the production of these osmolytes (22, 24). As osmolytes have general protective effects, the advantage seen in these transgenic plants may go beyond the increased tolerance to salt stress. For example, Holmstrom *et al.* (25) expressed bacterial glycine betaine synthesis genes in tobacco, a species that does not accumulate glycine betaine. They found that the transgenic plants accumulated glycine betaine, and that these plants are more tolerant to salt stress, high light photoinhibition, as well as to low-temperature stress.

Aside from genetic variations, the production of osmolytes in plants is likely regulated by stress signal transduction cascades. Although such signal transduction pathways are not understood at the moment (26), it is possible that these pathways

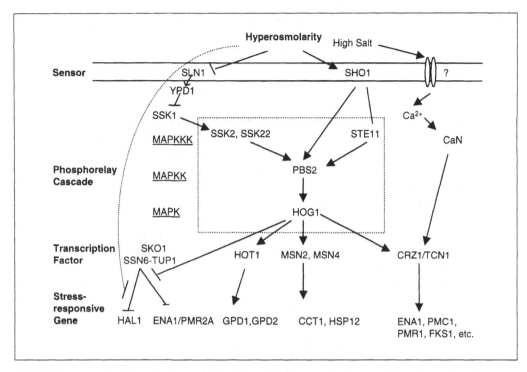

Fig. 1 Salt-stress signal transduction pathways in the yeast *Saccharomyces cerevisiae*. The major pathway, as depicted here, is the high osmolarity glycerol 1 (HOG1) pathway. In this pathway, osmolarity sensor is considered to be the membrane-bound two-component complex consisting of the osmosensor histidine kinase SLN1, an intermediate transmitter YPD1, and the response regulator SSK1. Under normal conditions, SLN1 is constitutively active and this results in the phosphorylation and inactivation of SSK1 to keep the HOG1 pathway in an 'off' state. Extracellular hyperosmolarity inhibits SLN1 and releases the repression on SSK1 and results in the activation of the MAPK cascade. Activated MAPK HOG1 in turn activates (or inhibits) transcription factors, presumably via phosphorylation, and leads to the expression of structural genes to survive osmotic stress. A parallel pathway with the transmembrane SH3-domain containing osmosensor SHO1 is also activated and the signal converges at the MAPKK PBS2. Under high salt conditions, a calcineurin (CaN) pathway is activated and this results in the induction of CaN-dependent gene expression. The MAPK module is dot-boxed. Arrows designate positive regulation and bars negative regulation. Dotted line indicates unknown pathways and a question mark indicates undefined component. The products of major stress-responsive genes showed are as follows: ENA1/PM2A, P-type ATPase; GPD, glycerol phosphate dehydrogenase; CCT1, cytosolic catalase; PMC1, putative vacuolar Ca^{2+}-ATPase; PMR1, Ca^{2+}/Mn^{2+}-ATPase; FKS1, 1,3-β-D-glucan synthase.

could be similar to those operating in yeast, i.e. the HOG1 pathway (Fig. 1). In addition, the production of osmolytes may also be modulated by feedback regulations. For example, the biosynthesis of proline in plants involves a feedback inhibitory regulation. Under salt stress, the P5CS protein, which is responsible for the limiting step in proline biosynthesis, may lose this negative regulation due to conformation changes in the protein. Tobacco plants expressing a mutated form of P5CS protein that disabled this negative regulation accumulated more proline under both stress and unstressed conditions and showed increased salt tolerance (23).

1.4 Role of Ca^{2+} and other ions in plant salt tolerance

Plant sensitivity to salt stress is also affected by other ions within the plants or outside in the immediate root environment. Ca^{2+} and K^+ are probably the most important ions that can alleviate the deleterious effects of toxic salts. It has long been observed that Ca^{2+} can increase salt tolerance both in soil-grown plants and in solution-cultured plants (e.g. reference 27), yet the exact mechanisms for these phenomena are not well understood. First, Ca^{2+} can regulate ion channel activities and restrict its permeability to Na^+. It was shown that Na^+ influx via VIC was inhibited by increased $[Ca^{2+}]$ in the external solution (28). As VIC is probably the major channel for Na^+ entry, this effect of Ca^{2+} may significantly reduce Na^+ accumulation in root cells. Second, there are many studies showing that Ca^{2+} increased the selectivity of these channels for K^+ against Na^+. Again, this increased selectivity is probably most relevant in controlling Na^+ uptake through VIC channels (e.g. 29, 30). Additionally, Ca^{2+} is an important signal molecule in transmitting salt-stress signals in plant cells.

Despite these early physiological studies and recent electrophysiological recording of channel activities, genetic evidence supporting the role of Ca^{2+} in plant salt tolerance was largely lacking till the isolation of the *Arabidopsis sos3* mutant (31). In addition to similar salt-sensitive phenotypes and increased requirement for K^+ nutrition as observed with *sos1* and *sos2* mutants, *sos3* mutant plants were unique in that the retarded growth on low K^+ culture medium was completely restored by increasing Ca^{2+} concentration up to 2 mM, and the salt sensitivity was partially alleviated by increased Ca^{2+} in external medium (31). This suggested that the defect in SOS3 might lead to the malfunction of K^+ transporters. It is likely that *sos3* mutation may primarily affect high-affinity K^+ uptake which in turn might affect Na^+ uptake. Yet, the failure to fully recover salt tolerance by increased Ca^{2+} in *sos3* implies that K^+ nutrition is not the sole determinant of salt sensitivity. The *SOS3* gene encodes a Ca^{2+}-binding protein with homology to yeast calcineurin B subunit and animal neural Ca^{2+} sensor (32). Further characterization of the SOS3 signalling pathway will illustrate how Ca^{2+} signalling through SOS3 or the SOS2/SOS3 complex might affect plant ion transport and salt compartmentation.

The ability of plants to maintain a high cytosolic K^+/Na^+ ratio is another determinant of salt tolerance (10, 28). The K^+/Na^+ ratio within plants is more important than the absolute Na^+ concentration. Genetic analysis of salt tolerance in *Arabidopsis* suggested that K^+ is a key determinant of Na^+ tolerance in this glycophyte (33). However, it is not clear which cellular processes or components have a critical requirement for a high K^+/Na^+ ratio. Although the possibility that K^+/Na^+ affects signalling cannot be ruled out, it is more likely that this ratio may be critical for the activity of some enzymes, or the selectivity of certain ion transporters (see Section 2.3). Genetic variations in K^+/Na^+ ratios among plant species may have their molecular basis. In particular, a subtle difference or mutation in some transporters may have a dramatic impact on the selectivity of the molecule. It was found that mutations in certain regions of the HKT1 transporter reduced its permeability to Na^+

and increased the selectivity of K^+ over Na^+ when expressed in yeast and in *Xenopus* oocytes. Thus, yeast cells expressing the mutated transporters were much more tolerant to salt stress (34–36).

2. Regulation of salt-tolerance responses

Each of the processes that control plant salt tolerance as outlined in the above section may be subjected to regulation in response to Na^+ stress. For example, plasma membrane ion channels can be regulated by transcription, post-transcription (e.g. phosphorylation), interaction with regulatory elements in the cytosolic medium, modulation on the extracellular side by the abundance of the transported ions or companion ions. Signal transduction networks activated by salt stress that finally modulate the channel activity are also envisioned. These hierarchies of regulation fine-tune the capacity of salt tolerance in plants.

2.1 Membrane potential

Under normal conditions, the plasma membrane potential (MP) of root cells is maintained at approximately -130 mv. Apparently, a more negative potential would favour the movement of the positively charged Na^+ into the cell. As mentioned before, the MP is generated and maintained by P-ATPases, which pump H^+ out of the cell and create an electrochemical potential facilitating the uptake of cations. Diminishing MP by using chemicals can greatly affect ion uptake. Meanwhile, the flux of ions also dynamically changes membrane potential.

Some transporters indirectly affect salt sensitivity by altering MP as a result of regulation in ion flux. The Pma1 H^+-ATPase in yeast is required to maintain membrane electrochemical potentials. Reduced Pma1 activity can reduce MP and increase salt tolerance (37). In yeast *trk1/trk2* mutants which are defective in K^+ transporters, the plasma membrane become hyperpolarized and this greatly enhanced the uptake of cations and rendered the mutants more sensitive to Na^+, Li^+, and low pH (38). An interesting observation was reported also in yeast, where a deletion in the *PMP3* gene resulted in hyperpolarization of the membrane as measured by methyl-ammonium uptake. The yeast mutant was more sensitive to salt stress due to increased uptake of Na^+ (39). As expected, this effect was not Na^+ specific. The *pmp3* mutation also suppressed the increased requirement for K^+ in the *trk1* and *trk2* mutants. Interestingly, Ca^{2+} can reverse the salt-sensitive growth in *pmp3* mutant. PMP3 is a small hydrophobic protein predicted to be localized on the plasma membrane. It is not known how this protein can regulate membrane potential. PMP3 is homologous to the plant protein RCI2A and RCI2B in *Arabidopsis* and BLT101 in barley. In fact, the *Arabidopsis* RCI2A can complement the yeast *pmp3* salt-sensitive phenotype (39). Interestingly, the expression of *RCI2A* and *RCI2B* genes in *Arabidopsis* and *BLT101* in barley was transiently induced by low-temperature treatment (40, 41). It is expected that the *RCI2* gene product in *Arabidopsis* may have similar functions in modulating ion uptake.

2.2 Acidity

Both extracellular and intracellular acidity may affect the function of membrane ion transporters. Among various transporters, Na^+/H^+ antiporters are probably the most vulnerable to pH changes with regard to their activities. In the yeast *pma1* mutant, it was thought that increased acidity in the cytosol may stimulate the activity of prevacuolar Na^+/H^+ antiporters and therefore increased the Na^+ compartmentation and salt tolerance in the mutant (14). In *Escherichia coli*, the activity of Na^+/H^+ antiporter NhaA is very sensitive to intracellular acidification. At pH 7.5 the activity of the antiporter is only 1% that of the activity at pH 8.0. In fact, this is probably a common property of all antiporters in this family. For example, the Na^+/H^+ antiporters NHE1, NHE2, and NHE3 are very sensitive to intracellular pH. At physiological internal pHs, these transporters are inactive, but they are rapidly activated when the intracellular pH drops. In mammalian cells, the way that a wide variety of hormones, integrins, and many growth factors regulate Na^+/H^+ antiporter activity is probably realized by binding to the cytoplasmic tail of the transporter and results in change of the sensitivity of the pH sensor in its transmembrane region (42). Through interaction with ligands at the cytosolic tails of the transporters, the sensitivity, and hence the activities, of these antiporters to H^+ can be modulated. Studies with the *E. coli* Na^+/H^+ antiporter NhaA suggested that pH induced a conformation change in the antiporter, which resulted in an alteration of the antiporter activity (43, 44).

2.3 Variety and concentration of salts

Apparently, different ions have quite different physiological effects on an organism. It is speculated that Na^+ sensors could be some ion transporters or regulatory factors of the transporters. The expression of *E. coli NhaA* Na^+/H^+ antiporter is regulated by the positive regulator NhaR, a LysR-type protein. Under normal conditions this protein binds with the *cis*-elements in the promoter of *NhaR* gene. Upon treatment with 100 mM Na^+, but not K^+, specific binding nucleotides were exposed. LysR was thus thought to be a Na^+ sensor and its conformational change upon Na^+ binding modifies its contact points with *NhaA* and induces *NhaA* expression. This Na^+ effect was also pH dependent, not observed at pH below 7.5 (45). In *Arabidopsis sos1*, *sos2*, and *sos3* mutants, the sensitivity was also only seen with Na^+ or Li^+ but not other monovalent ions (31, 33, 46). A specific signal role of intracellular Na^+ in plant cells has not been explored. In the nervous system, Na^+ regulates the activity of glutamate receptor *N*-methyl-D-aspartate (NMDA) Ca^{2+} channel and the sensitivity of the channel to Na^+ was set by an associated Src kinase (47). In *Arabidopsis*, there are also NMDA receptor homologues, yet their functions are not clear. Recently, it was reported that over-expression of an *Arabidopsis* glutamate receptor resulted in increased Ca^{2+} requirement and reduced tolerance to salt stress (48). The concentration of salts also affects the time course and amplitude of gene transcription. In yeast, high-salt treatment delays the induction of salt-regulated genes (49). The induction of these genes may also be activated via different signalling pathways under different salt concentrations. For example, in yeast, at relatively low salt concentration, the plasma membrane H^+-ATPase gene *ENA1* was activated by the HOG1 pathway, whereas at high concentration of salts, the activation of this gene was through a CaN-dependent pathway (Fig. 1).

2.4 Abscisic acid and other phytohormones

The plant hormone abscisic acid (ABA) is involved in many processes affected by salt stress. ABA level also increases during salt stress. Exogenous ABA was shown in some experiments to be able to increase salt tolerance of the treated plants or plant tissues. ABA may regulate plant salt responses at several levels. At the whole plant level, ABA can relieve the growth inhibition caused by decreased water potential (Ψ_w) under salt stress or drought conditions. Increased endogenous ABA level was showed to be important to maintain primary root elongation at low Ψ_w (50). This effect may have to do with the inhibition of ethylene production. Spollen *et al.* (51) reported that at a Ψ_w of -1.6 MPa, inhibition of maize primary root elongation by ABA synthesis inhibitor fluridone was prevented by applying ethylene synthesis or signal transduction inhibitors. As in ABA-deficient mutants, the production of ethylene in fluridone-treated seedlings was increased. Application of ABA again can reduce ethylene production back to the non-fluridone-treated control levels. Similar results were also found with the ABA-deficient mutant *vp5* (51). These results convincingly demonstrated that one role of ABA in promoting root growth at low Ψ_w is to restrict the production of ethylene and reduce the inhibitory effect of ethylene on root elongation. The interaction of ABA and ethylene in vegetative growth as well as in seed germination was further demonstrated by genetic analysis of ABA signal transduction in *Arabidopsis*. Characterization of mutants with altered responses to ABA during germination and analysis of ethylene response mutants showed that ethylene counteracts ABA signalling during seed germination, whereas it positively regulates ABA action in root growth (52, 53). At the molecular level, ABA induces the expression of numerous plant genes (for review, see reference 54) and most of these genes are also regulated by salt stress at the transcriptional level. Whereas some of these genes encode various signal transduction components such as putative receptors, protein kinases/phosphatases, and transcription factors that may participate in transducing salt-stress signals (see Section 3), others encode structural products that have been implicated to function in plant salt tolerances. These products include, for example, H^+-ATPase (55), P5CS (e.g. 56), and aquaporins (e.g. 57–59). In addition, ABA was shown to synergistically interact with salt in regulating the expression of salt-inducible genes (60).

Plant mutants defective in ABA and other hormone production or signal transduction are excellent tools to analyse the functions of these plant hormones in salt-stress tolerance. However, systemic analysis of gene expression or salt-stress tolerance in ABA mutants has not been reported. In a genetic analysis of abiotic stress signal transduction, we recovered a group of mutants that exhibit reduced salt induction of the luciferase reporter gene under control of the ABA and salt-stress-responsive promoter *RD29A* (61). Besides the endogenous *RD29A*, the expression patterns of many other stress-responsive genes are also changed in these mutants. Surprisingly, several of these mutants are found to be ABA deficient. Positional cloning of two of these affected genes identified that they are allelic to *ABA DEFICIENT 1* (*ABA1*) and *ABA3* respectively (L. Xiong and J.-K. Zhu, unpublished data). Both mutants are more tolerant to NaCl during germination. Interestingly, they showed different tolerance to salt stress at the seedling stage: relative to the wild-type plants, the *aba3* alleles are more sensitive to salt stress whilst the *aba1* allele shows the same sensitivity to salt as the wild-type. As ABA biosynthesis might still proceed via some minor shunt pathways in case of a block in the main pathway (62), a convincing assessment of the role of ABA in plant salt tolerance may require additional analysis of ABA responsive mutants or double mutants defective both in ABA biosynthesis and ABA responsiveness.

2.5 Transcriptional regulation

Transcriptional responses are among the most obvious changes within plant cells responding to hyperosmolarity. Through differential display and subtractive RNA analysis, the expression of many genes was found to be up-regulated or down-regulated by salt stress (for review, see references 4, 6, 63). In yeast, about 5% of the genes in the genome are regulated by salt stress (64). In another independent study, it was reported that around 7% of the genes in the yeast genome were induced more than fivefold after a brief osmotic shock (0.4 M NaCl, 10 minutes) (49). It is expected that comprehensive analysis of transcriptional responses under salt stress will be conducted shortly with *Arabidopsis*.

While the functions of many salt-inducible genes are unknown, some of them have clear roles in salt tolerance. These include various ion transporters, for example, the Na^+/H^+ antiporters SOS1 and AtNHX1 (17, 21). The modes of transcriptional regulation of salt-tolerant determinants also vary greatly. In yeast, salt-induced transcription can be realized by the activation of positive transcription factors, which results in the transcription of the target genes. It can also be realized by releasing transcription repression through the dissociation of the repressors from the promoter elements (Fig. 1).

2.6 Post-transcriptional regulation

Salt-stress signal transduction also results in the activation of many post-transcriptional events. These post-transcriptional modifications include, for example, modification of receptors, ion channels, ancillary proteins, and transcription factors by processes such as phosphorylation/dephosphorylation, prenylation, glycosylation, etc. Among these various modulation mechanisms, the most common one is protein phosphorylation.

One group of target proteins subjected to post-transcriptional modifications in response to salt stress are various transporters. For example, quite a few studies have shown that salt treatment increases tonoplast H^+-ATPase activity in plant cells (see reference 21 and references therein), whereas the transcription of the respective genes is not necessarily increased. To avoid internal acidification, animal cells use Na^+/H^+ exchanger to exclude H^+ in exchange of Na^+. There are at least six genes encoding Na^+/H^+ exchangers that have been identified in mammalian cells (42). A long stretch of C-termini of these transporters offers the possibility of multiple regulations. In mammalian cells, many hormonal regulations of Na^+/H^+ antiporters are achieved through the binding with C-termini of the respective transporters. Activation of NHE1 by growth factor correlates with increased NHE1 phosphorylation and cell alkalization. It was shown that p90 ribosomal S6 kinase is a key kinase that phosphorylates NHE1 and regulates the Na^+/H^+ exchange activity (65). For NHE3, hormone regulation is also through phosphorylation since inhibition of protein kinase activity completely or partially blocks the hormonal effects. Likewise, the inhibitory effect of cAMP on NHE3 is through protein kinase A instead of a direct effect of cAMP on the exchanger (for review see reference 66). Besides their role in the transcriptional regulation of stress-responsive genes, MAPK cascades are also shown to play roles in the phosphorylation of ion transporters. In rat myocardium, a MAPK and p90[rsk] are able to phosphorylate NHE1 *in vitro* (67). Besides direct phosphorylation of the transporter, phosphorylation of the regulatory factors can also modulate the activity of the transporters (see following sections). In plants, the hormone ABA regulates the expression of many stress-responsive genes (see Section 2.4). It is unclear whether ABA also regulates transporter activities similarly as animal hormone/growth factors. At least, the expression of the Na^+/H^+ antiporter *SOS1* in *Arabidopsis* does not appear to be regulated by ABA (21).

3. Salt-stress signal transduction modules and pathways

A generic signal transduction pathway in a living cell starts with signal perception, followed by the generation of the so-called second messengers, which are different from the primary extracellular signal in that it is a cell translation of the primary signal that can be directly coupled to subsequent cellular signalling. Second messengers usually result in the initiation of a phosphorelay cascade, with a major final target being transcriptional factors responsible for the transcription of specific sets of genes. We divide these signalling events, according to the order of signal flux, into various segments represented by modules, which are the relative independent functional units of the pathways.

3.1 Receptors

Salt stress initially may have an ionic, osmotic, or even a mechanical impact on the cell. Candidate receptors for ionic stress could be various ion transporters that contain ion-binding sites on either the extracelluar or intracellular side. Osmotic shock may result in a change of cell volume due to water flux along a water potential gradient. Ion channels that sense mechanical changes are found in various organisms. For example, plasma membrane mechanosensitive, non-selective cation channels were activated by membrane stretch caused by changes in cell volume in response to osmolarity alterations, which led to an increase in cytosolic Ca^{2+} (68–70). In *E. coli*, a change in osmolarity in the bath solution resulted in a rapid increase in K^+ influx through the Trk system. It was suggested that this rapid response might result from changes in turgor that acts as a regulator of these transporters (71). However, it is doubtful that these channels are the primary sensors, as their activities may be regulated by upstream protein phosphorylation. Previously, limited evidence suggested that the activation of NHE1 by mechanical stimuli, such as osmotic stress and cell spreading, was not regulated by phosphorylation (72). More recently, however, it was reported that in other cell types, this was actually controlled by phosphorylation. For example, shrinkage in endothelial cells leads to phosphorylation and activation of the Na/K/2Cl co-transporter (NKCC1). It was demonstrated that the c-Jun NH_2-terminal kinase (JNK) is responsible for this volume-dependent regulation of NKCC1 channel (73). In human polymorphonuclear leucocyte cells, Krump *et al.* (74) also showed that activation of NHE1 under shrinkage correlates with increased tyrosine phosphorylation activity. In *Arabidopsis sos1* mutants, the expression of some stress-responsive genes is affected by *sos1* mutation. It is thus likely that SOS1 Na^+/H^+ antiporter, which is homologous to NHE1, might have some signalling function as well (75). As plant cells possess a rigid cell wall, changes in cell volume may not be significant. On the other hand, alterations in turgor may generate a signal that could trigger conformation changes in membrane proteins and result in the initiation of a signalling cascade.

As the stress progresses, other secondary signals such as second messengers and stress hormones may arise from the primary signal and activate subsequent signalling pathways. Although the detailed mechanisms are unclear, it is thought that salt or osmotic stress may eventually be coupled with various receptors whose changes in conformation or energy status result in the activation of downstream phosphorelay cascades. These receptors include, for example, the receptor-like protein kinases, two-component receptors, and G-protein-coupled receptors (GPCRs). It seems that the latter two are more relevant to early events in salt signal transduction.

The two-component sensor-response regulator systems are used by both prokaryotes and eukaryotes for the perception of various environmental signals. When the extracellular sensor domain perceives a signal, the cytoplasmic histidine residue is autophosphorylated and the phosphoryl moiety is then passed on to a conserved aspartate receiver in a response regulator, which may constitute part of the sensor protein or a separate protein. The 'two-component' sensor may couple with a downstream MAP kinase cascade or directly phosphorylate specific targets to initiate cellular responses. In the yeast HOG1 (high-osmolarity glycerol 1) pathway, under normal osmolarity, the histidine kinase osmosensor SLN1 is constititutively active. As unphosphorylated response regulator SSK1 is the active form able to initiate the MAPK cascade, phosphorylation of the SSK1 by intermittent YPD1 protein under normal osmolarity prevents the activation of the MAP kinase cascade (see Section 3.4). Genetically, YPD1 is a negative regulator of the HOG1 pathway (Fig. 1). Extracellular hyperosmolarity inhibits SLN1 activity and thus activates the downstream MAPK cascade and results in the production of glycerol to survive osmotic stress (Fig. 1). A histidine kinase from *Arabidopsis*, AtHK1, which is structurally related to SLN1, has been implicated in salt signal transduction (76). AtHK1 can rescue the salt sensitivity of yeast mutants with deletions of SLN1 and SHO1 (another transmembrane osmosensor), implying that AtHK1 might have a similar function in plants. Expression of *AtHK1* was up-regulated by alterations in osmolarity of external solutions (76). Related to this work, it was also found that the expression of two genes that encode 'two-component' response regulator-like proteins in *Arabidopsis* was induced by low temperature, drought, and salt stress (77). Future genetic analysis should illustrate whether AtHK1 indeed plays a role in salt-stress signal transduction.

It is well documented that in plants salt stress rapidly activates the phosphoinositide signalling module where a key player is the phosphoinositide-specific phospholipase C (PLC) (see Section 3.2). In animal systems, PLC isoenzymes are activated by heterotrimeric G-protein subunits or, in a few cases, by receptor tyrosine kinases. More than 1000 hormone and neurotransmitter receptors that are structurally related to rhodopsin, the seven transmembrane (7TM) domain-containing G protein-coupled light receptor, are found in animal systems. They are collectively grouped as G-protein-coupled receptors (GPCRs). Similar receptors probably function in light perception in flagellate green alga (78). Surprisingly, it seems that orthologous GPCRs are under-represented in higher plant genomes. To date, there is only one sequence that was reported to have homology to animal GPCR (79, 80). However, there is a group of 7TM proteins encoded in the *Arabidopsis* genome that are homologous to the disease-resistant protein Mlo (81). The topology of these proteins is reminiscent of GPCR, yet they do not show significant homology with animal GPCR at the amino acid level. It is not known whether this group of proteins is the plant version of GPCR. As G-proteins are implicated to play roles in salt-stress signalling (see below), it can be envisaged that there are G-protein-associated receptors in plants that may participate in the perception of salt-stress-coupled signals.

3.2 Second messengers

Ca^{2+} is a ubiquitous signal involved in many intracellular signalling processes (for review, see reference 82) both in animals and in plants. At resting state, Ca^{2+} concentration in the cytosol is low. Upon stimulation, Ca^{2+} is released from intracellular storage or enters the cell via various Ca^{2+} channels. Ca^{2+} transients associated with abiotic stresses are well documented (for review, see reference 83). Studies with animal cells have shown that there are several paths that mediate transient Ca^{2+} increase in the cytoplasm. Ca^{2+} can enter cells from the outside by voltage-gated, receptor-operated or store-operated Ca^{2+} channels. Also, Ca^{2+} in intracellular stores can be released through ligand-sensitive Ca^{2+} channels. These ligands are second messengers such as inositol polyphosphates, cyclic ADP-ribose (cADPR) and nicotinic acid adenine dinucleotide phosphate (NAADP).

Phosphoinositides have been proven to play pivotal roles in signal transduction in many organisms. One prominent role of these phospholipids is that they serve as precursors for the production of soluble phosphoinositols or phosphatic acid in response to stimulation. Phosphoinositols, in particular, inositol 1,4,5-trisphosphate (IP_3), can induce Ca^{2+} transients in the cytoplasm. In higher plants and algae, phosphoinositide signalling has been implicated to be involved in responses to environmental stimuli (for review, see reference 84). Exogenous IP_3 was demonstrated to release Ca^{2+} from internal stores from various plant cells. Salt stress was shown to elicit a transient increase in IP_3 level in cultured plant cells (e.g. reference 85). The IP_3 precursor phosphatidylinositol 4,5-bisphosphate (PIP_2) is synthesized via phosphatidylinositol 4-phosphate 5-kinase. An *Arabidopsis* gene encoding this enzyme, *PIP5K*, was induced by water stress and ABA (86). IP_3 is generated by the hydrolysis of PIP_2 by PLC. In animal cells, there are at least three subfamilies of PLC and the regulation of each subfamily uses different signalling pathways, yet all PLCs strictly require Ca^{2+} as a cofactor for activation. In *Arabidopsis*, there are approximately ten genes encoding PLCs. An *Arabidopsis* PLC gene, *AtPLC1*, was strongly induced by salt and drought stress, and to a lesser extent by low temperature (87). Plant PLCs are similar to members of the animal PLCδ subfamily; they seem to be activated by G-proteins as evidenced by pathway activation with G-protein activator, mastoparan (e.g. reference 88). However, it should be noted that phosphoinositide signalling is subjected to multiple levels of regulation and interactions via PLC and many other enzymes in the pathway. For example, the expression of *AtPLC1* gene was negatively regulated by SOS2 (89), a protein kinase functioning in plant salt tolerance. Interestingly, it was reported that in mammalian cells the Na^+/H^+ exchanger regulatory factor 2 interacts with PLC-β3 and may regulate the activity of this isoform of PLC (90) (see Section 3.6).

Phospholipids are also hydrolysed by phospholipase D (PLD) to produce phosphatidic acid (PtdOH), which is a second messenger demonstrated in animal cells that may activate phosphatidylinositol 5-kinase (PI5K), phospholipase C (PLC) and protein kinase C (PKC). In the resurrection plant *Craterostigma plantagineum*, PLD activity was activated within minutes after dehydration. Additionally, PLD was activated by mastoparan, a G-protein agonist, suggesting the involvement of G-protein in early signal transduction (91). Phospholipid signalling via PLC or PLD also has 'cross-talk'. In animal cells, it was reported that PLD activity requires PIP_2 and hence PI5K for the production of this lipid (92).

Another group of second messengers are cyclic nucleotides. In the slime mould *Dictyostelium discoideum*, it was reported that osmotic stress induced cAMP via the hybrid histidine kinase DokA contributes to the survival of osmotic stress (93). cAMP also activates PKA, which can modulate the activities of many ion transporters (see Section 3.6). As in animal cells, cADPR is probably also a second messenger in plants that can trigger the release of Ca^{2+} from internal stores and initiates stress-induced gene expression. In plants, cADPR was demonstrated to release Ca^{2+} from vacuoles (94). Using the stress-responsive promoter of *RD29A* and *KIN1* fused with *GUS* reporter, Wu *et al.* (95) studied the transient expression of the reporter gene in tomato hypocotyl cells. They found that cADPR could induce the expression of the transgene and this process involved Ca^{2+} and protein phosphorylation/dephosphorylation (95). In animal cells, ryanodine receptor (RyR) Ca^{2+} channels are also regulated by NO, an endogenous oxidative signal (e.g. reference 96).

Reactive oxygen species (ROS) are potential second messengers that may initiate signalling cascades. Salt stress, like other abiotic stresses or biotic stresses, generates ROS. In rat cardic myocytes, hydrogen peroxide was shown to activate the MAPK pathway and stimulate Na^+/H^+ exchanger activity (97). In plants, exogenous hydrogen peroxide also activates Ca^{2+} channels in guard cells (98).

3.3 Phosphoproteins

Upon receiving a signal from membrane receptors, cells often utilize multiple phosphoprotein cascades to transduce and amplify the information. Protein phosphorylation regulates a wide range of cellular processes such as enzyme activation, assembly of macromolecules, protein localization, and degradation. In plants, many protein kinases and phosphatases are implicated to be involved in environmental stress responses based on pharmacological studies.

Virtually all plant protein kinases are serine/threonine kinases that play major roles in the protein phosphorelay. Although several protein kinases were implicated to function in salt-stress signal transduction, very few have been demonstrated genetically to play roles in the signalling. Phenotypic characterization of *Arabidopsis* *sos2* mutants demonstrated that SOS2 protein kinase is required for salt tolerance and signal transduction (33). The N-terminal catalytic domain of SOS2 showed homology to the yeast SNF1 protein kinase, which activates glucose-regulated gene expression under glucose starvation. SOS2 protein kinase activity is required for its function in salt tolerance (99), suggesting that SOS2 participates in salt signal transduction by phosphorylating target proteins.

Plant calcium-dependent protein kinases (CDPKs) are also implicated to play roles in stress responses. CDPKs consist of a serine/threonine kinase domain and a C-terminal calmodulin-like domain with up to four EF hand motifs. Most CDPKs have a myristoylation motif that potentially facilitates membrane association (see Section 3.6). Because these proteins can bind Ca^{2+}, they may couple changes in cytosolic Ca^{2+} to phosphoryl relay cascades. In fact, research has shown that CDPK may phosphorylate or regulate certain ion channels/transporters such as H^+-ATPase from maize roots (100) or beet root (101), a Cl^- channel (102) and the KAT1 potassium channel (103) in guard cells of *Vicia faba*. As these channels function in ion transport and osmotic adjustment in plants, CDPKs likely play some roles in salt-stress signal transduction in plants. Previous expression studies of two *Arabidopsis* CDPKs (*ATCDPK1* and *ATCDPK2*) indicated that these genes were rapidly induced by drought and salt treatments, yet their expression was not induced by cold, heat shock, or ABA, suggesting that they may participate in an ABA-independent pathway (104). By transiently expressing a chimeric gene consisting of a GFP reporter driven by a cold, salt, dark, and ABA-responsive promoter (*HVA1*) in maize leaf protoplasts, Sheen (105) showed that co-expression of a constitutively active ATCDPK1 activated the reporter gene, while a non-active CDPK1 mutant could not. Recently, Saijo *et al.* (106) reported that over-expression OsCDPK7 resulted in increased cold and salt/drought tolerance in transgenic rice plants.

In yeast and also in mammalian cells, cAMP-dependent kinase (PKA) regulates stress responses in a number of ways. PKA can negatively regulate the transcription of stress-responsive genes whose expression is either HOG1-dependent or HOG1-independent (Fig. 1). In addition, PKA can negatively modulate the activity of Na^+/H^+ exchangers through the Na^+/H^+ exchanger regulatory factor (NHERF)-mediated inhibition on the exchanger. Early *in vitro* studies suggested that PKA can phosphorylate both NHERF and the exchanger, and that phosphorylation of NHERF can increase the inhibitory effect of NHERF on the exchanger. However, recent studies suggested that PKA do not phosphorylate NHERF *in vivo* (e.g. reference 107). This is probably because within cells, NHERF is constitutively phosphorylated. Recently, work with human embryonic kidney HEK293 cells suggests that G-protein-coupled receptor kinase 6A is responsible for this constitutive phosphorylation of NHERF (108).

Inactivation of phosphoproteins is usually accomplished by dephosphorylation via protein phosphatases. There are four major subgroups of protein phosphatases: PP1, PP2A, PP2B (calcineurin), and PP2C. Among them, PP2B and PP2C are Ca^{2+}-dependent. As with kinases, studies using phosphatase inhibitors have indicated a role for phosphatases in stress signalling. The common heterotrimeric form of PP2A consists of a structural A subunit, a regulatory B subunit, and a catalytic C subunit. Each subunit has different interacting proteins. In yeast and animal cells, PP2A has been shown to affect cell cycling, vertebrate axis determination (Wnt signalling), and apoptosis. PP2A has also been implicated in low temperature signal transduction in plants (109). To identify PP2A targets or proteins associated with PP2A in plants, Harris *et al.* (110) used an *Arabidopsis* PP2A cDNA catalytic subunit as bait to conduct a yeast two-hybrid screen and obtained one protein, TAP46, that can associate with PP2A *in vivo*. This protein shows homology to yeast TAP42 and mammalian α4 proteins, both of which function in the TOR (target of rapamycin) pathway in response to nutrient availability. The yeast and mammalian proteins are involved in the activation of ribosomal S6 kinase and inactivation of a translation initiation inhibitor 4E-BP1. These activities positively regulate the translation initiation factor eIF-4E, and result in increased mRNA translation (111). Together with the previous finding that a ribosomal S6 kinase homologue was induced by low temperature and drought stress (112), it seems possible that a signalling branch using part of the TOR module involving MAPK/CDPK-PP2A/TAP42-S6K may be activated by low temperature and other stresses and regulates stress-responsive gene expression in plants.

In addition, PP2A may regulate MAPK cascade by dephosphorylating the core kinases, although other serine/threonine or dual specificity phosphatases are more common regulators of MAPK cascades. In *Arabidopsis*, the expression of *PTP1*, a gene encoding tyrosine-specific protein phosphatase, was significantly enhanced by NaCl but was down-regulated by short-term (6 hours or shorter) cold treatment (113). The significance of this differential regulation by two unrelated stresses is not clear. In a study of the interaction between low temperature and salt stress in the regulation of the stress-responsive gene *RD29A*, Xiong *et al.* (60) observed that when *Arabidopsis* seedlings were treated for 4.5 hours in the cold, the induction of *RD29A* by NaCl or ABA (in the cold) was completely blocked while longer cold treatment (2 days) did not block ABA induction. It is possible that this interaction between low temperature and salt stress occurs at the level of PTP1, as evidenced by similar regulation of *PTP1* expression by cold and salt stress (113). By analogy with other systems and also implicated in plants (e.g. 114), the role of this PTP1 in stress signalling may be to dephosphorylate a MAP kinase on a tyrosine residue. In yeast, the MAPK Hog1 upon activation (i.e. phosphorylation) will migrate to the nucleus to initiate gene transcription. The import and export of Hog1 depend on a number of factors. Interestingly, it was found that, among others, two protein tyrosine phosphatases, Ptp2 and Ptp3, affected the nucleocytoplasmic distribution of Hog1 (115). Through mutation and over-expression studies, it was found that Ptp2 is a nuclear tether that

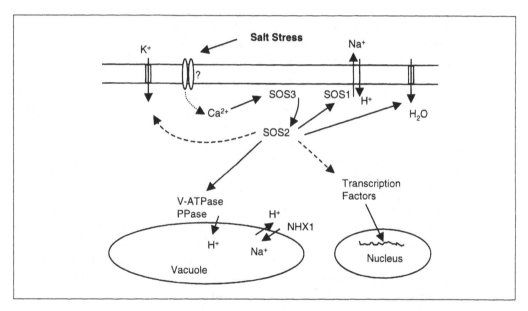

Fig. 2 A proposed salt signal transduction pathway in *Arabidopsis*. Salt stress induces cytosolic Ca^{2+} transients, which activate the Ca^{2+}-binding protein SOS3. SOS3 binds to the protein kinase SOS2 and activates SOS2. This SOS3/SOS2 protein complex may then activate transcription factors and induce stress-responsive gene expression. It is also likely that SOS3/SOS2 may directly phosphorylate certain transporters to regulate ion homeostasis (e.g. K^+ uptake and Na^+ extrusion or compartmentation). Potential effectors of the SOS3/SOS2 complex shown are K^+ channels, SOS1 Na^+/H^+ antiporters, plasma membrane or tonoplast water channels, V-type H^+-ATPase, and PPase. Arrows indicate the direction of signal or ion/water flux. Dotted lines indicate unknown pathways and question mark refers to undefined component.

helps to retain Hog1 in the nucleus whereas Ptp3 tends to keep Hog1 in cytoplasm (115).

The yeast calcineurin (CaN) B (CNB) (PP2B) is an important determinant of salt tolerance and it plays an important role in signal transduction (116). CaN affects many cellular processes both post-transcriptionally and transcriptionally. It has been shown that in animal cells, CaN can activate certain transcriptional factors, with the NFAT family transcription factors being the best studied. In human T cells, CaN dephosphorylates these NFATs, facilitates their nuclear localization, and promotes gene expression (for review see reference 117). The expression of a subset of ion transporters (i.e. PMC1, PMR1, PMR2A, FKS1) in yeast was demonstrated to be controlled by the zinc-finger transcription factor TCN1/CRZ1 in a CaN-dependent manner (118) (Fig. 1). Interestingly, over-expression of this transcription factor resulted in high induction of PMC1, FKS2, and PMR2A even at lower levels of Ca^{2+} (118). By functional complementation, plant transformation or expression studies, several homologous plant proteins were implicated to function in salt tolerance in a similar way as the yeast CNB (119–121). Plants expressing an active form of yeast calcineurin showed enhanced salt tolerance (119). In *Arabidopsis*, identification of the *SOS3* gene product as related to calcineurin B provided compelling evidence to support the role of calcineurin-like proteins in plant salt-tolerance (32) and signal transduction. However, unlike the yeast CNB, SOS3 does not seem to function through a protein phosphatase. Rather, SOS3 interacts with and activates a protein kinase encoded by SOS2 (99, 122). This SOS3/SOS2 protein kinase complex constitutes a core module controlling ion homeostasis and functions in salt tolerance, probably through transcriptional or post-transcriptional regulation of critical ion transporters (Fig. 2).

ABA signalling has an intimate relation with salt signalling. The *Arabidopsis abi1* and *abi2* are dominant mutants that show insensitivity to exogenous ABA during germination (123) and are also more resistant to NaCl during seed germination. The *ABI1* and *ABI2* genes encode homologous serine/threonine protein phosphatase 2C (124–126) and may have overlapping functions. Despite much interest in the ABI1 and ABI2 proteins, the *in vivo* substrates of these enzymes are still not known. In yeast two-hybrid screens, it was found that SOS2 interacts with ABI2 (J. K. Zhu, unpublished observation), suggesting that there probably exist connections between ABA-signalling pathways and the SOS module. Also, an alfalfa PP2C with homology to *Arabidopsis* ABI2 was shown to act as a wound-induced MAPK-specific phosphatase that regulates MAPK activity (127, 128). Thus, it is also possible that ABI proteins may connect to MAP kinase modules.

3.4 MAPK modules

The mitogen-activated protein (MAP) kinase pathways are intracellular signal modules that mediate signal transduction from the cell surface to the nucleus (for review see reference 129). MAPK cascades are likely conserved in all eukaryotes and seem to be widely used as osmolarity-signalling modules. The core MAPK cascades consist of three kinases that are activated sequentially by an upstream kinase. The MAPK kinase kinase (MAPKKK), upon activation, phosphorylates MAPK kinase (MAPKK) on serine and threonine residues. This dual specificity MAPKK in turn phosphorylates a MAP kinase (MAPK) on conserved tyrosine and threonine residues. The activated MAPK can then either migrate to the nucleus to activate transcription factors directly, or activate additional signalling components to regulate gene expression, cytoskeleton-associated proteins, enzyme activity, or target certain signal proteins for degradation.

The widespread adoption of MAPK modules by many cell-signalling cascades suggests some advantages of these modules in signal transduction. At a first glance, one may think that the use of three kinases in a cascade is costly in term of signalling speed. Yet, it was shown that MAPK signalling is very fast. This may be due to the fact that these kinases are very close in space due to scaffold proteins that pull these kinases together and thus speed up the phosphorelay process. More importantly, the cooperative activation of the three kinases results in an ultrasensitivity that makes the cascade operate in an on-or-off switch manner and prevents noise activation of the cascade (130). The incorporation of multiple kinases in the modules also offers the potential of multiple levels of regulation and integration of signals from different inputs. It has been shown that different MAPK pathways may share common components, yet activation of one pathway may not necessarily affect another pathway. The specificity is realized by scaffold proteins (such as Ste5 in yeast) that hold these kinases or by a specific component in the signalling cascade (131). The yeast HOG1 pathway is the best-studied MAP kinase pathway. Over 75% of the genes whose transcriptional induction are strongly regulated by osmotic stress are highly or fully dependent on the presence of Hog1 (49).

In plants, the MAPK cascade has been shown to participate in auxin and cytokinin signal transduction and cell-cycle regulation, and are implicated in wound and pathogenesis responses as well as in environmental stress signal transduction (for review, see reference 132). A salicylic acid-induced protein kinase belonging to the MAP kinase family that was activated by salicylic acid, pathogen attack, and wounding, was also activated within 5–10 minutes after osmotic stress (133). Similarly, in tobacco cells, the salicylic acid-induced protein kinase (SIPK, a MAPK) and another protein kinase, HOSAK, were activated by osmotic stress and this activation is independent of Ca^{2+} or ABA (134). In *Arabidopsis*, the transcription of a MAPK gene, *ATMPK3*, is induced by drought, low temperature, salinity, or touch (112). AtMPK3 is also activated by H_2O_2, and its activity is further enhanced by ectopically expressed ANP1, a MAPKKK (135). *ATMEKK1* (a MAPKKK), whose expression is induced within 5 minutes by NaCl treatment (136), can functionally complement the *ste11* (a *MAPKKK*) mutant of *S. cerevisiae* (112, 137). Yeast cells expressing *ATMEKK1* are more tolerant to osmotic stress and producing more glycerol under non-stressed conditions (136). Similarly, a pea PsMAP kinase is capable of rescuing morphological defects of yeast *hog1* mutants in hyperosmotic medium and partially restores the mutant growth (138). These results suggest functional similarity of plant MAPK components with their yeast counterparts.

The role of MAPK module in determining plant tolerance to abiotic stress was demonstrated by a recent study where a tobacco ANP orthologue, NPK1, which activates the H_2O_2-regulated gene expression in plants, was over-expressed in *Arabidopsis*. The engineered *Arabidopsis* plants showed increased tolerance to freezing, heat shock and salt stress (135). In this context, it is worth noting that, in addition to a signalling role of ROS as mentioned in Section 3.2, the accumulation of these compounds is toxic to plants. Detoxification of these compounds therefore contributes to increased tolerance to salt stress or other stresses. Genetic screening of salt-tolerant growth in *Arabidopsis* also resulted in the isolation of photoautotrophic salt tolerance (*pst*) mutants. The salt tolerance in *pst1* mutants is associated with enhanced active oxygen free radical detoxification (139).

3.5 Transcription factors

According to their conserved DNA-binding domains, common transcription factors in eukaryotes can be classified into several groups such as the basic region leucine zipper (bZip) proteins, MYB-like proteins (containing helix–turn–helix motifs), MADS-domain proteins, helix–loop–helix proteins, zinc-finger proteins and homeobox proteins. Many ABA-responsive genes have the ABA-responsive element (ABRE) with an ACGT core and additional coupling elements (140) to impart ABA induction. Proteins binding to this ABA-responsive complex contain bZIP motifs. Besides ABRE, the promoters of many cold- or drought-responsive genes contain another *cis*-regulatory element with the core sequence CCGAC that specifically responds to drought or cold signals. This element was termed C-repeat element (CRT) (141) or DRE (drought-responsive element) (142). A transcriptional activator containing AP2 domain that binds to this element, CBF1 (C-repeat binding element-1) was isolated (143). The expression of *CBF1* is itself cold up-regulated and this precedes the activation of other cold-regulated genes (144). Homologous genes are found clustered together with *CBF1* in the *Arabidopsis* genome, and they are named *CBF2/DREB1C* and *CBF3/DREB1A* (144, 145), respectively. Additionally, two similar genes encoding DRE-binding proteins, *DREB2A* and *DREB2B* were identified (145). *DREB2* genes were rapidly induced by dehydration and salt stress (145). Over-expression of *CBF/DREB* has been shown to induce stress-responsive gene expression and increase plant tolerance to cold and drought stresses (145–147).

The transcriptional regulatory mechanisms of other salt-tolerant determinants in plants are not well understood. Studies with the Na^+/H^+ antiporter SOS1 indicated that *SOS1* expression is up-regulated by NaCl stress and but not by low temperature or ABA (21). Reports on the regulation of Na^+/H^+ antiporter *NHX1* in *Arabidopsis* leaves are controversial. Whereas Apse *et al.* (16) did not observe induction of *NHX1* at the transcript level or protein level by either salt or ABA treatment, Quintero *et al.* (17) reported that *NHX1* transcript expression was induced both by salt stress and by ABA in *Arabidopsis* leaves but not in the roots. Transcription factors responsible for the expression of these and other related ion transporters have not been isolated. For the mammalian *NHE1* gene, an AP1 transcriptional factor is probably involved in the transcriptional regulation in cultured renal proximal tubule cells (148). An AP2-like transcription factor was found to bind to the promoter region of NHE1 and probably regulated the expression of this gene (149). In yeast, however, the transcriptional control of similar ion transporters that function in ion homeostasis during salt stress has been intensively studied. Many of the yeast stress-responsive genes contain a positive stress-responsive promoter element STRE with the consensus sequence CCCCT or AGGGG. This STRE is responsible for transcription in both HOG1-dependent and HOG1-independent pathways. Two zinc-finger transcription factors, Msn2p and Msn4p, were shown to bind to STRE and were required for STRE-dependent transcription (150). Some of the stress-responsive genes have a repressor element in the promoter and binding of this repressor element with a repressor protein can prevent the gene from being transcribed under non-stress conditions. It seems that the general repression complex Ssn6-Tup1 is responsible for this repression. Under salt-stress conditions, the association of this repressor complex with the repressor element is destroyed, resulting in gene induction (151). This repression is modulated by high levels of cAMP-dependent PKA. For example, the expression of the yeast plasma membrane ENA1/PMR2A gene is repressed under normal growth conditions, but osmotic stress can relieve the repression by a cAMP-response element binding factor (CREB) SKO1 through the general repression complex Ssn6p-Tup1p. As a consequence, *sko1* and *ssn6* mutants are more resistant to Na^+ or Li^+ stress (152). Ssn6-Tup1 complex may function through interaction with basic transcriptional machinery to halt transcription, and through interaction with histones to maintain the transcription-inaccessible state of the targeted DNA sequences.

3.6   Signal partners: protein modifiers, adapters, and scaffolds

Molecules described in the above sections, directly participate in signal transduction. However, there are others that may not directly mediate the flux of signals, but regulate the activity of the above-mentioned signalling molecules. These signalling partners are various protein modifiers, and adapters or scaffolds that physically support the assembly of signalling complexes. Besides phosphorylation, other common protein modifications include acetylation, methylation, ADP-ribosylation, glycosylation, and lipidation.

Protein lipidation is a common form of cotranslational or post-translational modification. Lipidation facilitates membrane association and this process can be very important for the transmission of extracellular signal into the cell. Common lipid modifications make use of fatty acids (myristate and palmitate) and isoprenoids (farnesol and geranylgeranol). Myristoylation is realized by cotranslational addition of myristic acid to the amino terminal glycine residue with the consensus sequence of MGXXXS/T(K). The enzyme responsible for this acylation process is *N*-myristoyltransferase. Studies in non-plant systems showed that there are approximately 100 proteins in lower and higher eukaryotes subjected to this modification, which are involved in oncogenesis, cellular signalling, and viral infection. These proteins include Src family tyrosine kinases, Ser/Thr kinases and anchoring proteins, calcineurin B, heterotrimeric G proteins, ADP ribosylation factors (ARFs), Ca^{2+}-binding proteins, membrane and cytoskeleton-bound structural proteins, and viral proteins (153). In plants, sequences that have potential myristoylation sites were also reported, but little is known of the impact of myristoylation on the functions of these proteins, such as CDPK, or, indeed, whether they are myristoylated. Plant mutants defective in myristoyltransferase have not been isolated. Like its homologues in animals and in yeast, the *Arabidopsis* SOS3 protein also contains a myristoylation consensus sequence (32). Recently, it was demonstrated that SOS3 can be myristoylated *in vitro* and, more importantly, it was found that myristoylation of SOS3 is required for its function in salt tolerance because mutated SOS3 with a destroyed myristoylation site failed to rescue the salt sensitivity of *sos3* mutant plants. The mutated SOS3 also resulted in less activation of SOS2 protein kinase (154). These data provide strong evidence indicating that myristoylation is critical for SOS3 function in salt signal transduction. Perhaps, besides its ability to bind Ca^{2+}, SOS3, upon myristoylation, is able to recruit other signal molecules and plays roles in the aggregation of signalling molecules such as SOS2 to particular membrane patches to regulate ion transporter activities.

The assembly of signalling complexes with the aid of adapter proteins and scaffolds is a common way to control signalling specificity. 'Piggybacking' of early signal components to membrane receptors by scaffolds and adapters also concentrates the signalling proteins to increase the steady signalling levels (155). Scaffolds and adapters play multiple roles at all steps of signal transduction, ranging from signal reception to the activation of transcription. They can be an 'anchor' that tethers recruited proteins to specific subcellular locations or they may have additional biochemical functions. Some scaffolds, such as Ste5 in the yeast pheromone pathway, have both functions (156). The role of scaffolds in signal transduction has been established in many systems. For example, A kinase anchoring proteins (AKAP) can simultaneously anchor protein kinase A and its phosphatase in a close proximity to ion channels to regulate ion flux (for review, see reference 157). The PDZ domain-containing Na^+/H^+ exchanger regulatory factor (NHERF) and E3KARP (NHE3 kinase A regulatory protein) (158), are interacting partners of Na^+/H^+ exchangers. They function as AKAP for NHE3 and are required for cAMP responsiveness of NHE3 (158), probably by anchoring PKA to NHE3 through association with erzin via the erzin/radixin/moesin (ERM)-binding domain (107). Recently, NHERF2/E3KAPP was shown to be associated with phospholipase Cβ3 and to potentiate its activity (90) by anchoring PLC-β3. This interaction may be important for forming a complex with other signal molecules in the PLC pathway. In another example, PKA, once anchored by AKAP, phosphorylated the proapoptotic protein, BAD, and bound with the 14-3-3 adapter protein to prevent apoptosis by blocking the association of BAD with pro-apoptotic factors. Disruption of AKAP anchoring prevented BAD phosphorylation and resulted in mitochondrial dysfunction and apoptosis (159).

Regarding the 14-3-3 protein family, a search of plant proteins in the database reveals that there are a number of signalling molecules containing 14-3-3 binding motifs (160). It was shown that 14-3-3 proteins interact with a CDPK (CPK-1) in *Arabidopsis* and increase the kinase activity of the protein (161). 14-3-3 proteins may be needed for the activation of H^+-ATPase by relieving the autoinhibition at the C-terminus. Recently, it was shown that the binding of 14-3-3 proteins to H^+-ATPase depends on the phosphorylation status of the H^+-ATPase in maize roots. A PP2A that seems to be able to dephosphorylate H^+-ATPase inhibited the binding between the ATPase and the 14-3-3 protein (162).

4. Outlook

Despite much information accumulated on salt-stress signal transduction as outlined above, our current understanding in plants is still at an early stage. To date, no single pathway of salt signal transduction has been worked out in plants. The majority of the information was obtained from biochemical or molecular studies, and the application of genetic analysis to salt-stress signal transduction is rather limited. It is clear that mutational analysis is key to the understanding of salt signal transduction pathways. For example, current knowledge on HOG1 pathway in yeast (Fig. 1) was built up step by step with mutational analysis. Conventional mutant screens using physiological criteria such as seed germination and root and seedling growth have been used to analyse plant salt-stress responses. For efficient genetic dissection of the salt-signalling network in plants, the use of reporter genes under the control of stress-responsive promoters as a screening approach can be an excellent alternative to overcome the shortage of salt-stress-specific responses in plants (61, 163).

One important prerequisite for genetic analysis of plant salt-stress responses is the use of a model system. Prior to recent development of the *Arabidopsis* model system, much information regarding salt signal transduction was obtained with the yeast system. Although it had been generally assumed that in response to salt stress yeast and plants may use similar signalling pathways, limited studies in plants have indicated that plant salt signalling has unique features (75). Despite the fact that many plant genes can increase yeast salt tolerance when expressed in yeast and vice versa, it is still not clear to what extent these heterologous expression results can reflect the native functions of the homologues in their own genetic background. Therefore, extrapolation of the findings from yeast to plants should be done with caution. With the complete sequencing of the *Arabidopsis* genome, this plant should find more use as a model system for studying plant salt signalling transduction. Current studies as described here and elsewhere have demonstrated the promises offered by this model system (75). This is because plants, whether glycophytes or halophytes, have similar signal transduction and salt tolerance machinery. The difference between salt tolerance and sensitivity likely results from changes in the threshold of some regulatory switches or mutations in some key determinants that are not vital for the hosts in their native habitats. Thus, using the *Arabidopsis* model system in the genetic analysis of salt signalling will contribute to building a blueprint of salt signal transduction networks applicable to all higher plants.

Once a framework has been set up, the application of the knowledge from

Arabidopsis to crop plants or, more specifically, the application of the knowledge to engineering salt-tolerant crop plants also requires specific information from crop plants and from other plants that are more tolerant than *Arabidopsis* to salts. In this regard, genetic analysis of salt signal transduction in halophytes should provide valuable information. Several salt-tolerant plants have been used as models for this purpose (164). These include the common ice plant *Mesembryanthemum crystallinum* and the resurrection plant *Craterostigma plantagineum* (164). One other potential halophyte model is *Thellungiella halophila* (75). This plant can grow in the presence of more than 300 mM NaCl, compared to around 75 mM NaCl for *Arabidopsis*. More importantly, *T. halophila* has many traits that are similar to *Arabidopsis*, such as small size, short life cycle, self-pollination, and a small genome. *T. halophila* can also be easily transformed via *Agrobacterium*-mediated flower-dipping transformation. One additional advantage is that at the cDNA level, homologous genes in this genome are over 90% identical to those in *Arabidopsis* (75). Therefore, genetic analysis of this halophyte plant can benefit from the availability of the *Arabidopsis* genome sequence.

Whilst genetic analysis or expression studies have yielded information on the functionality of individual components, many other studies also indicated that plant salt tolerance is a quantitative trait (e.g. reference 165). The fact that over-expressing individual components in a particular plant, e.g. *Arabidopsis*, increased salt tolerance is actually consistent with the notion of multiple determinants of salt tolerance, as it is possible that in each of these over-expressing studies there was a certain improvement in salt tolerance over the unmodified wild-type strain. However, each of these over-expression studies conducted in the laboratory was an isolated study and no comparison was made among transgenic plants engineered with different determinants. As a result, it is difficult to derive conclusions regarding the contribution of the individual determinants to the overall salt tolerance of the plants. Hopefully, with more information from both halophytes and crop plants, in the near future, it will be possible that a 'prescription' for engineering salt tolerance of crop plants in a particular environmental condition without compromising yield or other desired traits can be drawn, through comprehensive modelling of salt-signalling processes and the interaction between these cellular processes and the environment.

References

1. Postel, S. (1989) Water for agriculture: facing the limits. *Worldwatch Paper 93*. Worldwatch Institute, Washington DC.
2. Bohnert, H. J., Su, H. and Shen, B. (1999) Molecular mechanisms of salinity tolerance. In Shinozaki, K. and Yamaguchi-Shinozaki, K. (eds), *Molecular Responses to Cold, Drought, Heat, and Salt Stress in Higher Plants*. R. G. Landes Company, Austin, p. 29.
3. Hasegawa, P. M., Bressan, R. A., Zhu, J. K. and Bohnert, H. J. (2000) Plant cellular and molecular responses to high salinity. *Annu. Rev. Plant Physiol. Plant Mol. Biol.*, **51**, 463.
4. Shinozaki, K. and Yamaguchi-Shinozaki, K. (1997) Gene expression and signal transduction in water-stress response. *Plant Physiol.*, **115**, 327.
5. Yeo, A. (1998) Molecular biology of salt tolerance in the context of whole-plant physiology. *J. Exp. Bot.*, **49**, 913

6. Zhu, J. K., Hasegawa, P. M. and Bressan, R. A. (1997) Molecular aspects of osmotic stress in plants. *Crit. Rev. Plant Sci.*, **16**, 253.

7. Greenway, H. and Munns, R. (1980) Mechanisms of salt tolerance in non-halophytes. *Annu. Rev. Plant Physiol.* **31**, 149.

8. Amtmann, A. and Sanders, D. (1999) Mechanisms of Na^+ uptake by plant cells. *Adv. Bot. Res.*, **29**, 76.

9. Blumwald, E., Aharon, G., Apse, M. P. (2000) Sodium transport in plant cells. *Biochim. Biophy. Acta*, **1465**, 140.

10. Schachtman, D. and Liu, W. (1999) Molecular pieces to the puzzle of the interaction between potassium and sodium uptake in plants. *Trends Plant Sci.*, **4**, 281.

11. Tyerman, S. D. and Skerrett, I. M. (1999) Root ion channels and salinity. *Sci. Hort.*, **78**, 175.

12. Gassmann, W., Rubio, R. and Schroeder, J. I. (1996) Alkali cation selectivity of the wheat root high-affinity potassium transporter HKT1. *Plant J.*, **10**, 869.

13. White, P. J. (1999) The molecular mechanisms of sodium influx to root cells. *Trends Plant Sci.*, **7**, 245.

14. Nass, R., Cunningham, K. W. and Rao, R. (1997) Intracellular sequestration of sodium by a novel Na^+/H^+ exchanger in yeast is enhanced by mutations in the plasma membrane H^+-ATPase. *J. Biol. Chem.*, **272**, 26145.

15. Gaxiola, R. A., Rao, R., Sherman, A., Grisafi, P., Alper, S. L. and Fink, G. R. (1999) The *Arabidopsis thaliana* proton transporters, AtNhx1 and Avp1 can function in cation detoxification in yeast. *Proc. Natl Acad. Sci. USA*, **96**, 1480.

16. Apse, M. P., Aharon, G. S., Snedden, W. A. and Blumwald, E. (1999) Salt tolerance conferred by overexpression of a vacuolar Na^+/H^+ antiporter in *Arabidopsis*. *Science*, **285**, 1256.

17. Quintero, F. J., Blatt, M. R. and Pardo, J. M. (2000) Functional conservation between yeast and plant endosomal Na^+/H^+ antiporters. *FEBS Lett.*, **471**, 224.

18. Schumacher, K., Vafeados, D., McCarthy, M., Sze, H., Wilkins, T. and Chory, J. (1999) The *Arabidopsis det3* mutant reveals a central role for the vacuolar H^+-ATPase in plant growth and development. *Genes Dev.*, **13**, 3259.

19. Epstein, E. (1998) How calcium enhances plant salt tolerance. *Science*, **280**, 1906.

20. Gaymard, F., Pilot, G., Lacombe, B., Bouchez, D., Bruneau, D., Boucherez, J., Michaux-Ferrière, N., Thibaud, J.-B. and Sentenac, H. (1998) Identification and disruption of a plant shaker-like outward channel involved in K^+ release into the xylem sap. *Cell*, **94**, 647.

21. Shi, H., Ishitani, M., Kim, C. and Zhu, J. K. (2000) The *Arabidopsis thaliana* salt tolerance gene *SOS1* encodes a putative $Na+/H+$ antiporter. *Proc. Natl Acad. Sci. USA*, **97**, 6896.

22. Nuccio, M. L., Rhodes, D., McNeil, S. D. and Hanson, A. D. (1999) Metabolic engineering of plants for osmotic stress resistance. *Curr. Opin. Plant Biol.*, **2**, 128.

23. Hong, Z., Lakkineni, K., Zhang, Z. and Verma, D. P. S. (2000) Removal of feedback inhibition of Δ1-pyrroline-5-carboxylate synthetase results in increased proline accumulation and protection of plants from osmotic stress. *Plant Physiol.*, **122**, 1129.

24. Bohnert, H. J. and Sheveleva, E. (1998) Plant stress adaptations – making metabolism move. *Curr. Opin. Plant Biol.*, **1**, 267.

25. Holmstrom, K. O., Somersalo, S., Mandal, A., Palva, T. E. and Welin, B. (2000) Improved tolerance to salinity and low temperature in transgenic tobacco producing glycine betaine. *J. Exp. Bot.*, **51**, 177.

26. Hare, P. D., Cress, W. A. and van Staden, J. (1999) Proline synthesis and degradation: a model system for elucidating stress-related signal transduction. *J. Exp. Bot.*, **50**, 413.

27. LeHaye, P. A. and Epstein, E. (1971) Calcium and salt tolerance by bean plants. *Physiol. Plant.*, **25**, 213.

28. Maathuis, F. J. M. and Amtmann, A. (1999) K^+ nutrition and Na^+ toxicity: the basis of cellular K^+/Na^+ ratios. *Ann. Bot.*, **84**, 123.

29. Roberts, S. K., Tester, M. (1997) A patch clamp study of Na^+ transport in maize root. *J. Exp. Bot.*, **48**, 431.

30. Tyerman, S. D., Skerrett, M., Garrill, A., Findlay, G. P. and Leigh, R. A. (1997) Pathways for the permeation of Na^+ and Cl^- into protoplasts derived from the cortex of wheat roots. *J. Exp. Bot.*, **48**, 459.

31. Liu, J. and Zhu, J. K. (1997) An *Arabidopsis* mutant that requires increased calcium for potassium nutrition and salt tolerance. *Proc. Natl Acad. Sci. USA*, **94**, 14960.

32. Liu, J. and Zhu, J. K. (1998) A calcium sensor homolog required for plant salt tolerance. *Science*, **280**, 1943.

33. Zhu, J. K., Liu, J. and Xiong, L. (1998) Genetic analysis of salt tolerance in *Arabidopsis*. Evidence for a critical role of potassium nutrition. *Plant Cell*, **10**, 1181.

34. Rubio, F., Gassmann, W. and Schroeder, J. I. (1995) Sodium-driven potassium uptake by the plant potassium transporter HKT1 and mutations conferring salt tolerance. *Science*, **270**, 1660.

35. Rubio, F., Schwarz, M., Gassmann, W., Schroeder, J. I. (1999) Genetic selection of mutations in the high affinity K^+ transporter HKT1 that define functions of a loop site for reduced Na^+ permeability and increased Na^+ tolerance. *J. Biol. Chem.*, **274**, 6839.

36. Liu, W., Schachtman, D. P. and Zhang, W. (2000) Partial deletion of a loop region in the high affinity K^+ transporter HKT1 changes ionic permeability leading to increased salt tolerance. *J. Biol. Chem.*, **275**, 27924.

37. Withee, J. L., Sen, R., Cyert, M. S. (1998) Ion tolerance of *Saccharomyces cerevisiae* lacking the Ca^{2+}/CaM-dependent phosphatase (calcineurin) in improved by mutations in URE2 or PMA1. *Genetics*, **149**, 865.

38. Serrano, R., Mulet, J. M., Rios, G., Marguez, J. A., de Larrinoa, I. F., Leube, M. P., Mendizabal, I. M., Pascual-Ahuir, A., Proft, M., Ros. R. and Montesinos, C. (1999) A glimpse of the mechanisms of ion homeostasis during salt stress. *J. Exp. Bot.*, **50**, 1023.

39. Navarre, C. and Goffeau, A. (2000) Membrane hyperpolarization and salt sensitivity induced by deletion of PMP3, a highly conserved small protein of yeast plasma membrane. *EMBO J.*, **19**, 2515.

40. Goddard, N., Dunn, M., Zhang, L., White, A., Jack, P. and Hughes, M. (1993) Molecular analysis and spatial expression pattern of a low-temperature-specific barley gene, *blt101*. *Plant Mol. Biol.*, **23**, 871.

41. Capel, J., Jarrillo, J., Salinas, J. and Martinez-Zapater, J. (1997) Two homologous low-temperature genes from *Arabidopsis* encode highly hydrophobic proteins. *Plant Physiol.*, **115**, 569.

42. Counillon, L. and Pouyssegur, J. (2000) The expanding family of eucaryotic Na^+/H^+ exchangers. *J. Biol. Chem.*, **275**, 1.

43. Gerchman, Y., Rimon, A. and Padan, E. (1999) A pH-dependent conformational change of NhaA Na^+/H^+ antiporter of *Escherichia coli* involves loop VIII-IX, plays a role in the pH response of the protein, and is maintaned by the pure protein in dodecyle maltoside. *J. Biol. Chem.*, **274**, 24614.

44. Venturi, M., Rimon, A., Gerchman, Y., Hunte, C., Padan, E. and Michel, H. (2000) The monoclonal antibody 1F6 identifies a pH-dependent conformational change in the hydrophilic NH_2 terminus of NhaA Na^+/H^+ antiporter of *Escherichia coli*. *J. Biol. Chem.*, **275**, 4734.

45. Carmel, O., Rahav-Manor, O., Dover, N., Shaanan, B. and Padan, E. (1997) The Na^+-specific interaction between the LysR-type regulator, NhaR, and the Nha A gene encoding the Na^+/H^+ antiporter of *Escherichia coli. EMBO. J.*, **16**, 5922.

46. Wu, S. J., Ding, L. and Zhu, J. K. (1996) *SOS1*, a genetic locus essential for salt tolerance and potassium acquisition. *Plant Cell*, **8**, 617.

47. Yu, X. M. and Salter, M. W. (1998) Gain control of NMDA-receptor currents by intracellular sodium. *Nature*, **396**, 469.

48. Kim, S. A., Wang, M. H., Jae, S. K., Kwak, J. M., Schroeder, J. I. and Nam, H. G. (2000) Overexpression of AtGluR2, an *Arabidopsis* homolog of mammalian ionotropic glutamate receptor renders transgenic plants inefficient in calcium utilization and hypersensitive to ionic stresses. In *Plant Biology 2000 Abstract*. American Society of Plant Physiologists, Baltimore, p. 1030

49. Posas, F., Chamber, J. R., Heyman, J. A., Hoeffler, J. P., de Nadal, E. and Arino, J. (2000) The transcriptional response of yeast to saline stress. *J. Biol. Chem.* **275**, 17249.

50. Saab, I. N., Sharp, R. E., Pritchard, J. and Voetberg, G. S. (1990) Increased endogenous abascisic acid maintains primary root growth and inhibits shoot growth of maize seedlings at low water potential. *Plant Physiol.*, **93**, 1329.

51. Spollen, W. G., LeNoble, M. E., Samuel, T. D., Bernstein, N. and Sharp, R. E. (2000) Abscisic acid accumulation maintains maize primary root elongation at low water potentials by restricting ethylene production. *Plant Physiol.*, **122**, 967.

52. Beaudoin, N., Serizet, C., Gosti, F. and Giraudat, J. (2000) Interactions between abscisic acid and ethylene signaling cascades. *Plant Cell*, **12**, 1103.

53. Ghassemian, M., Nambara, E., Cutler, S., Kawaide, H., Kamiya, Y. and McCourt, P. (2000) Regulation of abscisic acid signaling by the ethylene response pathway in *Arabidopsis. Plant Cell*, **12**, 1117.

54. Rock, C. D. (2000) Pathways to abscisic acid-regulated gene expression. *New Phytol.*, **148**, 357.

55. Barkla, B. J., Vera-Estrella, R., Maldonado-Gama, M. and Pantoja, O. (1999) Abscisic acid induction of vacuolar H^+-ATPase activity in *Mesembryanthemum crystallinum* is developmentally regulated. *Plant Physiol.*, **120**, 811.

56. Strizhov, N., Abraham, E., Okresz, L., Blicking, S., Ziberstein, A., Schell, J., Koncz, C. and Szabados, L. (1997) Differential expression of two P5CS genes controlling proline accumulation during salt-stress requires ABA and is regulated by ABA1, ABI1, and AXR2 in *Arabidopsis. Plant J.*, **12**, 557.

57. Kaldenholf, R., Kolling, A. and Ritchter, G. (1993) A novel blue light-inducible and abscisic acid-inducible gene of *Arabidopsis thaliana* encoding an intrinsic membrane protein. *Plant Mol. Biol.*, **23**, 1187.

58. Tyerman, S. D., Bohnert, H. J., Maurel, C., Steudle, E. and Smith, J. A. C. (1999) Plant aquaporins: their molecular biology, biophysics and significance for plant water relations. *J. Exp. Bot.*, **50**, 1055.

59. Pih, K. T., Kabilan, V., Lim, J. H., Kang, S. G., Piao, H. L., Jin, J. B. and Hwang, I. (1999) Characterization of two new channel protein genes in *Arabidopsis. Mol. Cell*, **28**, 84.

60. Xiong, L., Ishitani, M. and Zhu, J. K. (1999) Interaction of osmotic stress, temperature, and abscisic acid in the regulation of gene expression in *Arabidopsis. Plant Physiol.*, **119**, 205.

61. Ishitani, M., Xiong, L., Stevenson, B. and Zhu, J. K. (1997) Genetic analysis of osmotic and cold stress signal transduction in *Arabidopsis*: interactions and convergence of abscisic acid-dependent and abscisic acid-independent pathways. *Plant Cell*, **9**, 1935.

62. Rock, C. D., Heath, T. G., Gace, D. A. and Zeevaart, J. A. D. (1991) Abscisic alcohol is an intermediate in abscisic acid biosynthesis in a shunt pathway from abscisic aldehyde. *Plant Physiol.* **97,** 670.

63. Bray, E. A. (1993) Molecular responses to water deficit. *Plant Physiol.,* **103,** 1035.

64. Rep, M., Krantz, M., Thevelein, J. M. and Hohmann, S. (2000) The transcriptional response of *Saccharomyces cerevisiae* to osmotic shock. *J. Biol. Chem.,* **275,** 8290.

65. Takahashi, E., Abe, J. I., Gallis, B., Aebersold, R., Spring, D. J., Krebs, E. G. and Berk, B. (1999) p90RSK is a serum-stimulated Na^+/H^+ exchanger isoform-1 kinase. *J. Biol. Chem.,* **274,** 20206.

66. Moe, O. W. (1999) Acute regulation of proximal tubule apical membrane Na/H exchanger NHE-3: role of phosphorylation, protein trafficking, and regulatory factors. *J. Am. Nephrol.,* **10,** 2412.

67. Moor, A. N. and Fliegel, L. (1999) Protein kinase-mediated regulation of the Na^+/H^+ exchanger in the rat myocardium by mitogen-activated protein kinase-dependent pathways. *J. Biol. Chem.,* **274,** 22985.

68. Urbachm V., Leguen, I., O'Kelly, I. and Harvey, B. J. (1999) Mechanosesitive calcium entry and mobilization in renal A6 cells. *J. Membr. Biol.,* **168,** 29.

69. Christensen, O. (1987) Mediation of cell volume regulation by Ca^{2+} flux through stretch-activated channels. *Nature,* **330,** 66.

70. Strotmann, R., Harteneck, C., Nunnenmacher, K., Schultz, G. and Plant, T. D. (2000) OTRPC4, a nonselective cation channel that confers sensitivity to extracellular osmolarity. *Nature Cell Biol.* **2,** 695.

71. Epstein, W., Buurman, E., McLaggan, D., Naprstek, J. (1993) Multiple mechanisms, roles and controls of K^+ transport in *Escherichia coli. Biochem. Soc. Trans.,* **21,** 1006.

72. Grinstein, S., Woodside, M., Sardet, C., Pourssegur, J. and Rotin, D. (1992) Activation of the Na^+/H^+ antiporter during cell volume regulation. Evidence for a phosphorylation-dependent mechanism. *J. Biol. Chem.,* **267,** 23823.

73. Klein, J. D., Lamitina, S. T., O'Neill, W. C. (1999) JNK is a volume-sensitive kinase that phosphorylates the Na-K-2Cl cotransporter in vitro. *Am. J. Physiol.,* **277,** 425.

74. Krump, E., Nikitas, K. and Grinstein, S. (1997) Induction of tyrosine phosphorylation and Na^+/H^+ exchanger activation during shrinkage of human neutrophilis. *J. Biol. Chem.,* **272,** 17303.

75. Zhu, J. K. (2000) Genetic analysis of plant salt tolerance using *Arabidopsis. Plant Physiol.,* **124,** 941.

76. Urao, T., Yakubov, B., Satoh, R., Yamaguchi-Shinozakia, K., Seki, B., Hirayama, T. and Shinozaki, K. (1999) A transmembrane hybrid-type histidine kinase in *Arabidopsis* functions as an osmosensor. *Plant Cell,* **11,** 1743.

77. Urao, T., Katagiri, T., Mizoguchi, T., Yamaguchi-Shinozaki, K., Hayashida, N. and Shinozaki, K. (1994) Two genes that encode Ca^{2+}-dependent protein kinases are induced by drought and high salt stresses in *Arabidopsis thaliana. Mol. Gen. Genet.,* **224,** 331.

78. Calenberg, M., Brohsonn, U., Zedlacher, M. and Kreimer, G. (1998) Light- and Ca^{2+}-modulated heterotrimeric GTPases in the eyespot apparatus of a flagellate green alga. *Plant Cell,* **10,** 91.

79. Josefsson, L. G. and Rask, L. (1997) Cloning of a putative G-protein-coupled receptor from *Arabidopsis thaliana. Eur. J. Biochem.,* **249,** 415.

80. Plakidou-Dymock, S., Dymock, D. and Hooley, R. (1998) A higher plant seven-trans-membrane receptor that influences sensitivity to cytokinins. *Curr. Biol.,* **8,** 315.

81. Devoto, A., Piffanelli, P., Nilsson, I., Wallin, E., Panstruga, R., von Heijne, G. and Schulze-Lefert, P. (1999) Topology, subcellular localization, and sequence diversity of the Mlo family in plants. *J. Biol. Chem.*, **274**, 34993.

82. Berridge, M. J., Lipp, P. and Bootman, M. D. (2000) The versatility and university of calcium signalling. *Nature Rev. Mol. Cell Biol.*, **1**, 11.

83. Sanders, D., Brownlee, C. and Harper, J. F. (1999) Communicating with calcium. *Plant Cell*, **11**, 691.

84. Munnik, T., Irvine, R. F. and Musgrave, A. (1998), Phospholipid signaling in plants. *Biochim. Biophy. Acta*, **1389**, 222.

85. Drobak, B. K. and Watkins, P. A. C. (2000) Inositol (1,4,5) trisphosphate production in plant cells: an early response to salinity and hyperosmotic stress. *FEBS Lett.*, **481**, 240.

86. Mikami, K., Katagiri, T., Luchi, S., Yamaguchi-Shinozaki, K. and Shinozaki, K. (1998) A gene encoding phosphatidylinositol 4-phosphate 5-kinase is induced by water stress and abscisic acid in *Arabidopsis thaliana*. *Plant J.*, **15**, 563.

87. Hirayama, T., Ohto, C., Mizoguchi, T. and Shinozaki, K. (1995) A gene encoding a phosphatidylinositol-specific phospholipase C is induced by dehydration and salt stress in *Arabidopsis thaliana*. *Proc. Natl Acad. Sci. USA*, **92**, 3903.

88. Quarmby, L. M., Yueh, Y. G., Cheshire, J. L., Keller, L. R., Snell, W. J. and Crain, R. C. (1992) Inositol phospholipid metabolism may trigger flagellar excision in *Chlamydomonas reinhardtii*. *J. Cell Biol.*, **116**, 737.

89. Zhu, J. K., Liu, J. and Xiong, L. (1998) Genetic analysis of salt tolerance in *Arabidopsis*. Evidence for a critical role of potassium nutrition. *Plant Cell*, **10**, 1181.

90. Hwang, J. I., Heo, K., Shin, K.J, Kin, E. and Yun, C. H. C. (2000) Regulation of phospholipase C-β3 activity by Na^+/H^+ exchanger regulation factor 2. *J. Biol. Chem.* **275**, 16632.

91. Frank, W., Munnik, T., Kerkmann, K., Salamini, F. and Bartels, D. (2000) Water deficit triggers phospholipase D activity in the resurrection plant *Craterostigma plantagineum*. *Plant Cell*, **12**, 111.

92. Divecha, N., Roefs, M., Halstead, J. R., D'Andrea, S., Fernandez-Borga, M., Oomen, L., Saqib, K. M., Wakelam, M. J. O. and D'Santos, C. (2000) Interaction of the type Iα PIPkinase with phospholipase D: a role for the local generation of phosphatidylinositol 4,5-bisphosphate in the regulation of PLD2 activity. *EMBO J.*, **19**, 5440.

93. Ott, A., Oehme, F., Keller, H. and Schuster, S. C. (2000) Osmotic stress response in *Dictyostelium* is mediated by cAMP. *EMBO J.*, **19**, 5782.

94. Allen, G.J, Muir, S. R. and Sanders, D. (1995) Release of Ca^{2+} form individual plant vacuoles by both InsP$_3$ and cyclic ADP-ribose. *Science*, **268**, 735.

95. Wu, Y., Kuzma, J., Marechal, E., Graeff, R., Lee, H. C., Foster, R. and Chua, N. H. (1997) Abscisic acid signaling through cyclic ADP-ribose in plants. *Science*, **278**, 2126.

96. Eu, J. P., Sun, J., Xu, L., Stamler, J. S. and Meissner, G. (2000) The skeletal muscle calcium release channel: coupled O_2 sensor and NO signaling functions. *Cell*, **102**, 499.

97. Sabri, A., Byron, K. L., Samarel, A. M., Bell, J. and Lucchesi, P. A. (1998) Hydrogen peroxide activates mitogen-activated protein kinases and Na^+/H^+ exchanger in neonatal rat cardiac myocytes. *Circul. Res.*, **82**, 1053.

98. Pei, Z. M., Murata, Y., Benning, G., Thomine, S., Klusener, B., Allen, G. J., Grill, E. and Schroeder, J. I. (2000) Calcium channels activated by hydrogen peroxide mediate abscisic acid signaling in guard cells. *Nature*, **406**, 731.

99. Liu, J., Ishitani, M., Halfter, U., Kim, C. S. and Zhu, J. K. (2000) The *Arabidopsis thaliana* SOS2 gene encodes a protein kinase that is required for salt tolerance. *Proc. Natl Acad. Sci. USA*, **97**, 3730.

100. Camoni, L., Fullone, M. R., Marra, M. and Aducci, P. (1998) The plasma membrane H+-ATPase from maize roots is phosphorylated in the C-terminal domain by a calcium-dependent protein kinase. *Physiol. Plant.*, **104**, 549.

101. Lino, B., Baizabal-Aguirre, V. M. and Gonzalez de la Vara, L. E. (1998) The plasma membrane H+-ATPase from beet root is inhibited by a calcium-dependent phosphorylation. *Planta*, **204**, 352.

102. Pei, Z. M., Ward, J. M., Harper, J. F. and Schroeder, J. I. (1996) A novel chloride channel in *Vici faba* guard cell vacuoles activated by the serine/threonine kinase, CDPK. *EMBO J.*, **15**, 6564.

103. Li, J., Lee, Y. R. J. and Assmann, S. M. (1998) Guard cells posses a calcium-dependent protein kinase that phosphorylates the KAT1 potassium channel. *Plant Physiol.*, **116**, 785.

104. Urao, T., Yahubov, B., Yamaguchi-Shinozaki, K. and Shinozaki, K. (1998) Stress-responsive expression of genes for two-component response regulator-like proteins in *Arabidopsis thaliana FEBS Lett.*, **427**, 175.

105. Sheen, J. (1996) Ca^{2+} dependent protein kinases and stress signal transduction in plants. *Science* **274**, 1900.

106. Saijo, Y., Hata, S., Kyozuka, J., Shimamoto, K. and Izui, K. (2000) Over-expression of a single Ca^{2+}-dependent protein kinase confers both cold and salt/drought tolerance on rice plants. *Plant J.*, **23**, 319.

107. Lamprecht, G., Weinman, E. J. and Yun, C. H. C. (1998) The Role of NHERF and E3KARP in the cAMP-mediated Inhibition of NHE3. *J. Biol. Chem.*, **273**, 29972.

108. Hall, R. A., Spurney, R. F., Premont, R. T., Rahman, N., Blitzer, J. T., Pitcher, J. A. and Lefkowitz, R. J. (1999) G protein-coupled receptor kinase 6A phosphorylates the Na^+/H^+ exchanger regulatory factor via a PDZ domain-mediated interaction. *J. Biol. Chem.*, **274**, 24328.

109. Monroy, A. F., Sangwan, V. and Dhindsa, R. S. (1998) Low temperature signal transduction during cold acclimation: protein phosphatase 2A as an early target for cold-inactivation. *Plant J.*, **13**, 653.

110. Harris, D. M., Myrick, T. L. and Rundle, S. J. (1999) The *Arabidopsis* homolog of yeast TAP42 and mammalian α4 binds to the catalytic subunit of protein phosphatase 2A and is induced by chilling. *Plant Physiol.*, **121**, 609.

111. Gingras, A. C., Gygi, S. P., Raught, B., Polakiewicz, R. D., Abraham, R. T., Hoekstra, M. F., Aebersold, R. and Sonenberg, N. (1999) Regulation of 4E-BP1 phosphorylation: a novel two-step mechanism. *Genes Dev.*, **13**, 1422.

112. Mizoguchi, T., Irie, K., Hirayam, T., Hayashida, N., Yamaguchi-Shinozaki, K., Matsumoto, K. and Shinozaki, K. (1996) A gene encoding a MAP kinase kinase kinase is induced simultaneously with genes for a MAP kinase and an S6 kinase by touch, cold and water stress in *Arabidopsis thaliana*. *Proc. Natl Acad. Sci. USA*, **93**, 765.

113. Xu, Q., Fu, H. H., Gupta, R. and Luan, S. (1998) Molecular characterization of a tyrosine-specific protein phosphatase encoded by a stress-responsive gene in *Arabidopsis*. *Plant Cell*, **10**, 849.

114. Knetsch, M. L. W., Wang, M., Snaar-Jagalska, B. E. and Heimovaara-Dijkstra., S. (1996) Abscisic acid induces mitogen-activated protein kinase activation in barley aleurone protoplasts. *Plant Cell*, **8**, 1061.

115. Mattison, C. P. and Ota, I. M. (2000) Two protein tyrosine phosphatases, Ptp2 and Ptp3, modulate the subcellular localization of the Hog1 MAP kinase in yeast. *Genes Dev.* **14**, 1229.

116. Mendoza, I., Rubio, F., Rodriguez-Navarro, A. and Pardo, J. M. (1994) The protein phosphatase calcineurin is essential for NaCl tolerance of *Saccharomyces cerevisiae*. *J. Biol. Chem.*, **269**, 8792.

117. Rao, A., Luo, C. and Hogan, P. G. (1997) Transcriptional factors of the NFAT family: regulation and function. *Annu. Rev. Immunol.*, **15**, 707.

118. Matheos, D. P., Kingsbury, T. J., Ashan, U. S., Cunningham, K. W. (1997) Tcn1p/Crz1p, a calcineurin-dependent transcription factor that differentially regulates gene expression in *Saccharomyces cerevisiae*. *Genes Dev.*, **11**, 3445.

119. Pardo, J. M., Reddy, M. P., Yang, S., Maggio, A., Huh, G. H., Matsumoto, T., Coca, M. A., Paino-D'Urzo, M., Koiwa, H., Yun, D. J., Watad, A. A., Bressan, R. A. and Hasegawa, P. M. (1998) Stress signaling through Ca^{2+}/calmodulin-dependent protein phosphatase calcineurin mediates salt adaptation in plants. *Proc. Natl Acad. Sci. USA*, **95**, 9681.

120. Kudla, J., Xu, Q., Harter, K., Gruissem, W. and Luan, S. (1999) Genes for calcineurin B-like proteins in *Arabidopsis* are differentially regulated by stress signals. *Proc. Natl Acad. Sci. USA*, **96**, 4718.

121. Piao, H. L., Pih, K. T., Lim, J. H., Kang, S. G., Jin, J. B., Kim, S. H. and Hwang, I. (1999) An *Arabidopsis* GSK3/shaggy-like gene that complements yeast salt stress-sensitive mutants in induced by NaCl and abscisic acid. *Plant Physiol.*, **119**, 1527.

122. Halfter, U., Ishitani, M. and Zhu, J. K. (2000) The *Arabidopsis* SOS2 protein kinase physically interacts with and is activated by the calcium-binding protein SOS3. *Proc. Natl Acad. Sci. USA*, **97**, 3735.

123. Koornneef, M., Reuling, C. and Karssen, C. M. (1984) The isolation and characterization of abscisic acid-insensitive mutants of *Arabidopsis thaliana*. *Physiol. Plant.*, **61**, 377.

124. Leung, J., Bouvier-Durand, M., Morris, P. C., Guerrier, D., Chefdor, F. and Giraudat, J. (1994) *Arabidopsis* ABA response gene ABI1: features of a calcium modulated phosphatase. *Science*, **264**, 1448.

125. Meyer, K., Leube, M. P. and Grill, E. (1994) A protein phosphatase 2C involved in ABA signal transduction in *Arabidopsis thaliana*. *Science*, **264**, 1452.

126. Leung, J., Merlot, S. and Giraudat, J. (1997) The *Arabidopsis* ABSCISIC ACID-INSENSITIVE 2 (*ABI2*) and *ABI1* genes encode homologous protein phosphatase 2C involved in abscisic acid signal transduction. *Plant Cell*, **9**, 759.

127. Meskiene, I., Boge, L., Glaser, W., Balog, J., Brandstotter, M., Zwerger, K., Ammerer, G. and Hirt, H. (1998) MP2C, a plant protein phosphatase 2C, functions as a negative regulator of mitogen-activated protein kinase pathways in yeast and plants. *Proc. Natl Acad. Sci. USA*, **95**, 1938.

128. Baudouin, E., Meskiene, I. and Hirt, H. (1999) Unsaturated fatty acids inhibit MP2C, a protein phosphatase 2C involved in the wound-induced MAP kinase pathway regulation. *Plant J.*, **20**, 343.

129. Robinson, M. J. and Cobb, M. H. (1997) Mitogen-activated protein kinase pathways. *Curr. Opin. Cell Biol.*, **9**, 180.

130. Huang, C. Y. F. and Ferrell. J. E.Jr. (1996) Ultrasensitivity in the mitogen-activated protein kinase cascade. *Proc. Natl Acad. Sci. USA*, **92**, 10098.

131. O'Rourke, S. M. and Herskowitz, I. (1998) The Hog1 MAPK prevents cross talk between the HOG and pheromone response MAPK pathways in *Saccharomyces cerevisiae*. *Genes Dev.* **12**, 2874.

132. Meskiene, I. and Hirt, H. (2000) MAP kinase pathways: molecular plug-and-play chips for the cell. *Plant Mol. Biol.*, **42**, 791.

133. Mikolajczyk, M., Awotunde, O. S., Muszynska, G. and Klessig, D. F. (2000) Osmotic stress induces rapid activation of a salicylic acid-induced protein kinase and a homolog of protein kinase ASK1 in tobacco cells. *Plant Cell*, **12**, 165.

134. Hoyos, M. E. and Zhang, S. (2000) Calcium-independent activation of salicylic acid-induced protein kinase and a 40-kilodalton protein kinase by hyperosmotic stress. *Plant Physiol.*, **122**, 1355.

135. Kovtun, Y., Chiu, W. L., Tena, G. and Sheen, J. (2000) Functional analysis of oxidative stress-activated mitogen-activated protein kinase cascade in plants. *Proc. Natl Acad. Sci. USA*, **97**, 2940.

136. Covic, L., Silva, N. F. and Lew, R. R. (1999) Functional characterization of ARAKIN (ATMEKK1): a possible mediator in an osmotic stress response pathway in higher plants. *Biochim. Biophys. Acta*, **1451**, 242.

137. Covic, L. and Lew, R. R. (1996) *Arabidopsis thaliana* cDNA isolated by functional complementation shows homology to serine/threonine protein kinases. *Biochim. Biophys. Acta.*, **1305**, 125.

138. Popping, B., Gibbons, T. and Watson, M. D. (1996) The *Pisum sativum* MAP kinase homolog (PsMAPK) rescues the *Saccharomyces cerevisiae* HOG1 deletion mutant under conditions of high osmotic stress. *Plant Mol. Biol.* **31**, 355.

139. Tsugane, K., Kobayashi, K., Niwa, Y., Ohba, Y. and Wada, K. (1999) A recessive *Arabidopsis* mutant that grows photoautotrophically under salt stress shows enhanced active oxygen detoxification. *Plant Cell*, **11**, 1195.

140. Shen, Q. and Ho, T. H. D. (1995) Functional dissection of an abscisic acid (ABA)-inducible gene reveals two independent ABA-responsive complexes each containing a G-box and a novel *cis*-acting element. *Plant Cell*, **7**, 295.

141. Baker, S. S., Wilhelm, K. S. and Thomashow, M. F. (1994) The 5′-region of *Arabidopsis thaliana cor15a* has *cis*-acting elements that confer cold-, drought-, and ABA-regulated gene expression. *Plant Mol. Biol.* **24**, 701.

142. Yamaguchi-Shinozaki, K. and Shinozaki, K. (1994) A novel cis-acting element in an *Arabidopsis* gene is involved in responsiveness to drought, low-temperature, or high-salt stress. *Plant Cell*, **6**, 251.

143. Stockinger, E. J., Gilmour, S. J. and Thomashow, M. F. (1997) *Arabidopsis thaliana CBF1* encodes an AP2 domain-containing transcriptional activator that binds to the C-repeat/DRE, a *cis*-acting DNA regulatory element that stimulates transcription in response to low temperature and water deficit. *Proc. Natl Acad. Sci. USA*, **94**, 1035.

144. Gilmour, S. J., Zarka, D. G., Stockinger, E. J., Salazar, M. P., Hougton, J. M. and Thomashow, M. F. (1998) Low temperature regulation of the *Arabidopsis* CBF family of AP2 transcriptional activators as an early step in cold-induced *COR* gene expression. *Plant J.*, **16**, 433.

145. Liu, Q., Kasuga, M., Sakuma, Y., Abe, H., Miura, S., Yamaguchi-Shinozaki, K. and Shinozaki, K. (1998) Two transcription factors, DREB1 and DREB2, with an EREBP/AP2 DNA binding domain separate two cellular signal transduction pathways in drought- and low-temperature-responsive gene expression, respectively, in *Arabidopsis*. *Plant Cell*, **10**, 1391.

146. Jaglo-Ottosen, K. R., Gilmour, S. J., Zarka, D. G., Schabenberger, O. and Thomashow, M. F. (1998) *Arabidopsis CBF1* overexpression induces *COR* genes and enhances freezing tolerance. *Science*, **280**, 104.

147. Kasuga, M., Liu, Q., Miura, S., Yamaguchi-Shinozaki, K. and Shinozaki, K. (1999) Improving plant drought, salt, and freezing tolerance by gene transfer of a single stress-inducible transcriptional factor. *Nature Biotech.*, **17**, 287.

148. Horie, S., Moe, O., Yamaji, Y., Cano, A., Miller, R. T. and Alpern, R. J. (1992) Role of protein kinase C and transcription factor AP-1 in the acid-induced increase in Na/H antiporter activity. *Proc. Natl Acad. Sci. USA*, **89**, 5236.

149. Dyck, J. R. B., Silva, N. L. C. and Fliegel, L. (1995) Activation of the Na^+/H^+ exchanger gene by the transcription factor AP-2. *J. Biol. Chem.* **270**, 1375.

150. Martinez-Pastor, M. T., Marcher, G., Schuller, C., Marcher-Bauer, A., Ruis, H. and Estruch, F. (1996) The *Saccharomyces cerevisiae* zinc finger proteins Msn2p and Msn4p are required for transcriptional induction through the stress-response element (STRE) *EMBO. J.*, **15**, 2227.

151. Marquez, J. A., Pascual-Ahuir, A., Proft, M. and Serrano, R. (1998) The Ssn6-Tup1 repressor complex of *Saccharomyces cerevisiae* is involved in the osmotic induciton of HOG-dependent and -independent genes. *EMBO J.*, **17**, 2543.

152. Proft, M. and Serrano, R. (1999) Repressors and upstream repressing sequences of the stress-regulated ENA1 gene in *Saccharomyces cerevisiae*: bZIP protein Sko1p confers HOG-dependent osmotic regulation. *Mol. Cell. Biol.*, **19**, 537.

153. Resh, M. D. (1999) Fatty acylation of proteins: new insights into membrane targeting of myristoylated and palmitoylated proteins. *Biochim. Biophys. Acta*, **1451**, 1.

154. Ishitani, M., Liu, J., Halfter, U., Kim, C. S., Shi, W. and Zhu, J. K. (2000) SOS3 function in plant salt tolerance requires N-myristoylation and calcium binding. *Plant Cell*, **12**, 1667.

155. Kholodenko, B. N., Hoek, J. B. and Westerhoff, H. V. (2000) Why cytoplasimic signalling proteins should be recruited to cell membranes. *Trends Cell Biol.*, **10**, 173.

156. Mahanty, S. K., Wang, Y. M., Farley, F. W. and Elion, E. A. (1999) Nuclear shuttling of yeast scaffold Ste5 is required for its recruitment to the plasma membrane and activation of the mating MAPK cascade. *Cell*, **98**, 501.

157. Fraser, I. D. and Scott, J. D. (1999) Modulation of ion channels: a 'current' view of AKAPs. *Neuron*, **23**, 423.

158. Yun, C. H. C., Oh, S., Zizak, M., Steplock, D., Tsao, S., Tse, C. M., Weinman, E. J. and Donowitz, M. (1997) cAMP-mediated inhibition of the epithelial brush border Na^+/H^+ exchanger, NHE3, requires an associated regulatory protein. *Proc. Natl Acad. Sci. USA*, **94**, 3010.

159. Harada, H., Becknell, B., Wilm, M., Mann, M., Huang, L. J., Taylor, S. S., Scott, J. D. and Korsmeyer, S. J. (1999) Phosphorylation and inactivation of BAD by mitochondria-anchored protein kinase A. *Mol. Cell*, **3**, 413.

160. Finnie, C., Borch, J. and Collinge, D. B. (1999) 14-3-3 proteins: eukaryotic regulatory proteins with many functions. *Plant Mol. Biol.* **40**, 545.

161. Camoni, L., Harper, J. F. and Palmgren, M. G. (1998) 14-3-3 proteins activate a plant calcium-dependent protein kinase (CDPK). *FEBS Lett.*, **430**, 381.

162. Camoni, L., Iori, V., Marra, M. and Aducci, P. (2000) Phosphorylation-dependent interaction between plant plasma membrane H^+-ATPase and 14-3-3 proteins. *J. Biol. Chem.*, **275**, 9919.

163. Xiong, L., David, L., Stevenson, B. and Zhu, J. K. (1999) High throughput screening of signal transduction mutants with luciferase imaging. *Plant Mol. Biol. Rep.*, **17**, 159.

164. Cushman, J. C. and Bohnert, H. J. (2000) Genomic approaches to plant stress tolerance. *Curr. Opinion Plant Biol.*, **3**, 117.

165. Foolad, M. R., Jones, R. A. (1993) Mapping salt-tolerance genes in tomato (*Lycopersicon esculentum*) using trait-based marker analysis. *Theor. Applic. Gen.*, **87**, 184.

9 | Recognition and defence signalling in plant/bacterial and fungal interactions

JONG HYUN HAM AND ANDREW BENT

1. Introduction

Plants, like other organisms, have defence systems that can be induced following infection by diverse microorganisms with a variety of pathogenic traits (1). Interest in disease resistance remains high among both scientists and plant breeders because disease outbreaks are one of the major limiting factors for crop yield. The development of disease-resistant plant lines has been an effective strategy for disease control (2–4). Genetic and physiological aspects of plant disease resistance have been widely studied, and there are an increasing number of opportunities to apply these findings to the development of disease-resistant plants.

Recent efforts have led to a dramatic increase in what is known about plant defence signal transduction. In particular, intensive genetic dissection of the model plant *Arabidopsis thaliana* has added many new features to integrated models of the signalling pathways for disease resistance.

This chapter discusses the molecular basis of pathogen recognition and the subsequent activation of defence responses in plants. Much of the focus is on the signal transduction pathways for gene-for-gene resistance, local and systemic acquired resistance, and induced systemic resistance. Attention is drawn to the methodological approaches used to study defence signal transduction. These subjects have also been covered recently in a number of brief reviews (e.g. 5–9).

2. Gene-for-gene resistance and the initiation of defence signal transduction

This opening section discusses some of the known signals from bacterial and fungal pathogens that are recognized by host plants, and which trigger their defence responses. Also discussed are the structure and function of plant resistance gene

(R gene) products that perceive these signals from pathogens, and the evolution of R genes and the generation of novel pathogen-recognition specificities.

2.1 A molecular foundation for the gene-for-gene hypothesis

Soon after Mendel's pioneering genetic studies were rediscovered in the early twentieth century, many cases of single genes that control plant disease resistance were described (2). In the 1940s, Flor suggested the 'gene-for-gene hypothesis' to explain his detailed genetic study of both host and pathogen in the interaction between flax (*Linum ultissimum*) and flax rust fungi (*Melampsora lini*). Flor found that individual host R genes only functioned to repel infections by fungal strains that carried a corresponding dominant avirulence (*avr*) gene of matched specificity. This gene-for-gene hypothesis has now been shown to describe numerous host–pathogen interactions involving fungal, bacterial, or viral diseases (2, 10, 11). In many cases, both R and *avr* genes behave as single dominant genes (2, 10, 11). Disease resistance in a gene-for-gene interaction is often accompanied by the rapid local collapse of the infected region, the so-called hypersensitive response (HR), which is known to be a programmed cell death response (12).

Receptor–ligand binding has been a likely physical model for the gene-for-gene hypothesis, with the receptor encoded by the plant R gene perceiving the signal encoded by a corresponding pathogen *avr* gene (10, 11, 14). The HR and associated defence reactions are thought to be triggered by this receptor–ligand-mediated pathogen recognition event. Multiple *avr* and R genes have now been cloned, making it possible to study the biological and biochemical nature of the products of these genes. However, direct physical interaction between R gene products and *avr* gene products has only been shown in a limited number of cases (15–18), and the detailed mechanisms of recognition and signal delivery to the subsequent defence pathway remain to be elucidated.

2.2 Bacterial signals

Bacterial *avr* genes have been extensively studied to understand the molecular interaction between plants and bacterial pathogens (14, 19). Since an *avr* gene from the incompatible interaction between soybean cultivars and *Pseudomonas syringae* pv. *glycinea* was first isolated in 1984, more than 30 bacterial *avr* genes have been cloned and characterized (14). Introduction of cloned *avr* genes into virulent pathogen strains can make the 'compatible' (disease-causing) interaction 'incompatible' (disease-blocking) if the host plant possesses the cognate R gene, yet leave the bacteria fully virulent on host plants that do not express the appropriate R gene.

The molecular activity of *avr* gene products, beyond their macroscopic defence-eliciting phenotype, is in most cases not clearly understood. Sequence data for *avr* genes have generally not suggested functions. However, a number of *avr* genes, such as *avrE*, *avrBs2*, and *avrRpm1*, are required either for full virulence or for multiplication within plants susceptible to the pathogen, suggesting positive functions of *avr* genes in pathogen virulence (e.g. 14, 19–22). *Erwinia amylovora dspA/E*, a homologue of *avrE* of *Pseudomonas syringae* pv. *tomato*, is a gene originally identified as a pathogenicity determinant that was subsequently shown to have avirulence activity when placed in a *Pseudomonas syringae* strain (23, 24). Nevertheless, most *avr* genes do not show obvious virulence phenotypes, although this may due to virulence assays that bypass the process where *avr* genes are required.

Clues regarding the mode of action of Avr proteins have been uncovered for AvrBs3-family products. *avrBs3* was cloned from *Xanthomonas campestris* pv. *vesicatoria*, and the gene encodes 17.5 copies of a 34 aa direct tandem repeat (25). The number of repeats, and/or subtle sequence variation, was shown to determine race-cultivar specificity among pepper and tomato cultivars. Deletion of repeat units generated new avirulence specificities. This work suggested that the physical structure of the Avr protein directly confers defence-eliciting activity.

Other bacterial genes with structures similar to *avrBs3* (including the 102 bp repeats) include *pth*, the pathogenicity factor of citrus pathogen *Xanthomonas citri*, and several *avr* genes of cotton pathogen *Xanthomonas campestris* pv. *malvacearum* and the rice pathogen *Xanthomonas oryzae* pv. *oryzae* (19, 26–29). As in *avrBs3*, the102 bp repeat region determined the specificity of those *avr* genes and *pth* function (26, 28, 30). In addition to the tandem repeat regions, AvrBs3 homologues contain a putative leucine zipper region and three nuclear localization signals in the C-terminus of each protein. Intriguingly, Avr/Pth proteins expressed in plant cells are localized to nuclei, and this localization is important for avirulence function (31, 32). An active DNA-binding domain is also present in many of these AvrBs3-family gene products, suggesting a direct role for these proteins in control of host gene transcription (19, 29).

Although most Avr proteins are thought to encode defence-eliciting structures, *avrD* of *Pseudomonas syringae* pv. *glycinea* encodes an enzyme responsible for the production of a syringolide elicitor compound (33).

2.2.1 Type III secretion

Gram-negative plant pathogenic bacteria, such as *Pseudomonas syringae*, *Erwinia amylovora*, *Xanthomonas campestris*, *Ralstonia solanacearum*, and *Erwinia chrysanthemi*, possess the type III (Hrp) secretion machinery that is also an important pathogenesis determinant in animal pathogenic bacteria (34, 35). Interestingly, it was found that a *Pseudomonas syringae* Hrp secretion system, together with single *Pseudomonas syringae avr* genes, was sufficient to allow *Escherichia coli* to elicit an HR in soybean or *Arabidopsis* carrying a corresponding *R* gene (36, 37). The dependence of *avr* gene function on the Hrp secretion system strongly suggests that Avr proteins are secreted through the type III (Hrp) secretion machinery. Recently, secretion of Avr proteins through Hrp secretion machinery has been demonstrated (38–40).

Avr signal delivery by the Hrp secretion system is apparently very similar to virulence protein delivery by type III secretion systems of animal pathogenic bacteria, which inject virulence proteins directly into host cells in a contact-dependent manner (34, 35). The model of direct transfer of Avr proteins from bacterial cell to plant cell through the Hrp system strongly suggests, together with the deduced cytoplasmic location of many *R* gene products, that the site of these particular Avr–R interactions is inside the plant cell. Although there is not yet direct evidence for the localization of Avr proteins in plant cells by bacterial injection, transient expression of *avr* genes in plants containing corresponding *R* genes causes plant cell death, strongly indicating that Avr–R interactions take place inside of plant cells (34, 35). Avr proteins can undergo further processing within the plant host (39, 41).

As a parallel to the above avirulence studies, a current paradigm for bacterial pathogenesis in plants is that effector proteins (virulence factors), including Avr proteins of plant pathogenic bacteria, are entering into plant cells through the type III (Hrp) secretion pathway. The resulting bacterial proliferation presumably results in part from suppression of plant defence and/or from release of nutrients from plant cells (35).

2.2.2 Non-Avr bacterial elicitors

In addition to the extensive work that has been done on bacterial Avr proteins, bacterial products have been identified that elicit plant defence responses but are not known *avr* gene products that mediate race-specific resistance (e.g. 42–44). It is intriguing that a plant receptor for flagellin elicitor has been identified and discovered to encode a protein resembling the plant *R* gene products that are discussed below (45).

2.3 Fungal signals

A limited number of Avr proteins have been isolated from fungal pathogens (46). Avr9 and Avr4 from *Cladosporium fulvum* carry elicitor activities on tomato cultivars carrying *Cf-9* and *Cf-4* resistance genes, respectively. It is known that *Avr9* and *Avr4* encode small cysteine-rich peptides – 28 amino acids and 86 amino acids, respectively – processed from preproteins. Because the pathogen grows extracellularly in the plant apoplast and the resistance genes (*Cf-9* and *Cf-4*) have an extracellularly oriented leucine-rich repeat (LRR) region that determines the resistance specificity, it is very likely that the recognition of Avr proteins by Cf-resistance gene products occurs in the plant apoplastic region (46–48). This is in contrast to the bacterial Avr proteins discussed above. However, the presence of a high-affinity binding site for Avr9 in plasma membranes of $Cf-9^+$ and $Cf-9^-$ tomato mesophyll cells implies that Cf-9 is not the primary receptor of AVR9 peptide. Instead, in analogy to the complex encoded by the *CLAVATA* (*CLV*) genes *CLV1, 2,* and *3* that are involved in *Arabidopsis* development, at least three molecules may be involved in the *Cf-9*-dependent signal transduction. (47).

The barley pathogen *Rhyncosporium secalis* produces NIP1 protein, another small secreted cysteine-rich protein with Avr activity, albeit from an entirely different host–pathogen interaction (49).

An Avr protein from the rice blast fungus, *Magnaporthe grisea*, is encoded by *Avr-Pita*, formerly *Avr2-YAMO*. Avr-Pita, a putative zinc metalloprotease, directly interacts with the product of the cognate rice resistance gene, *Pita* (17). The cytoplasmic nature of the Pita protein, together with transient expression experiments, suggests that recognition occurs inside the plant cell. However, the mechanism of Avr-Pita translocation is not clear (17).

Several fungal proteinaceous and non-proteinaceous elicitors are known to induce defence responses of host plants in a non-gene-for-gene fashion (not host-genotype-specific; or host species-specific) (9, 44, 50). For example, oligo-*N*-acetylglucosamine and oligogalacturonides are known to have broad elicitor activities in several plants, whereas elicitins, a type of protein secreted by oomycete pathogens, are known to have species-specific elicitor activities, as are a glycoprotein and a heptaglucan from the oomycete pathogen, *Phytophthora sojae* (9, 50). These elicitors are not known to participate in gene-for-gene pathogen recognition/defence activation systems, but they do elicit defence responses and may serve as important defence-eliciting determinants *in vivo*. These elicitors have been used in many biochemical studies of defence signal transduction (discussed below).

2.4 Structure and function of *R* genes

The molecular cloning of resistance genes in the mid-1990s opened the way for many types of studies of the molecular interaction between plant and pathogen. Excitingly, many *R* gene products from different plant species share similar structural patterns despite their function against diverse viruses, bacteria or fungi. These patterns include almost universally leucine-rich repeat (LRR) (10, 11). Nucleotide-binding sites (NBS) are also common. Known *R* genes can be classified into several groups based on their encoded protein structures (Fig. 1).

2.4.1 Serine/threonine kinases

The tomato resistance gene *Pto* was the first cloned gene-for-gene resistance gene. It confers resistance against a bacterial pathogen, *Pseudomonas syringae* pv. *tomato*, carrying the *avrPto* gene. *Pto*, which encodes a serine/threonine protein kinase, lies in a gene cluster with other highly similar kinase genes (51). For resistance function, *Pto* requires *Prf*, which encodes an NBS-LRR class protein with an amino-terminal leucine zipper (LZ) domain like other resistance genes such as *RPS2* and *RPM1* (52). The discovery of *Pto* demonstrates the importance of protein phosphorylation events in gene-for-gene defence signal transduction systems.

Pto protein interacts with the cognate avirulence gene product, AvrPto, in yeast two-hybrid systems (15, 16). This interaction correlates with the resistance function of Pto, in that AvrPto does not interact with the kinase encoded by the close *Pto* homologue, *Fen*, which also requires *Prf* for activity (15, 16). The physical interaction between Pto and AvrPto provides a molecular model for the mechanism of gene-for-gene resistance, in which the resistance gene product and the avirulence gene product act as a receptor and a ligand, respectively. Besides AvrPto, plant host factors that interact with Pto kinase have been identified and a number of other structure–function studies have been completed using this system, as described in detail in Section 4.3.

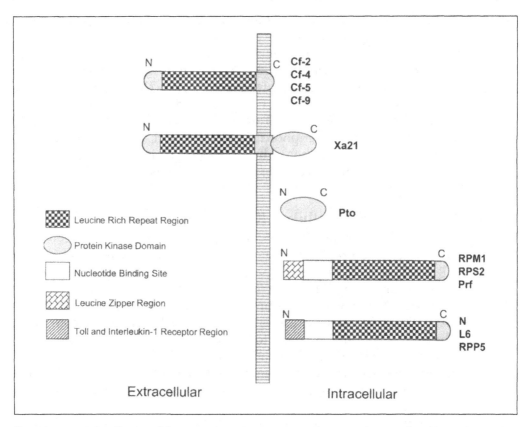

Fig. 1 Structural classification of *R* gene products. Representative *R* gene products are listed for each example.

2.4.2 NBS-LRR proteins

Most other *R* genes characterized to date encode leucine-rich repeat (LRR) domains, and many also encode nucleotide binding sites (NBS) (10, 11). The biochemical role of the NBS in resistance processes is not clear, but models can be built based on similar proteins of animal cells. In Ras and other guanosine-nucleotide-binding proteins, signal transduction is off in the GDP-bound state but on in the GTP-bound state. Apaf-1 and CED-4, which have NBS domains and other regions of similarity to *R* gene products, are activators of mammalian and nematode apoptosis proteases, caspase-9 and CED-3, respectively (53, 54). Nod1, which has both NBS and LRR domains, was found to regulate apoptosis and NF-κB activation pathways, and this required a functional NBS domain (55). Additional similarities between defence signal transduction and cell death in plant and animal defence systems have been noted (54, 56, 57).

The NBS-LRR resistance genes can be divided roughly into three classes based on the structure of the amino-terminus (10, 11). *RPS2*, *RPM1*, *Prf*, and other *R* genes encode proteins that have a leucine zipper (LZ) domain in their amino-terminus (10,

11). LZ are commonly known to function in protein dimerization, but it is not known if *R* gene products form homo- or hetero-dimers. Other NBS-LRR resistance genes encode within the amino-terminus a domain with similarity to Toll or interleukin-1 receptor(s) (TIR). This class includes, for example, the flax rust resistance genes (*L6* and *M*), the tobacco mosaic virus resistance gene (*N*), and the *Arabidopsis* downy mildew resistance gene (*RPP5*) (53). The third class encodes neither LZ nor TIR domains. It is noteworthy that Toll and IL-1R function through ligand binding to activate Rel-family transcriptional factors of insect and mammalian cells, respectively, resulting in antimicrobial host responses. It is very likely that different taxa also share other common components and mechanism for defence responses (10, 56, 57).

2.4.3 Proteins with extracellular LRR domains

Tomato resistance genes that function against *Cladosporium fulvum* (*Cf-9, Cf-2, Cf-4*, and *Cf-5*) encode proteins that consist largely of amino-terminal extracytoplasmic LRR domains with very small putative transmembrane/cytoplasmic domains for membrane anchoring (11). *Cf-9* and *Cf-4* are tightly linked and are very similar each other except for a few deletions and amino acid substitutions in the amino-terminal half of *Cf-4*. *Cf-2*, which exists as two functionally redundant genes at a locus, is tightly linked to and also similar to *Cf-5*. Cf-2 and Cf-5 proteins differ in their amino-terminal portions, where Cf-5 has a deletion of six LRRs and several amino acid changes. The diversity in LRR domains and the high homology in the C-termini of Cf proteins suggest that the extracytoplasmic LRR domains are responsible for recognition of pathogens (11), a point that is discussed below.

The rice *Xa21* gene, which confers resistance to the bacterial leaf blight pathogen *Xanthomonas campestris* pv. *oryzae*, encodes a 1025 amino acid protein with an extracytoplasmic LRR domain, a transmembrane domain, and a cytoplasmic serine/threonine kinase domain (58). Because Xa21 possesses both an extracytoplasmic LRR domain and a Pto-like cytoplasmic serine/threonine kinase, this protein provides clues for the potential linkage between an LRR-mediated recognition event and the downstream activation of intracellular gene-for-gene signal transduction pathways.

2.5 The evolution and specificity of *R* genes

R genes are highly polymorphic and many of them reside in complex loci as members of multigene families (59, 60). Tandem arrays of closely related genes are found at the loci of *Cf* genes, *Pto*, and *RPP8*, *Rp1*, *M*, *Dm*, and other *R* genes (59, 60). Novel *R* gene specificities apparently arise from several genetic events that can occur at these complex loci, including gene duplication, equal- or unequal-exchange during meiotic recombination, single base changes, and small insertion/deletion events (11, 60). Retroelements, which provide sites for recombination and translocation events, and transposable elements may also be important factors for the *R* gene evolution (59). Characterization of many *R* gene loci indicates that although other domains can also contribute, LRR regions play a central role in determination of pathogen specificity (59, 60, 154, 155). Although most genes in an organism are under conservative selection that retains amino acid sequence, diversifying selection apparently occurs within the LRR-encoding domains of R gene loci (59–62). It is particularly exciting that future work may allow the generation of R genes with new specificity using *in vitro* recombinant DNA methods (e.g. 154, 155).

3. What actually stops pathogen growth?

Plants carry out a variety of defence reactions at local and systemic levels in response to microbial attacks. In strong defence responses such as those mediated by many R/*avr* interactions, the defence response at the site of interaction is often rapid, robust, and multifaceted. Some of the same responses may be induced more slowly and/or less dramatically in interactions involving 'quantitative', 'horizontal', or 'partial' resistance that can slow or partially limit disease development (63, 64). More delayed but long-lasting systemic responses such as systemic acquired resistance (SAR) or induced systemic resistance (ISR) can also be induced by pathogen attack (9, 65, 66). This section briefly outlines some of the defence responses that apparently comprise local and systemic defences. Components of the signal transduction pathways that induce these defence responses are described in Section 4.

Cell wall fortification by protein cross-linking, lignification, and papilla formation are among the most common and intuitively logical host responses against the pathogen attack, producing a physical blockade that can restrict pathogen spread (64). Fragments of plant cell walls, cuticles and other extracellular components may also have significant defence-eliciting (signal transduction) activity (44).

Expression of an array of PR ('pathogenesis-related') proteins becomes elevated following infection, especially in cells surrounding the site of infection but also to a lesser extent in distal tissues (9, 67). Some of the classically defined PR proteins are listed in Table 1. With the introduction of microarray expression profiling capabilities, the list of *PR* genes can be expected to grow substantially (68). Some PR proteins may have a direct role in disease resistance, based on their antimicrobial activities *in vitro* or on the partially enhanced resistance of transgenic plants expressing some *PR* genes (64, 69). The profile of *PR* gene expression varies based on the plant species and the inducing agent. For example, PR-1 protein is produced abundantly by tobacco and *Arabidopsis* during SAR but is only weakly produced by cucumber (70). In addition, the systemic resistance induced by certain rhizobacteria results in a different pattern of PR protein expression from that induced by SAR (65).

Production of phytoalexins such as phenylpropanoid or flavonoid antimicrobial compounds is another typical response to pathogen infection or elicitor treatment (71). Phytoalexins have long been studied as a resistance determinant. Their relative role in restricting pathogen growth is probably more quantitative than qualitative in most host–pathogen interactions (71).

Reactive oxygen intermediates (ROI) such as hydrogen peroxide or superoxide are also important defence components. In addition to their role in chemically activating

Table 1 The families of pathogenesis-related (PR) proteins[a]

Family	Type Member	Properties
PR-1	Tobacco PR-1a	Unknown
PR-2	Tobacco PR-2	β-1,3-Glucanase
PR-3	Tobacco P, Q	Chitinase type I, II, IV, V, VI, VII
PR-4	Tobacco 'R'	Chitinase type I, II
PR-5	Tobacco 'S'	Thaumatin-like
PR-6	Tomato inhibitor I	Proteinase inhibitor
PR-7	Tomato P_{69}	Endoproteinase
PR-8	Cucumber chitinase	Chitinase type III
PR-9	Tobacco 'lignin-forming peroxidase'	Peroxidase
PR-10	Parsley 'PR1'	Ribonuclease-like
PR-11	Tobacco class V chitinase	Chitinase type I
PR-12	Radish Rs-AFP3	Defensin
PR-13	*Arabidopsis* THI2.1	Thionin
PR-14	Barley LTP4	Lipid-transfer protein

[a] Adapted from reference 67.

cell wall reinforcement, and their key role in defence signal transduction (discussed below), ROI can also be directly antimicrobial.

The hypersensitive response (HR) is frequently associated with gene-for-gene resistance responses (12, 72). The HR is a programmed cell death (PCD) process driven by *de novo* protein synthesis and active metabolism (12). The HR is also associated with the 'non-host' resistance response that some plants develop when infected by a plant pathogen that is not pathogenic on that host species (12), although this may be driven by *R/avr* (gene-for-gene) interactions (73). Even though HR is a hallmark of gene-for-gene resistance, the direct role of HR in disease resistance is not clear (12). Death of the infected cell could clearly be beneficial in slowing the growth of virus or biotrophic fungus pathogens. Cell death may also serve to release antimicrobial compounds, as well as signal molecules that induce or enhance resistance in adjacent or distal uninfected cells. However, a number of recent studies have shown that cell death *per se* is not required for all gene-for-gene defence responses (74–76).

Plant defence responses vary greatly depending on the plant and pathogen involved, and it is important to keep in mind that the differences between these defence responses can be as significant as the overlaps and similarities.

4. Defence signal transduction pathways

Plants have a complex network of signal transduction pathways for defence against a variety of microbial attack (Fig. 2; see also ref. 5–9 and 64). Genetic analysis has been a powerful tool for identifying these defence pathways and their components (7, 8). Biochemical, pharmacological, and molecular biological methods have provided valuable information that cannot be found by genetic analysis (9, 77). This section

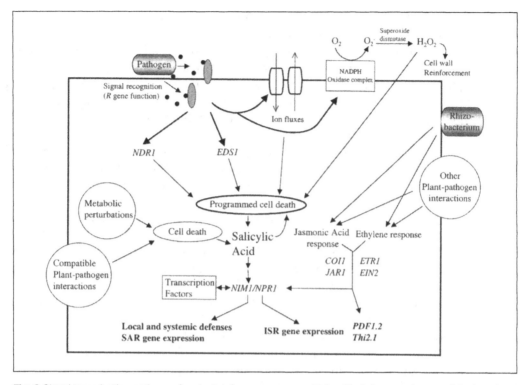

Fig. 2 Signal transduction pathways for plant defence responses. A simplified diagram of some of the key signal transduction components identified in recent studies. Names in italics are gene names. Note that many other known components of plant defence signal transduction are omitted.

discusses defence signal transduction pathways and the use of different strategies to reveal relevant components of those pathways.

4.1 Identification and dissection of defence signal transduction components by genetic analysis

4.1.1 Gene-for-gene resistance pathways

Mutations that abolish or impair gene-for-gene disease resistance can arise in individual *R* genes, or in other genes that participate in gene-for–gene signal transduction pathways. *R* genes, which to date have been identified almost exclusively by genetic strategies, were discussed in a previous section. The present section discusses some of those 'other' genes that are likely to act downstream of *avr*/*R* interactions. Genes that are known to participate both in gene-for-gene signal transduction and in other types of defence responses are discussed in later sections.

The *Arabidopsis* gene *NDR1* (for *non-race-specific disease resistance*) was identified using *ndr1* mutants, which are susceptible to avirulent *Pseudomonas syringae* pv.

tomato DC3000 expressing one of four bacterial avirulence genes (*avrB*, *avrRpm1*, *avrRpt2*, or *avrPph3*) (78). Mutation of *ndr1* also caused susceptibility to several avirulent isolates of *Peronospora parasitica*, but the *ndr1* mutant retained an HR-like phenotype in response to some avirulent *Pseudomonas syringae* pv. *tomato* strains (78). NDR1 was isolated and found to encode a 219 amino acid protein with two putative transmembrane domains (79). Elevated expression of *NDR1* was observed following pathogen infection, indicating that it is a defence response gene. The biochemical function of NDR1 remains unclear.

EDS1 (for *enhanced disease susceptibility*) of *Arabidopsis* was originally identified using *eds1* mutants that showed susceptibility to *Peronospora parasitica* strains that are avirulent on the wild-type parent (80). *EDS1* encodes a protein with amino-terminal homology to eukaryotic lipases (81).

Further study with *eds1* and *ndr1* mutants revealed preferred usage of the wild-type gene products, in particular *R* gene pathways (82). *EDS1* is strongly required for *RPP2*, *RPP4*, *RPP5*, and *RPP21*(*R* genes for *Peronospora parasitica*) and *RPS4* (for bacterial *avrRps4*) but not *RPS2*, *RPM1*, or *RPS5*. Conversely, *NDR1* is necessary for *RPS2*, *RPM1*, and *RPS5* but not *RPP2*. *RPP4*- and *RPP5*-mediated resistances were impaired partially by *ndr1*, indicating some degree of cross-talk between the two pathways. However, a hypothesis was advanced that the structure of the *R* gene product, rather than the pathogen species, determines the dependence on *EDS1* or *NDR1* for defence signal transduction (82). The *R* genes that have the amino-terminal leucine zipper domains were proposed to be dependent on *NDR1* for their functions, whereas those with amino-terminal TIR domains are dependent on *EDS1* (82). Subsequently, McDowell and colleagues found that resistance mediated by the *R* genes *RPP7* and *RPP8* is only weakly impacted in a double-mutant *eds1⁻ ndr1⁻* double mutant, implying the existence of at least one additional pathway (83). The action of *RPP7* was not compromised in *coi1⁻ npr1⁻* double mutants (discussed below), suggesting that for *R* genes such as *RPP7* those additional pathways operate independent of salicylic acid or jasmonic acid (83).

Additional components responsible for gene-for-gene resistance have been identified in *Arabidopsis* (7, 8). For example, the *pbs1*, *pbs2*, and *pbs3* mutations (for *avrPphB* susceptible) identify genes that are required for *RPS5*–*avrPphB*-mediated resistance (84). Interestingly, whilst *pbs1* blocked resistance only for *RPS5*, *pbs2* reduced resistance mediated by *RPS5* or *RPM1*, and *pbs3* partially suppressed all four resistance genes tested (*RPS5*, *RPM1*, *RPS2*, *RPS4*). It should prove very interesting to learn the structures of the products encoded by these *pbs* genes.

Turning to species other than *Arabidopsis*, *Prf* is a tomato gene essential for the function of *Pto*, whose protein kinase product confers resistance against *Pseudomonas syringae* pv. *tomato* containing the *avrPto* gene (52). Interestingly, *Prf* meets the classical definition of a resistance gene, and encodes an NBS-LRR protein similar to many other *R* gene products (52). Work with *Pto* and *Prf* demonstrates that more than one 'resistance gene' can participate in defence activated due to a single avirulence gene. As mentioned previously, this work also provides an example in which an NBS-LRR protein and a protein kinase function together to confer gene-for-gene resistance. It is interesting that *Prf* is very tightly linked to *Pto* in the tomato genome.

The barley gene *Rar1* is a convergence point for the gene-for-gene resistance pathways triggered by a number of distinct barley *Mla R* genes (85, 86). The sequence of the deduced 25.5 kDa protein reveals two copies of a 60 amino acid domain, CHORD, that is conserved in tandem organization in protozoa, plants, and metazoa. CHORD defines a novel eukaryotic Zn^{2+}-binding domain. The molecular function of this type of protein in defence signal transduction remains to be discovered, although there are hints from work on other CHORD proteins that *Rar1* may influence protein ubiquitination and proteolysis.

The above sample of genes represents the tip of the iceberg of genes important for gene-for-gene defence signal transduction. When studying any biological process there is concern that mutational approaches will fail to identify all relevant components. A mutation might not be isolated if disruption of the gene is lethal, a particular concern given that gene-for-gene resistance often involves HR cell death. It may also be difficult to identify mutations in a relevant gene if there is functional redundancy, or if the mutation causes only a subtle alteration in phenotype. Nevertheless, mutagenesis and other standard genetic strategies are likely to remain productive for the identification of components of gene-for-gene resistance (8, 87) A number of the genes discussed below influence gene-for-gene resistance as well as other forms of disease resistance.

4.1.2 Systemic acquired resistance (SAR)

SAR is a type of systemic and broad-spectrum disease resistance activated upon primary infection by some pathogens (66). SAR causes elevated constitutive expression of many pathogenesis-related (PR) proteins. SAR also potentiates defence signal transduction such that defence responses are induced more rapidly and strongly in response to subsequent infections. Salicylic acid (SA) plays an essential role in SAR signal transduction (66).

Mutants that show altered SAR phenotypes were initially divided into two groups, those with impaired SAR (*npr1/nim1/sai1*) and those with constitutive SAR (*cpr*, *cim*, and other mutants) (66, 87). *npr1* (non-expresser of *PR* genes), *nim1* (non-inducible *im*munity) and *sai1* (salicylic *a*cid *i*nducible) were identified independently in different laboratories by screening for mutants that, in response to SAR elicitors, failed to induce pathogenesis-related genes or effective resistance. These mutations all define the same gene (87). *npr1/nim1* mutants show significantly increased susceptibility to virulent or avirulent *P. syringae* strains and allow sporulation of many avirulent *Peronospora parasitica* isolates.

NPR1/NIM1 encodes a protein carrying putative ankyrin repeat protein–protein interaction domains, as are found for example in the mammalian transcription factor inhibitor IκB (88, 89). NPR1/NIM1 protein is not required for SA accumulation but does mediate *PR* gene induction in response to SA, indicating that NPR1/NIM1 acts downstream of SA in the SAR signalling pathway (Fig. 2) (87). However, interesting feedback loops have been identified in the SA-*NPR1* pathway (66). NPR1/NIM1 protein has recently been shown to localize to the nucleus (13). Interaction of NPR1/NIM1 with transcription factors is discussed below.

Mutants that exhibit the converse phenotype, constitutive expression of *PR* genes, include those called *cpr* (constitutive expresser of *PR* genes) or *cim* (constitutive *im*munity) (87, 90). These mutants exhibit elevated levels of SA, constitutive expression of genes that encode PR proteins such as PR1, PR5, and β-glucanases, and increased resistance against virulent bacterial and fungal pathogens. In many of these mutants, constitutive expression of *PR* genes is SA-dependent but *NPR1/NIM1*-independent (87, 90). For example, the *cpr6-1* mutation is dominant and its phenotype for increased resistance against *Pseudomonas syringae* pv. *maculicola* ES4326 is controlled by *NPR1/NIM1* even though constitutive *PR* gene expression is not (90). This and other work has indicated that there are *NPR1/NIM1*-dependent antimicrobial responses that are distinct from the induction of known *PR* genes (90).

4.1.3 Ethylene, jasmonate and alternative defence pathways

The SA-independent systemically inducible expression of antimicrobial peptides thionin (Thi2.1) and defensin (PDF1.2) led to genetic studies that revealed alternative defence pathways for disease resistance (Fig. 2) (91). *PDF1.2* is induced by methyl jasmonic acid (JA), ethylene, and the non-host fungal pathogen *Alternaria brassicola* but not by SA (92). The induction of *PDF1.2* is not affected in *npr1* and *cpr1* mutants or *NahG* transgenic plants but is greatly reduced in *ein2* and *coi1* mutants, which are ethylene- and JA-insensitive, respectively (92). *Thi2.1* also shows the similar expression pattern to that of *PDF1.2* and plays a role in disease resistance (93). This defines a systemic resistance pathway(s) involved in Thi2.1 and PDF1.2 production that is JA- and ethylene-dependent but distinct from the SA-independent SAR pathway (Fig. 2) (91). Penninckx *et al.* reported that concomitant rather than sequential activation of JA and ethylene signalling pathways is required for the induction of PDF1.2 (94).

The induced systemic resistance (ISR) induced by plant growth-promoting rhizobacteria (PGPR) also requires ethylene and JA but not SA. However, *NPR1* is reportedly essential for ISR, indicating that ISR incorporates a novel JA- and ethylene-dependent pathway that shows some features of NPR1 dependence beyond the pathways controlling PDF1.2 and Thi2.1 production (65, 95). Genetic strategies are being actively pursued to identify the additional genes and pathways suggested by the above work.

4.1.4 Other genes responsible for altered disease resistance and cell death

Many 'lesion-mimic' mutants have been identified in *Arabidopsis* and other plants, and many of them have the associated phenotype of enhanced disease resistance (72, 87). *Rp1* of maize encodes an NBS-LRR-type resistance gene product that confers resistance to the rust pathogen *Puccinia sorghi*, and several mutant alleles of the *Rp1* locus cause a lesion-mimic phenotype. In barley, the stronger mutant alleles of *mlo*, which confer resistance to the powdery mildew pathogen *Erysiphe graminis*, are responsible for spontaneous lesion formation as well as cell wall appositions (96). *Mlo* encodes a seven-pass transmembrane protein of unknown function (97). Although lesion formation has been suggested to play a role in activating or propagating resistance (66), it may be more appropriate to say that lesion formation is a downstream outcome or symptom of physiological conditions that also promote strong defence activation.

In *Arabidopsis*, several *acd* (*accelerated cell death*) and *lsd* (*lesions simulating disease*) mutants have been identified. In many of these mutants, lesion formation is accompanied elevated SA levels, constitutive expression of PR genes, and resistance to virulent pathogens (87). One of those genes, *LSD1*, is thought to be a sensor that blocks the feedforward and amplification of the cell-death signal derived from extra-cellular superoxide (98). Recently, *LSD1* was shown to encode a novel zinc finger protein that may either repress transcription for a cell-death pathway or activate transcription for an anti-cell-death pathway (99).

Arabidopsis dnd (*defence, no-death*) mutants are like *cpr, cim, lsd, acd*, and other mutants in exhibiting elevated SA levels, constitutive expression of PR genes, a dwarfed growth habit, and resistance to virulent pathogens, but with an additional twist: the *dnd* mutants produce little or no HR in response to avirulent *Pseudomonas syringae* (76). The *dnd* mutants can still carry out gene-for-gene resistance, providing evidence that cell death is separable from other aspects of gene-for-gene defence signal transduction (76). Recently, *DND1* was cloned and characterized as a cyclic nucleotide-gated ion channel. This supports long-standing evidence that ion fluxes play an important role in defence and HR signalling (100).

The gain-of-function *acd6* mutant of *Arabidopsis* shows an elevated level of defences and increased resistance to *Pseudomonas syringae* as well as spontaneous patches of cell death (101). In addition, the phenotypes of *acd6* mutant was solely SA-dependent (101).

Although most constitutive defence *Arabidopsis* mutants studied to date are dwarfed and have reduced seed set, mutants showing a quantitatively less effective constitutive defence have also been isolated and these lines are not dwarfed (102).

Other mutants that show the altered disease resistance include *pad* (for phytoalexin-deficient) and *eds* (for enhanced disease susceptibility) mutants in *Arabidopsis* (7, 87). It is very interesting that both *EDS1*, which is a key component for the resistance pathway mediated by TIR-NBS-LRR class *R* genes, and *PAD4*, which is responsible for multiple defence responses including phytoalexin production, encode the putative eukaryotic lipases (81, 103). This suggests that lipids and lipases may play an important role in plant defence signalling (103).

The growing family of *eds* mutants, isolated because of enhanced disease susceptibility to *Pseudomonas syringae*, include known genes such as *NPR1* and *PAD4* as well as previously unidentified genes (7, 87). Mutant screens for enhanced disease susceptibility or for enhanced disease resistance are being carried out with other pathogens as well (104, 105). The resulting mutants often show pathogen species-specific alteration in disease resistance, and will be a valuable resource for identification of additional defence-associated genes.

From the above, it should be clear that genetic analysis of *Arabidopsis* is a powerful tool for identifying novel defence signal transduction components (7, 87). Key paradigms have been defined using other plant species, but new defence-associated genes are being identified in *Arabidopsis* at an increasing rate. By working with the same species, diverse research groups can more effectively build on the findings of others and generate integrated models (e.g. Fig. 2) for the function of defence signal transduction components.

4.2 Biochemical/pharmacological methods

Biochemical and pharmacological methods have also been used very productively to study early signal transduction events in the plant defence responses (9). This work has examined plant responses both to pathogens and to purified defence-eliciting compounds. Phenomena implicated by these methods as playing a role in defence signal transduction have included ion fluxes, protein kinases, and protein phosphatases, and the production of reactive oxygen species and nitric oxide.

4.2.1 Ion fluxes

Changes in the permeability of the plasma membrane, followed by calcium and proton influx and potassium and chloride efflux, are among the earliest events in the reaction of plant cells to elicitors or pathogens (9, 64, 77). For example, physiological changes in membrane permeability are observed in parsley suspension cells between 2 and 5 minutes following treatment with a fungal elicitor (106). These ion fluxes are typically caused by activation of ion channels and pumps, although late in the progression of some plant responses (such as the hypersensitive response) a more general ion leakage can occur due to failed regulation of membrane potential as the cell dies. Ca^{2+} in particular is known as very important for defence signalling (9, 77, 107). Ca^{2+} channel blockers have been shown to inhibit the HR in tobacco and soybean systems (108–111). Ca^{2+} channel blockers also disrupt other defence responses to fungal and bacterial elicitors (109, 112, 113). Ion fluxes are thought to be required for the activation of a MAP kinase specific to defence responses, which may serve to activate gene expression following its translocation to the nucleus (114). Significantly, Ca^{2+} influx and the transient increase in cytosolic Ca^{2+} levels following elicitor treatment have been shown to be necessary and sufficient for the induction of the oxidative burst and subsequent induction of defence responses (115, 116). The finding (discussed above) that mutation of the *DND1/AtCNGC2* cyclic nucleotide-gated ion channel can strongly influence defence activation identified a specific ion channel that may contribute to Ca^{2+} influx during plant responses to pathogens (100).

4.2.2 Oxidative burst (ROI production)

ROIs not only have direct toxicity to pathogens, but also are central components of plant defence signal transduction pathways that lead to the HR, cell wall reinforcement, and defence gene expression (9, 12, 64, 117, 118). For example, an oxidative burst can be essential for the activation of PR production and phytoalexin production

(115). It is intriguing that a two-phase oxidative burst is observed in incompatible or non-host interactions (117, 118), the first of which is a generic response to virulent or avirulent pathogens and the second of which is dependent on gene-for-gene interaction.

ROI are known to control HR cell death. Inhibition of oxidative burst can cause a decrease in HR cell death, whereas the suppression of the genes for ROI scavengers (ascorbate peroxidase and catalase) has caused an increased HR (117, 119, 120). In addition, superoxide was reported to be a necessary and sufficient signal for cell death in the *lsd1* (for *l*esions *s*imulating *d*isease resistance *1*) mutant of *Arabidopsis* (99).

In strongly reacting tissues, SA may cause elevated H_2O_2 levels by direct inhibition of catalase activity (121). ROIs also play an important role in the induction of salicylate-mediated systemic resistance, which may be propagated in part by micro-bursts of ROI generation in tissues distant from the initial immunizing signal (122).

4.2.3 Nitric oxide

Nitric oxide (NO), an important factor for innate immune and inflammatory responses in animals, was recently reported to function as a critical signal for disease resistance in plants (123–125). NO enhances the ROI-mediated induction of hyper-sensitive cell death by avirulent bacterial pathogens in both soybean cell culture and in *Arabidopsis* (123). Inhibitors of NO synthesis not only blocked hypersensitive cell death but also caused increased susceptibility in incompatible interactions (123). Cyclic nucleotides may act as second messengers for NO signalling (124). In tobacco, an NO synthase (NOS)-like activity was strongly correlated with the expression of *PR-1* (126).

4.2.4 Protein phosphorylation/dephosphorylation

Following a ubiquitous theme in signal transduction, GTP-binding proteins and protein phosphorylation/dephosphorylation are involved in transferring signals from receptors to downstream pathways in defence signal transduction (9, 12, 117). Mastoparan, a G-protein-activating peptide, acts like a defence elicitor in plant cells, activating ion fluxes, ROI production, and phytoalexin accumulation (9). Using tomato cells, evidence has been generated that a G-protein mediates the signals generated by the *R*-gene-specific elicitors (117).

Inhibitors have been used in a number of systems to show that activation of the oxidative burst is dependent on the state of phosphorylation/dephosphorylation of unknown proteins in the signal transduction machinery (117). In soybean cells, inhibition of protein phosphatases resulted in the induction of oxidative burst in the absence of elicitors and potentiated the response to the avirulent *Pseudomonas syringae* pv. *glycinea*, whereas inhibition of kinase caused the opposite effects, suggesting that the regulation of protein phosphorylation/dephosphorylation is important for the tight control of induction, duration, and magnitude of the oxidative burst (117, 127).

The resistance genes *Pto* and *Xa21* can activate very strong defence responses, and both have been shown to encode active protein kinases, as discussed above.

Mitogen-activated protein kinases (MAPKs) have been shown to participate in plant defence responses in tobacco plants (113, 128, 129). They include wound-induced protein kinases (WIPK) (130, 131) and SA-induced protein kinases (SIPK) (113, 128, 131, 132). In parsley cultured cells, a MAP kinase is involved in the defence signal transduction in response to fungal elicitor (133). This kinase is translocated to the nucleus upon receptor-mediated activation and acts downstream of ion influxes and upstream or independent of oxidative burst (133).

4.3 Immunoprecipitation, interaction cloning, site-directed mutagenesis

Many important components for plant defence pathways that would not have been found by genetic or pharmacological approaches have been identified by other bio-chemical and molecular biological methods. Recently, interaction between AvrRpt2 and RPS2, a bacterial avirulence protein and the cognate resistance gene product, was shown using an immunoprecipitation assay (18). Interestingly, an unknown 75 kDa protein was co-precipitated along with the AvrRpt2–RPS2 complex (18). Unfortunately, *R* gene products and related components are typically expressed at extremely low levels in plant cells, making application of immunuoprecipitation methods technically challenging.

Yeast two-hybrid interaction cloning methods have contributed substantially to study of defence signal transduction, especially for the tomato defence pathways activated by Pto. Pto and AvrPto have been shown to interact in yeast two-hybrid assays, providing the first evidence for direct interaction between *R* and *avr* gene products (15, 16). Tomato cDNA libraries have been screened using Pto as bait. Pti1 (for Pto-interacting) is a serine/threonine kinase that is phosphorylated specifically by the Pto kinase (134). Transgenic tobacco expressing *Pti1* show an enhanced HR to a *Pseudomonas syringae* pv. *tabaci* strain carrying the avirulence gene, *avrPto* (134). Additional Pto-interacting proteins, Pti4, Pti5, and Pti6, are apparent plant tran-scription factors (135) that are discussed in the next section. Yeast two-hybrid work using *Arabidopsis* NPR1/NIM1 also identified interaction with transcription factors (141–143). The identification of additional factors that interact with *R* gene products and other defence components will be a ripe area for future study.

Site-directed mutagenesis is another tool that is providing valuable information on defence signal transduction. Returning to the Pto example, substitution of Thr-204 into the related Fen kinase allowed that kinase to interact with AvrPto and to confer an AvrPto-specific defence response in tobacco leaves (136). Thus, simple mutations appear capable of giving rise to new resistance gene specificities. Rathjen *et al.* have made dominant negative mutants that identify key activation domains of Pto (137). Constitutively active *Pto* mutants required kinase capability for activity, and were unable to interact with proteins previously shown to bind to wild-type Pto. The constitutive gain-of-function phenotype was dependent on a functional *Prf* gene, demonstrating activation of the cognate disease resistance pathway and precluding a role for Prf upstream of Pto (137).

4.4 Transcriptional control of defence-related genes in plants

Pathogen infection brings about active defence responses of plants in large part by reprogramming gene expression. In particular, genes involved in the production of antimicrobial secondary metabolites (e.g. phytoalexins), degradative enzymes (e.g. chitinases, glucanases) and an array of other pathogenesis-related (*PR*) genes are strongly induced (138). The promoter regions of these pathogen-induced genes comprise an important end point in defence signal transduction cascades.

Several *cis*-elements and corresponding *trans*-elements have been identified from pathogen-responsive genes of various plants (138). Ethylene-responsive element-binding proteins (EREBPs) and WRKY transcription factors are unique to plants and bind to the GCC box (ethylene-responsive element) and the W box, respectively, in the promoter regions of defence-associated genes (138, 139).

It is noteworthy that the recently found protein components that interact with *Pto*, a resistance gene product, or NPR1/NIM1, a central component for SAR, are *trans*-elements for the *cis*-element of *PR* genes. Pti4/5/6, the Pto-interacting proteins, are EREBPs and bind to the GCC boxes of *PR* genes (135). Recently, it was shown that *Pti4* expression is induced by ethylene and by bacterial infection, and that Pti4 is phosphorylated by Pto kinase, and that this phosphorylation enhances binding of Pti4 to the *PR* gene promoter sequence (140). These results show that relatively direct signal transduction pathways can exist between *R–avr* gene product interaction and the activation of at least some defence responses.

TGA2, a NPR1/NIM1-interacting protein, is a member of bZIP family and binds to the *as-1* element and the *LS7* element of *PR1* promoter (141–143). NPR1/NIM1 enhanced the binding of TGA2 to the promoter region of the *PR1* gene, and the activity of NPR1/NIM1 structural variants in two-hybrid interaction with TGA2 correlated with their defence-activating capacity in plants (141). Here again, these results complete at least a rough-draft outline of an important plant defence signal transduction cascade.

4.5 Microarrays and global gene expression patterns

Defence signal transduction research, like many other disciplines, is undergoing significant evolution with the recent development of DNA microarrays that can monitor mRNA levels for thousands of genes at once. Two of the earliest plant microarray papers have addressed *Arabidopsis* defence-associated gene expression (68, 144). Schenk *et al.* (68) monitored the expression of 2375 genes plants responding to avirulent pathogen, salicylic acid, methyl jasmonate, or ethylene; 705 genes showed changes in mRNA abundance in at least one of the treatments. Both groups noted substantial overlap in gene expression patterns among different signalling pathways, including pathways that were formerly thought to act in a largely antagonistic fashion (68, 144). An important contribution was recently made by Malek *et al.*, who used microarrays to characterize the response of Arabidopsis to 14 different SAR-inducing or repressing conditions, including pathogen and chemical treatments and use of plant mutants (156). This study was also used to discover that the genes exhibiting a similar regulatory pattern to PR-1 across the various treatments often contained a particular W-box sequence motif in their upstream promoter sequences (156). Microarrays are now available with more than 8000 *Arabidopsis* genes represented, and substantial progress can be expected upon the use of these resources with previously defined plant mutants or physiological treatments.

5. Application of defence signal transduction findings

Knowledge of defence signal transduction provides an opportunity to develop new strategies for disease control (69, 145–147). Many of the findings discussed above are relatively recent and few applied strategies have reached commercialization, but this is an active area of research and development. A detailed discussion of practical applications is beyond the scope of this chapter (see 69, 145–147), but a few general examples can highlight this important segment of defence signal transduction research.

Based on the important role of *R* genes in initiating defence signal transduction cascades, transgenic use of *R* genes from sexually incompatible species is a promising first step (e.g. 148). Engineering of *R* genes for novel specificity, activation by a broader spectrum of pathogens, or more stable resistance is another primary goal. The LRR domains of many *R* genes are particularly important in determining pathogen specificity, while protein regions holding the NBS, kinase, and other domains seem important for downstream signalling (53, 149). Either can be manipulated for enhanced resistance, but more sophisticated modification of R proteins would be possible if the structure/function aspects of R-Avr recognition and subsequent signalling were understood in greater detail.

Genes for the major components of defence signalling pathways (e.g. *EDS1*, *NDR1*, and *NPR1/NIM1*) are also strong candidates for development of disease resistant crops, due to their broad-spectrum effects against different pathogens. As a promising example, constitutive elevated expression of *NPR1/NIM1* has been shown to enhance disease resistance with few side-effects in a laboratory/greenhouse setting (150).

Studies of downstream regulatory elements for defence, including transcription factors and *cis*-elements, may allow engineering of plants to express their defence arsenals in a more tightly controlled manner. This could minimize the negative effects such as lesion formation, reduced plant growth, and yield reduction that arise in plants undergoing excessive defence activation.

6. Conclusions

Plant defence systems incorporate a very complex network of signal transduction and metabolic pathways. Diverse pathogen-associated inputs activate subsets of these pathways, many of which can also be activated by other biotic and abiotic stimuli. Signal transduction intermediaries, such as salicylic acid and jasmonic acid, sometimes work independently, sometimes in concert, and sometimes in opposition with respect to activation of a given response. There are certainly common signal transduction cascades and stereotypical defence genes that are activated in response to diverse pathogen signals, but there are also important differences. Temporal and spatial as well as quantitative control of these defence responses determine the ultimate outcome of the defence responses.

Knowledge of plant signal transduction has increased steadily through work on disease resistance systems. Importantly, this work is also fostering development of more effective and stable disease resistance traits in crop plants. Studies of disease resistance are carried out in many important plant species, and this must continue. Many biological particulars can only be learned by study of diverse plant–pathogen interactions, and a number of resources are available for economically important crop species. However, study of *Arabidopsis* has proven to be especially productive, and accelerated progress is likely given the completed *Arabidopsis* genome sequence, the ease of transformation, the potential for reverse genetics using gene knockout collections, and other experimental advantages (151–153). Work with *Arabidopsis* can expedite discovery and help to provide an integrated picture of defence signal transduction, but it will remain important to relate *Arabidopsis* discoveries back to the specifics of defence signal transduction in crop plants.

References

1. Jackson, A. O. and Taylor, C. B. (1996) Plant-microbe interactions: life and death at the interface. *Plant Cell*, **8**, 1651.
2. Crute, I. R. and Pink, D. A. C. (1996) Genetics and utilization of pathogen resistance in plants. *Plant Cell*, **8**, 1747.
3. Wenzel, G. (1985) Strategies in unconventional breeding for disease resistance. *Annu. Rev. Phytopathol.*, **23**, 149.
4. Borlaug, N. E. (1983) Contributions of conventional plant breeding to food production. *Science*, **219**, 689.
5. McDowell, J. M. and Dangl, J. L. (2000) Signal transduction in the plant immune response. *Trends Biochem. Sci.*, **25**, 79.
6. Yang, Y., Shah, J. and Klessig, D. F. (1997) Signal perception and transduction in plant defense responses. *Genes Dev.*, **11**, 1621.
7. Glazebrook, J. (1999) Genes controlling expression of defense responses in *Arabidopsis*. *Curr. Opin. Plant Biol.*, **2**, 280.
8. Innes, R. R. (1998) Genetic dissection of R gene signal transduction pathways. *Curr. Opin. Plant Biol.*, **1**, 229.
9. Scheel, D. (1998) Resistance response physiology and signal transduction. *Curr. Opin. Plant Biol.*, **1**, 305.
10. Bent, A. (1996) Plant disease resistance genes: function meets structure. *Plant Cell*, **8**, 1757.
11. Hammond-Kosack, K. E. and Jones, J. D. G. (1997) Plant disease resistance genes. *Annu. Rev. Plant Physiol.*, **48**, 575.
12. Greenberg, J. T. (1997) Programmed cell death in plant-pathogen interactions. *Annu. Rev. Plant Physiol. Plant Mol. Biol.*, **48**, 525.
13. Kinkema, M., Fan, W., Dong, X. (2000) Nuclear localization of NPR1 is required for activation of PR gene expression. *Plant Cell*, **12**, 2339.
14. Leach, J. E. and White, F. F. (1996) Bacterial avirulence genes. *Annu. Rev. Phytopathol.*, **34**, 153.

15. Tang, X., Frederick, R. D., Zhou, J., Halterman, D. A., Jia, Y. and Martin, G. B. (1996) Initiation of plant disease resistance by physical interaction of AvrPto and Pto kinase. *Science*, **274**, 2060.

16. Scofield, S. R., Tobias, C. M., Rathjen, J. P., Chang, J. H., Lavelle, D. T., Michelmore, R. W. and Staskawicz, B. J. (1996) Molecular basis of gene-for-gene specificity in bacterial speck disease of tomato. *Science*, **274**, 2063.

17. Jia, Y., McAdams, S. A., Bryan, G. T., Hershey, H. P. and Valent, B. (2000) Direct interaction of resistance gene and avirulence gene products confers rice blast resistance. *EMBO J.*, **19**, 4004.

18. Leister, R. T. and Katagiri, F. (2000) A resistance gene product of the nucleotide binding site-leucine rich repeats class can form a complex with bacterial avirulence proteins *in vivo*. *Plant J.*, **22**, 345.

19. White, F. F., Yang, B. and Johnson, L. B. (2000) Prospects for understanding avirulence gene function. *Curr. Opin. Plant Biol.*, **3**, 291.

20. Kearney, B. and Staskawicz, B. J. (1990) Widespread distribution and fitness contribution of *Xanthomonas campestris* avirulence gene *avrBs2*. *Nature*, **346**, 385.

21. Lorang, J. M., Shen, H., Kobayashi, D., Cooksey, D. and Keen, N. T. (1994) *avrA* and *avrE* in *Pseudomonas syringae* pv. *tomato* PT23 play a role in virulence on tomato plants. *Mol. Plant–Microbe Interact.*, **7**, 508.

22. Ritter, C. and Dangl, J. L. (1995) The *avrRpm1* gene of *Pseudomonas syringae* pv. *meculicola* is required for virulence on *Arabidopsis*. *Mol. Plant–Microbe Interact.*, **8**, 444.

23. Bogdanove, A. J., Kim, J. F., Wei, Z., Kolchinski, P., Charkowski, A. O., Colin, A. K., Collmer, A. and Beer, S. V. (1998) Homology and functional similarity of an *hrp*-linked pathogenicity locus, *dspEF*, of *Erwinia amylovora* and the avirulence locus *avrE* of *Pseudomonas syringae* pathovar tomato. *Proc. Natl Acad. Sci. USA*, **95**, 1325.

24. Gaudriault, S., Malandrin, L., Paulin, J. P. and Barny, M. A. (1997) DspA, and essential pathogenicity factor of *Erwinia amylovora* showing homology with AvrE of *Pseudomonas syringae*, is secreted via Hrp secretion pathway in a DspB-dependent way. *Mol. Microbiol.*, **26**, 1057.

25. Herbers, K., Conrads-Strauch, J. and Bonas, U. (1992) Race-specificity of plant resistance to bacterial spot disease determined by repetitive motifs in a bacterial avirulence protein. *Nature*, **356**, 172.

26. Feyter, R. D., Yang, Y. and Gabriel, D. W. (1993) Gene-for-gene interactions between cotton *R* genes and *Xanthomonas campestris* pv. *malvacearum avr* genes. *Mol. Plant–Microbe Interact.*, **6**, 225.

27. Swarup, S., Feyter, R. D., Brlansky, R. H. and Gabriel, D. W. (1991) A pathogenicity locus from *Xanthomonas citri* enables strains from several pathovars of *X. campestris* to elicit cankerlike lesions on citrus. *Phytopathology*, **81**, 802.

28. Yang, Y., Feyter, R. D. and Gabriel, D. W. (1994) Host-specific symptoms and increased release of *Xanthomonas citri* and *X. campestris* pv. *malvacearum* from leaves are determined by the 102-bp tandem repeats of *pthA* and *avr6*, respectively. *Mol. Plant–Microbe Interact.*, **7**, 345.

29. Yang, B., Zhu, W., Johnson, L. B. and White, F. F. (2000) The virulence factor AvrXa7 of *Xanthomonas oryzae* pv. *oryzae* is a type III secretion pathway-dependent nuclear-localized double-stranded DNA- binding protein. *Proc. Natl Acad. Sci. USA*, **97**, 9807.

30. Yang, Y. and Gabriel, D. W. (1995) Intragenic recombination of a single plant pathogen gene provides a mechanism for the evolution of new host specificities. *J. Bacteriol.*, **177**, 4963.

31. Yang, Y. and Gabriel, D. W. (1995) *Xanthomonas* avirulence/pathogenicity gene family encodes functional plant nuclear targeting signals. *Mol. Plant–Microbe Interact.*, **8**, 627.
32. van den Ackerveken, E. A., Marois, E. and Bonas, U. (1996) Recognition of the bacterial avirulence protein AvrBs3 occurs inside the host plant cell. *Cell*, **87**, 1307.
33. Murillo, J., Shen, H., Gerhold, D., Sharma, A., Cooksey, D. A. and Keen, N. T. (1994) Characterization of pPT23B, the plasmid involved in syringolide production by *Pseudomonas syringae* pv. *tomato* PT23. *Plasmid*, **31**, 275.
34. Alfano, J. R. and Collmer, A. (1996) Bacterial pathogens in plants: life up against the wall. *Plant Cell*, **8**, 1683.
35. Alfano, J. R. and Collmer, A. (1997) The type III (Hrp) secretion pathway of plant pathogenic bacteria: trafficking Harpins, Avr proteins, and death. *J. Bacteriol.*, **179**, 5655.
36. Gopalan, S., Bauer, D. W., Alfano, J. R., Loniello, A. O., He, S. Y. and Collmer, A. (1996) Expression of the *Pseudomonas syringae* avirulence protein AvrB in plant cells alleviates its dependence on the hypersensitive response and pathogenicity (Hrp) secretion system in eliciting genotype specific hypersensitive cell death. *Plant Cell*, **8**, 1095.
37. Pirhonen, M. U., Lidell, M. C., Rowley, D. L., Lee, S. W., Jin, S., Liang, Y., Siverstone, S., Keen, N. T. and Hutcheson, S. W. (1996) Phenotypic expression of *Pseudomonas syringae* avr genes in *E. coli* is linked to the activities of the *hrp*-encoded secretion system. *Mol. Plant–Microbe Interact.*, **9**, 252.
38. Rossier, O., Wengelnik, K., Hahn, K. and Bonas, U. (1999) The Xanthomonas Hrp type III system secretes proteins from plant and mammalian bacterial pathogens. *Proc. Natl Acad. Sci. USA*, **96**, 9368.
39. Mudgett, M. B. and Staskawicz, B. J. (1999) Characterization of the *Pseudomonas syringae* pv. *tomato* AvrRpt2 protein: demonstration of secretion and processing during bacterial pathogenesis. *Mol. Microbiol.*, **32**, 927.
40. Ham, J. H., Bauer, D. W., Fouts, D. E. and Collmer, A. (1998) A cloned *Erwinia chrysanthemi* Hrp (type III protein secretion) system functions in *Escherichia coli* to deliver *Pseudomonas syringae* Avr signals to plant cells and to secrete Avr proteins in culture. *Proc. Natl Acad. Sci. USA*, **95**, 10206.
41. Nimchuk, Z., Marois, E., Kjemtrup, S., Leister, R. T., Katagiri, F. and Dangl, J. L. (2000) Eukaryotic fatty acylation drives plasma membrane targeting and enhances function of several type III effector proteins from *Pseudomonas syringae*. *Cell*, **101**, 353.
42. Arlat, M., Van Gijsegem, F., Huet, J. C., Pernollet, J. C. and Boucher, C. A. (1994) PopA1, a protein which induces a hypersensitivity-like response on specific Petunia genotypes, is secreted via the Hrp pathway of Pseudomonas solanacearum. *Embo J.*, **13**, 543.
43. Wei, Z.-M., Sneath, B. and Beer, S. (1992) Expression of *Erwinia amylovora hrp* genes in response to environmental stimuli. *J. Bacteriol.*, **174**, 1875.
44. Hahn, M. G. (1996) Microbial elicitors and their receptors in plants. *Annu. Rev. Phytopathol.*, **34**, 387.
45. Gomez-Gomez, L. and Boller, T. (2000) FLS2: an LRR receptor-like kinase involved in the perception of the bacterial elicitor flagellin in *Arabidopsis*. *Mol. Cell*, **5**, 1003.
46. Hutcheson, S. W. (1998) Current concepts of active defense in plants. *Annu. Rev. Phytopathol.*, **36**, 59.
47. deWit, P. J. G. M. and Joosten, M. H. A. J. (1999) Avirulence and resistance genes in the *Cladosporium fulvum*-tomato interaction. *Curr. Opin. Microbiol.*, **2**, 368.
48. Piedras, P., Rivas, S., Droge, S., Hillmer, S. and Jones, J. D. (2000) Functional, c-myc-tagged Cf-9 resistance gene products are plasma- membrane localized and glycosylated. *Plant J.*, **21**, 529.

49. Rohe, M., Gierlich, A., Hermann, H., Hahn, M., Schmidt, B., Rosahl, S. and Knogge, W. (1995) The race-specific elicitor, NIP1, from the barley pathogen, *Rhynchosporium secalis*, determines avirulence on host plants of the *Rrs1* resistance genotype. *EMBO J.*, **14**, 4168.

50. Knogge, W. (1996) Fungal infection of plants. *Plant Cell*, **8**, 1711.

51. Martin, G. B., Brommonschenkel, S. H., Chunwongse, J., Frary, A., Ganal, M. W., Spivey, R., Earle, E. D. and Tanksley, S. D. (1993) Map-based cloning of a protein kinase gene conferring disease resistance in tomato. *Science*, **262**, 1432.

52. Salmeron, J. M., Oldroyd, G. E., Rommens, C. M., Scofield, S. R., Kim, H. S., Lavelle, D. T., Dahlbeck, D. and Staskawicz, B. J. (1996) Tomato *Prf* is a member of the leucine-rich repeat class of plant disease resistance genes and lies embedded within the Pto kinase gene cluster. *Cell*, **86**, 123.

53. Ellis, J. and Jones, D. (1998) Structure and function of proteins controlling strain-specific pathogen resistance in plants. *Curr. Opin. Plant Biol.*, **1**, 288.

54. van der Biezen, E. A. and Jones, J. D. (1998) The NB-ARC domain: a novel signalling motif shared by plant resistance gene products and regulators of cell death in animals [letter]. *Curr. Biol.*, **8**, R226.

55. Inohara, N., Koseki, T., Peso, L., Hu, Y., Yee, C., Chen, S., Carrio, R., Merino, J., Liu, D., Ni, J., *et al.* (1999) Nod1, an Apaf-1-like activator of caspase-9 and nuclear factor-kappa B. *J. Biol. Chem.*, **274**, 14560.

56. Medzhitov, R. and Janeway, C. A. J. (1998) An ancient system of host defense. *Curr. Opin. Immunol.*, **10**, 12.

57. Medzhitov, R. and Janeway, C. (2000) Innate immune recognition: mechanisms and pathways. *Immunol. Rev.*, **173**, 89.

58. Song, W. Y., Wang, G. L., Chen, L. L., Kim, H. S., Pi, L. Y., Holsten, T., Gardner, J., Wang, B., Zhai, W. X., Zhu, L. H., *et al.* (1995) A receptor kinase-like protein encoded by the rice disease resistance gene, *Xa21*. *Science*, **270**, 1804.

59. Ronald, P. C. (1998) Resistance gene evolution. *Curr. Opin. Plant Biol.*, **1**, 294.

60. Ellis, J., Dodds, P. and Pryor, T. (2000) Structure function and evolution of plant disease resistance genes. *Curr. Opin. Plant Biol.*, **3**, 273.

61. Michelmore, R. W. and Meyers, B. C. (1998) Clusters of resistance genes in plants evolve by divergent selection and a birth-and-death process. *Genome Research*, **8**, 1113.

62. McDowell, J. M., Dhandaydham, M., Long, T. A., Aarts, M. G., Goff, S., Holub, E. B. and Dangl, J. L. (1998) Intragenic recombination and diversifying selection contribute to the evolution of downy mildew resistance at the *RPP8* locus of *Arabidopsis*. *Plant Cell*, **10**, 1861.

63. Nelson, R. R. (1978) Genetics of horizontal resistance to plant diseases. *Annu. Rev. Phytopathol.*, **16**, 359.

64. Hammond-Kosack, K. E. and Jones, J. D. G. (1996) Resistance gene-dependent plant defense reponses. *Plant Cell*, **8**, 1773.

65. van Loon, L. C., Bakker, P. A. H. M. and Pieterse, C. M. J. (1998) Systemic resistance induced by rhizosphere bacteria. *Annu. Rev. Phytopathol.*, **36**, 453.

66. Ryals, J. A., Neuenschwander, U. H., Willits, M. G., Molina, A., Steiner, H.-Y. and Hunt, M. D. (1996) Systemic aquired resistance. *Plant Cell*, **8**, 1809.

67. van Loon, L. C. and van Strien, E. A. (1999) The families of pathogenesis-related proteins, their activities, and comparative analysis of PR-1 type proteins. *Physiol. Mol. Plant Pathol.*, **55**, 85.

68. Schenk, P. M., Kazan, K., Wilson, I., Anderson, J. P., Richmond, T., Somerville, S. C. and Manners, J. M. (2000) Coordinated plant defense responses in arabidopsis revealed by microarray analysis [In Process Citation]. *Proc. Natl Acad. Sci. USA*, **97**, 11655.

69. Bent, A. F. and Yu, I.-c. (1999) Applications of molecular biology to plant disease and insect resistance. *Adv. Agron.*, **66**, 251.

70. Sticher, L., Mauch-Mani, B. and Metraux, J. P. (1997) Systemic aquired resistance. *Annu. Rev. Phytopathol.*, **35**, 235.

71. Hammerschmidt, R. (1999) Phytoalexins: What have we learned after 60 years? *Annu. Rev. Phytopathol.*, **37**, 285.

72. Dangl, J. L., Dietrich, R. A. and Richberg, M. H. (1996) Death don't have no mercy: Cell death programs in plant-microbe interactions. *Plant Cell*, **8**, 1793.

73. Whalen, M. C., Stall, R. E. and Staskawicz, B. J. (1988) Characterization of a gene from a tomato pathogen determining hypersensitive resistance in non-host species and genetic analysis of this resistance in bean. *Proc. Natl Acad. Sci.*, **85**, 6743.

74. Bendahmane, A., Kanyuka, K. and Baulcombe, D. C. (1999) The *Rx* gene from potato controls separate virus resistance and cell death responses. *Plant Cell*, **11**, 781.

75. Schiffer, R., Gorg, R., Jarosch, B., Beckhove, U., Bahrenberg, G., Kogel, K.-H. and Schulze-Lefert, P. (1997) Tissue dependence and differential cordycepin sensitivity of race-specific resistance responses in the barley-powdery mildew interaction. *Mol. Plant-Microbe Interact.*, **10**, 830.

76. Yu, I.-c., Parker, J. and Bent, A. F. (1998) Gene-for-gene disease resistance without the hypersensitive response in *Arabidopsis dnd1* mutant. *Proc. Natl Acad. Sci. USA*, **95**, 7819.

77. Ebel, J. and Scheel, D. (1997) Signals in host-parasite interactions. In Carroll, B. C. and Tudzynski, P. (eds), *The Mycota*, Vol. V, *Plant Relationships, Part A.* Springer-Verlag, Berlin, p. 85.

78. Century, K. S., Holub, E. B. and Staskawicz, B. J. (1995) *NDR1*, a locus of *Arabidopsis thaliana* that is required for disease resistance to both a bacterial and a fugal pathogen. *Proc. Natl Acad. Sci. USA*, **92**, 6597.

79. Century, K. S., Shapiro, A. D., Repetti, P. P., Dahlbeck, D., Holub, E. and Staskawicz, B. J. (1997) *NDR1*, a pathogen-induced component required for *Arabidopsis* disease resistance. *Science*, **278**, 1963.

80. Parker, J. E., Holub, E. B., Frost, L. N., Falk, A., Gunn, N. D. and Danniels, M. J. (1996) Characterization of *eds1*, a mutation in *Arabidopsis* supressing resistance to *Peronospora parasitica* specified by several different *RPP* genes. *Plant Cell*, **8**, 2033.

81. Falk, A., Feys, B. J., Frost, L. N., Jones, J. D. G., Danniels, M. J. and Parker, J. E. (1999) *EDS1*, an essential component of *R* gene-mediated disease resistance in *Arabidopsis* has homology to eukaryotic lipases. *Proc. Natl Acad. Sci. USA*, **96**, 3292.

82. Aart, N., Metz, M., Holub, E., Staskawicz, B. J., Daniels, M. J. and Parker, J. E. (1998) Different requirements for *EDS1* and *NDR1* by disease resistance genes defines at least two *R* gene-mediated signaling pathways in *Arabidopsis*. *Proc. Natl Acad. Sci. USA*, **95**, 10306.

83. McDowell, J. M., Cuzick, A., Can, C., Beynon, J., Dangl, J. L. and Holub, E. B. (2000) Downy mildew (*Peronospora parasitica*) resistance genes in *Arabidopsis* vary in functional requirements for *NDR1*, *EDS1*, *NPR1*, and salicylic acid accumulation. *Plant J.*, **22**, 523.

84. Warren, R. F., Merritt, P. M., Holub, E. and Innes, R. W. (1999) Identification of three putative signal transduction genes involved in *R* gene-specified disease resistance in *Arabidopsis*. *Genetics*, **152**, 401.

85. Shirasu, K., Lahaye, T., Tan, M. W., Zhou, F., Azevedo, C. and Schulze-Lefert, P. (1999) A novel class of eukaryotic zinc-biding proteins is required for disease resistance signaling in barley and development in *C. elegans*. *Cell*, **99**, 355.

86. Lahaye, T., Shirasu, K. and Schulze-Lefert, P. (1998) Chromosome landing at the barley *Rar1* locus. *Mol. Gen. Genet*, **260**, 92.

87. Glazebrook, J., Rogers, E. E. and Ausubel, F. M. (1997) Use of *Arabidopsis* for genetic dissection of plant defense response. *Annu. Rev. Genet.*, **31**, 547.

88. Cao, H., Glazebrook, J., Clarke, J. D., Volko, S. and Dong, X. (1997) The *Arabidopsis NPR1* gene that controls systemic aquired resistance encodes a novel protein containing ankyrin repeats. *Cell*, **88**, 57.

89. Ryals, J., Weymann, K., Lawton, K., Friedrich, L., Ellis, D., Steiner, H.-Y., Johnson, J., Delaney, T. P., Jesse, T., Vos, P., *et al.* (1997) The *Arabidopsis* NIM1 protein shows homology to the mammalian transcription factor inhibitor I kappa B. *Plant Cell*, **9**, 425.

90. Clarke, J. D., Liu, Y., Klessig, D. F. and Dong, X. (1998) Uncoupling PR gene expression from NPR1 and bacterial resistance: characterization of the dominant *Arabidopsis cpr6-1* mutant. *Plant Cell*, **10**, 557.

91. Dong, X. (1998) SA, JA, ethylene, and disease resistance in plants. *Curr. Opin. Plant Biol.*, **1**, 316.

92. Penninckx, I. A. M. A., Eggermont, K., Terras, F. R. G., Thomma, B. P. H. J., Samblanx, G. W. D., Buchala, A., Metraux, J.-P., Manners, J. M. and Broekaert, W. F. (1996) Pathogen-induced systemic activation of a plant defensin gene in Arabidopsis follows a salicylic acid-independent pathway. *Plant Cell*, **8**, 2309.

93. Epple, P., Apel, K. and Bohlmann, H. (1997) Overexpression of an endogenous thionine enhances resistance of *Arabidopsis* against *Fusarium oxysporum*. *Plant Cell*, **9**, 509.

94. Penninckx, I. A. M. A., Thomma, B. P. H. J., Buchala, A., Metraux, J.-P. and Broekaert, W. F. (1998) Concomitant activation of jasmonate and ethylene response pathways is required for induction of a plant gene in *Arabidopsis*. *Plant Cell*, **10**, 2103.

95. Pieterse, C. M., Van Wees, S. C. M., Van Pelt, J. A., Knoester, M., Laan, R., Gerrits, H., Weisbeek, P. J. and Van Loon, L. C. (1998) A novel signaling pathway controlling induced systemic resistance in *Arabidopsis*. *Plant Cell*, **10**, 1571.

96. Wolter, M., Hollricher, K., Salamini, F. and Schulze-Lefert, P. (1993) The *mlo* resistance alleles to powdery mildew infection in barley trigger a developmentally controlled defence mimic phenotype. *Mol. Gen. Genet.*, **239**, 122.

97. Buschges, R., Hollricher, K., Panstruga, R., Simons, G., Wolter, M., Frijters, A., van Daelen, R., van der Lee, T., Diergaarde, P., Groenendijk, J., *et al.* (1997) The barley *Mlo* gene: a novel control element of plant pathogen resistance. *Cell*, **88**, 695.

98. Jabs, T., Dietrich, R. A. and Dangl, J. L. (1996) Initiation of runaway cell death in an *Arabidopsis* mutant by extracellular superoxide. *Science*, **273**, 1853.

99. Dietrich, R. A., Richberg, M. H., Schmidt, R., Dean, C. and Dangl, J. L. (1997) A novel zinc finger protein is encoded by the *Arabidopsis LSD1* gene and functions as a negative regulator of plant cell death. *Cell*, **88**, 685.

100. Clough, S. J., Fengler, K. A., Yu, I., Lippok, B., Jr., R. K. S. and Bent, A. F. (2000) The *Arabidopsis dnd1* 'defense, no death' gene encodes a mutated cyclic nucleotide-gated ion channel. *Proc. Natl Acad. Sci. USA*, **97**, 9323.

101. Rate, D. N., Cuenca, J. V., Bowman, G. R., Guttman, D. S. and Greenberg, J. T. (1999) The gain-of-function *Arabidopsis acd6* mutant reveals novel regulation and function of the salicylic acid signaling pathway in controlling cell death, defense, and cell growth. *Plant Cell*, **11**, 1695.

102. Yu, I., Fengler, K. A., Clough, S. J. and Bent, A. F. (2000) Identification of *Arabidopsis* mutants exhibiting an altered hypersensitive response in gene-for-gene disease resistance. *Mol. Plant–Microbe Interact.*, **13**, 277.

103. Jirage, D., Tootle, T. L., Reuber, T. L., Frost, L. N., Feys, B. J., Parker, J. E. and Ausubel, F. M. (1999) *Arabidopsis thaliana PAD4* encodes a lipase-like gene that is important for salicylic acid signaling. *Proc. Natl Acad. Sci. USA*, **96**, 13583.

104. Vogel, J. and Somerville, S. (2000) Isolation and characterization of powdery mildew-resistant *Arabidopsis* mutants. *Proc. Natl Acad. Sci. USA*, **97**, 1897.

105. Frye, C. A. and Innes, R. W. (1998) An *Arabidopsis* mutant with enhanced resistance to powdery mildew. *Plant Cell*, **10**, 947.

106. Hahlbrock, K., Scheel, D., Logemenn, E., Nurnberger, T., Papniske, M., Reinhold, S., Sacks, W. R. and Schmelzer, E. (1995) Oligopeptide elicited defense gene activation in cultured parsley cells. *Proc. Natl Acad. Sci. USA*, **92**, 4150.

107. Gabriel, D. and Rolfe, B. (1990) Working models of specific recognition in plant-microbe interactions. *Annu. Rev. Phytopathol.*, **28**, 365.

108. Atkinson, M. M., Keppler, L. D., Orlandi, E. W., Baker, C. J. and Mischke, C. F. (1990) Involvement of plasma membrane calcium influx in bacterial induction of the K^+/H^+ and hypersensitive responses in tobacco. *Plant Physiol.*, **92**, 215.

109. Atkinson, M. M., Midland, S. L., Sims, J. J. and Keen, N. T. (1996) Syringolide 1 triggers Ca^{2+} influx, K^+ efflux, and extracellular alkalization in soybean cells carrying the disease resistance gene *Rpg4*. *Plant Physiol.*, **112**, 297.

110. He, S.-Y., Huang, H.-C. and Collmer, A. (1993) *Pseudomonas syringae* pv. *syringae* harpin-Pss: a protein that is secreted via the *hrp* pathway and elicits the hypersensitive response in plants. *Cell*, **73**, 1255.

111. Levine, A., Pennell, R. I., Alvarez, M. E., Palmer, R. and Lamb, C. (1996) Calcium-mediated apoptosis in a plant hypersensitive disease resistance response. *Curr. Biol.*, **6**, 427.

112. Nurnberger, T., Nennstiel, D., Jabs, T., Sacks, W. R., Hahlbrock, K. and Scheel, D. (1994) High affinity binding of a fungal elicitor to parsley plasma membranes. *Cell*, **78**, 449.

113. Romeis, T., Piedras, P., Zhang, S., Klessig, D. F., Hirt, H. and Jones, J. D. G. (1999) Rapid Avr9- and Cf-9-dependent activation of MAP kinases in tobacco cell cultures and leaves: convergence of resistance gene, elicitor, wound, and salicylate responses. *Plant Cell*, **11**, 273.

114. Ligternik, W., Kroj, T., zur Nieden, U., Hirt, H. and Scheel, D. (1997) Receptor-mediated activation of a MAP kinase in pathogen defense of plants. *Science*, **276**, 2054.

115. Jabs, T., Tschope, M., Colling, C., Hahlbrock, K. and Scheel, D. (1997) Elicitor-stimulated ion fluxes and O_2^- from the oxidative burst are essential components in triggering defense gene activation and phytoalexin synthesis in parsley. *Proc. Natl Acad. Sci. USA*, **94**, 4800.

116. Blume, B., Nurnberger, T., Nass, N. and Scheel, D. (2000) Receptor-mediated increase in cytoplasmic free calcium required for activation of pathogen defense in parsley. *Plant Cell*, **12**, 1425.

117. Lamb, C. and Dixon, R. (1997) The oxidative burst in plant disease resistance. *Annu. Rev. Plant Physiol. Plant Mol. Biol.*, **48**, 251.

118. Baker, C. J. and Orlandi, E. W. (1995) Active oxygen in plant pathogenesis. *Annu. Rev. Phytopathol.*, **33**, 299.

119. Mittler, R., Herr, E. H., Orvar, B. L., Camp, W., Willwkens, H., Inze, D. and Ellis, B. E. (1999) Transgenic tobacco plants with reduced capability to detoxify reactive oxygen intermediates are hyperresponsive to pathogen infection. *Proc. Natl Acad. Sci. USA*, **96**, 14165.

120. Chamnongpol, S., Willekens, H., Moeder, W., Langebartels, C., Sandermann Jr, H., Van Montagu, M., Inze, D. and Van Camp, W. (1998) Defense activation and enhanced pathogen tolerance induced by H_2O_2 in transgenic tobacco. *Proc. Natl Acad. Sci. USA*, **95**, 5818.

121. Chen, Z., Silva, H. and Klessig, D. F. (1993) Active oxygen species in the induction of plant systemic acquired resistance by salicylic acid. *Science*, **262**, 1883.

122. Alvarez, M. E., Pennell, R. I., Meijer, P.-J., Ishikawa, A., Dixon, R. A. and Lamb, C. (1998) Reactive oxygen intermediates mediate a systemic signal network in the establishment of plant immunity. *Cell*, **92**, 773.

123. Delledonne, M., Xia, Y., Dixon, R. A. and Lamb, C. (1998) Nitric oxide functions as a signal in plant disease resistance. *Nature*, **394**, 585.

124. Durner, J., Wendehenne, D. and Klessig, D. F. (1998) Defense gene induction in tobacco by nitric oxide, cyclic GMP, and cyclic ADP-ribose. *Proc. Natl Acad. Sci. USA*, **95**, 10328.

125. Durner, J. and Klessig, D. F. (1999) Nitric oxide as a signal in plants. *Curr. Opin. Plant Biol.*, **2**, 369.

126. Klessig, D. F., Durner, J., Noad, R., Navarre, D. A., Wendehenne, D., Kumar, D., Zhou, J. M., Shah, J., Zhang, S., Kachroo, P., *et al.* (2000) Nitric oxide and salicylic acid signaling in plant defense. *Proc. Natl Acad. Sci. USA*, **97**, 8849.

127. Levine, A., Tenhaken, R., Dixon, R. and Lamb, C. (1994) H_2O_2 from the oxidative burst orchestrates the plant hypersensitive disease response. *Cell*, **79**, 583.

128. Zhang, S., Du, H. and Klessig, D. F. (1998) Activation of the tobacco SIP kinase by both a cell wall-derived carbohydrate elicitor and purified proteinaceous elicitins from *Phytophthora* spp. *Plant Cell*, **10**, 435.

129. Zhang, S. and Klessig, D. F. (1998) Resistance gene N-mediated *de novo* synthesis and activation of a tobacco mitogen-activated protein kinase by tobacco mosaic virus infection. *Proc. Natl Acad. Sci. USA*, **95**, 7433.

130. Seo, S., Okamoto, M., Seto, H., Ishizuka, K., Sano, H. and Ohashi, Y. (1995) Tobacco MAP kinase: a possible mediator in wound signal transduction pathways. *Science*, **270**, 1988.

131. Zhang, S. and Klessig, D. F. (1998) The tobacco wounding-activated mitogen-activated protein kinase is encoded by *SIPK*. *Proc. Natl Acad. Sci. USA*, **95**, 7225.

132. Zhang, S. and Klessig, D. F. (1997) Salicylic acid activates a 48-kD MAP kinase in tobacco. *Plant Cell*, **9**, 809.

133. Ligterink, W., Kroj, T., zur Nieden, U., Hirt, H. and Scheel, D. (1997) Receptor-mediated activation of a MAP kinase in pathogen defense of plants. *Science*, **276**, 2054.

134. Zhou, J., Loh, Y. T., Bressan, R. A. and Martin, G. B. (1995) The tomato gene *Pti* encodes a serine/threonine kinase that is phosphorylated by Pto and is involved in the hypersensitive response. *Cell*, **83**, 925.

135. Zhou, J., Tang, X. and Martin, G. B. (1997) The Pto kinase conferring resistance to tomato bacterial speck disease interacts with proteins that bind a *cis*-element of pathogenesis-related genes. *EMBO J.*, **16**, 3207.

136. Frederick, R. D., Thilmony, R. L., Sessa, G. and Martin, G. B. (1998) Recognition specificity for the bacterial avirulence protein AvrPto is determined by Thr-204 in the activation loop of the tomato Pto kinase. *Mol. Cell*, **2**, 241.

137. Rathjen, J. P., Chang, J. H., Staskawicz, B. J. and Michelmore, R. W. (1999) Constitutively active *Pto* induces a *Prf*-dependent hypersensitive response in the absence of *avrPto*. *EMBO J.*, **18**, 3232.

138. Rushton, P. J. and Somssich, I. E. (1998) Transcriptional control of plant genes responsive to pathogens. *Curr. Opin. Plant Biol.*, **1**, 311.

139. Eulgem, T., Rushton, P. J., Schmelzer, E., Hahlbrock, K. and Somssich, I. E. (1999) Early nuclear events in plant defense signaling: rapid gene activation by WRKY transcription factors. *EMBO J.*, **18**, 4689.

140. Gu, Y.-Q., Yang, C., Thara, V. K., Zhou, J. and Martin, G. B. (2000) *Pti4* induced by ethylene and salicylic acids, and its product is phosphorylated by the Pto kinase. *Plant Cell*, **12**, 771.

141. Zhang, Y., Fan, W., Kinkema, M., Li, X. and Dong, X. (1999) Interaction of NPR1 with basic leucine zipper protein transcription factors that bind sequences required for salicylic acid induction of the PR-1 gene. *Proc. Natl Acad. Sci. USA*, **96**, 6523.

142. Despres, C., DeLong, C., Glaze, S., Liu, E. and Fobert, P. R. (2000) The *Arabidopsis* NPR1/NIM1 protein enhances the DNA binding activity of a subgroup of the TGA family of bZIP transcription factors. *Plant Cell*, **12**, 279.

143. Zhou, J. M., Trifa, Y., Silva, H., Pontier, D., Lam, E., Shah, J. and Klessig, D. F. (2000) NPR1 differentially interacts with members of the TGA/OBF family of transcription factors that bind an element of the PR-1 gene required for induction by salicylic acid. *Mol. Plant–Microbe Interact.*, **13**, 191.

144. Reymond, P., Weber, H., Damond, M. and Farmer, E. E. (2000) Differential gene expression in response to mechanical wounding and insect feeding in *Arabidopsis* [see comments]. *Plant Cell*, **12**, 707.

145. Salmeron, J. M. and Vernooij, B. (1998) Transgene approaches to microbial disease resistance in crop plants. *Curr. Opin. Plant Biol.*, **1**, 347.

146. Shah, D. M. (1997) Genetic engineering for fungal and bacterial diseases. *Curr. Opin. Biotechnol.*, **8**, 208.

147. Dempsy, D. A., Silva, H. and Klessig, D. F. (1998) Engineering disease and pest resistance in plants. *Trends Microbiol.*, **6**, 54.

148. Rommens, C. M. T., Salmeron, J. M., Oldroyd, G. E. D. and Staskawicz, B. J. (1995) Intergeneric transfer and functional expression of the tomato disease resistance gene *Pto*. *Plant Cell*, **7**, 1537.

149. Ellis, J. G., Lawrence, G. J., Luck, J. E. and Dodds, P. N. (1999) Identification of regions in alleles of the flax rust resistance gene *L* that determine differences in gene-for-gene specificity. *Plant Cell*, **11**, 495.

150. Cao, H., Li, X. and Dong, X. (1998) Generation of broad-spectrum disease resistance by overexpression of an essential regulatory gene in systemic acquired resistance. *Proc. Natl Acad. Sci. USA*, **95**, 6531.

151. Krysan, P. J., Young, J. C. and Sussman, M. R. (1999) T-DNA as an insertional mutagen in *Arabidopsis*. *Plant Cell*, **11**, 2283.

152. Tissier, A. F., Marillonnet, S., Klimyuk, V., Patel, K., Torres, M. A., Murphy, G. and Jones, J. D. (1999) Multiple independent defective suppressor-mutator transposon insertions in *Arabidopsis*: a tool for functional genomics. *Plant Cell*, **11**, 1841.

153. Somerville, C. and Somerville, S. (1999) Plant functional genomics. *Science*, **285**, 380.

154. Van Der Hoorn, R. A., Roth, R. and De Wit, P. J. (2001) Identification of distinct specificity determinants in resistance protein Cf-4 allows construction of a Cf-9 mutant that confers recognition of avirulence protein Avr4. *Plant Cell*, **13**, 273.

155. Wulff, B. B., Thomas, C. M., Smoker, M., Grant, M. and Jones, J. D. (2001) Domain Swapping and Gene Shuffling Identify Sequences Required for Induction of an Avr-Dependent Hypersensitive Response by the Tomato Cf-4 and Cf-9 Proteins. *Plant Cell*, **13**, 255.

156. Maleck, K., Levine, A., Eulgem, T., Morgan, A., Schmid, J., Lawton, K. A., Dangl, J. L. and Dietrich, R. A. (2000) The transcriptome of *Arabidopsis thaliana* during systemic acquired resistance. *Nat. Genet.*, **26**, 403.

10 | Signalling in plant–virus interactions

STEVEN A. WHITHAM AND S. P. DINESH-KUMAR

1. Introduction

Plant viruses are obligate parasites that are dependent on host cells for multiplication and completion of their life cycles. Viral genomes typically encode 4–12 proteins that facilitate replication, encapsidation, local and systemic movement within a host, and transmission to new hosts. In susceptible hosts, viral proteins may promote replication and movement by directly inducing host encoded susceptibility factors and/or suppressing host defence responses. Mammalian virology contains many examples of viral pathogens promoting their infections by altering or highjacking host signal transduction pathways (1–4). There is recent evidence to suggest that plant viruses carry on in a similar fashion (5). Plant viruses have acquired at least three ways to alter the host cellular environment in order to ensure that infection will proceed. First, members of the Geminiviridae cause the plant cell cycle to proceed past critical checkpoints so that necessary DNA replication machinery is present in infected cells. Second, many plant viruses suppress post-transcriptional gene silencing to their RNA genomes from degradation. Third, viruses cause alterations in plasmodesmata that facilitate movement of viral nucleoprotein complexes from cell to cell. Another influence viral proteins have in susceptible hosts is the induction of symptoms that are usually associated with a systemic infection. The onset of symptoms may be a direct result of viruses interfering with hosts to promote infection. In resistant plants, viral proteins elicit defence responses in hosts endowed with cognate resistance (R) genes resulting in inhibition of viral spread. Thus, viruses exert a variety of effects on host signal transduction in both compatible and incompatible interactions. This chapter discusses how viruses interface with their hosts to bring about subsequent responses in both susceptible and resistant plants, with emphases on:

- induction of host DNA synthesis factors for viral replication;
- viral suppression of post-transcriptional gene silencing
- alteration of host gene expression during virus infection;
- induction of R gene-mediated defence responses.

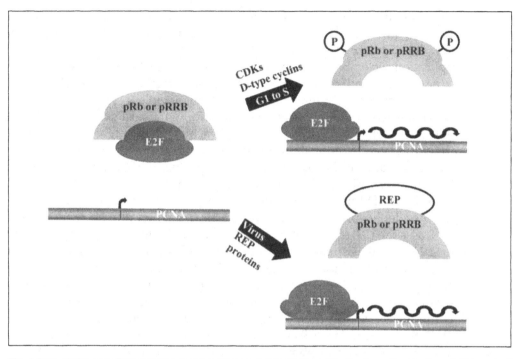

Fig. 1 Role of pRb and pRRB proteins in cell cycle and viral replication. Mammalian retinoblastoma (pRb) or plant retinoblastoma-related (pRRB) proteins bind E2F. During G1 to S transition, they become phosphorylated by cyclin-dependent kinases (CDKs) and D-type cyclins, resulting in release of E2F and transcription of genes such as proliferating cell nuclear antigen (PCNA). During virus infection, REP protein binding substitutes for phosphorylation, resulting in release of E2F and transcription of genes such as PCNA.

2. Induction of host DNA synthesis factors for viral replication

If a host lacks factors required for a virus to replicate or move, then the infection cannot proceed. In some instances, a host may possess the necessary factors, but they might not be expressed in the appropriate cells. To overcome this problem, geminiviruses and nanoviruses induce host cells to produce the appropriate factors required for infection. Geminiviruses are small, single-stranded DNA (ssDNA) viruses with circular genomes of 2.7–3.0 kb. Within the geminivirus family, the mastreviruses and curtoviruses have monopartite genomes that encode all proteins needed to direct replication, movement, encapsidation, and plant-to-plant transmission; and the begomoviruses have bipartite genomes that encode replication, encapsidation, and plant-to-plant transmission on the A component and movement on the B component. Nanoviruses also have circular, ssDNA genomes with replication and movement proteins that can be distributed among 6–10 ssDNAs of 1 kb or less. Geminiviruses and nanoviruses are transmitted from plant to plant by insect vectors that feed on phloem contents.

Geminiviruses and nanoviruses possess multifunctional proteins, termed RepA, AL1, C1, and Rep, that control viral gene expression and facilitate genome replication. In addition, they require host DNA synthesis enzymes to complete their replication cycles. Based on this fact, we might predict these viruses to be dead if they enter quiescent cells, but this is not the case. RepA, AL1, C1, or Rep (for simplicity these are generically referred to as REP proteins unless specified) creates conducive environments for viral replication by overriding the host's control over the cell cycle. This alteration in cell cycle control is achieved through interaction with plant proteins related to retinoblastoma (pRRB) (Fig. 1) (6).

In mammalian cells, the retinoblastoma protein (pRb) is a key regulator of the cell division cycle. pRb is phosphorylated late in G1 by cyclin-dependent kinases in concert with D-type cyclins, is increasingly phosphorylated in S and G2, and then becomes dephosphorylated during mitosis (7). When pRb becomes phosphorylated it releases the E2F transcription factor, which activates expression of genes necessary for DNA synthesis, such as proliferating cell nuclear antigen (PCNA) and other genes required for the G1 to S transition (Fig. 1). Thus, pRb prevents cells from making the G1 to S transition in the absence of appropriate stimuli. pRb is targeted by small DNA tumour viruses, such as adenoviruses and simian virus (SV) 40, that require host cells to provide enzymes for genome replication and gene expression (2). Both the SV40 large T antigen and adenovirus E1A protein bind to pRb. These interactions mimic phosphorylation of pRb by causing E2F to dissociate and mediate expression of PCNA and other genes and induce cells to complete the G1 to S transition (7). The ability of large T antigen and E1A to interact with pRb overrides regulation by phosphorylation and allows each virus to promote DNA replication by inducing host cells to produce the necessary factors.

The first indications that plant viruses might use similar mechanisms to control replication came from studies of geminivirus REP proteins. The Rep protein of wheat dwarf mastrevirus (WDV) was found to possess an LXCXE motif, which is a signature for viral (large T antigen and E1A) and host proteins that interact directly with pRb in mammalian cells. Mutational analysis of the LXCXE motif in WDV Rep demonstrated that it was necessary for interaction with pRb (p130Rbr2) in the yeast two-hybrid system, and it was required for efficient replication of WDV in wheat cell culture (8). The LXCXE motif is also present in other mastrevirus and nanovirus REP proteins (9–13), and in many of these cases it mediates an *in vitro* or yeast two-hybrid interaction with pRb. Since these initial discoveries, several publications have described the identification of plant RRB proteins (14–19), plant cyclin D homologues that are highly conserved with mammalian counterparts and possess the LXCXE Rb-binding motif (20), and plant E2F transcription factor homologues (21, 22). The identification of these cell-cycle regulators in plants suggests that the Rb pathway and certain aspects of cell cycle control are conserved between plants and mammals. Furthermore, geminiviruses alter the expression of markers for the plant cell cycle. Accumulation of PCNA was observed within differentiated plant cells infected by tomato golden mosaic geminivirus (TGMV). PCNA is an accessory factor for DNA polymerase, and normally, it is only expressed in actively dividing cells. The AL1 protein of TGMV is sufficient to induce the accumulation of PCNA in transgenic plants (23). These observations suggested that geminivirus REP proteins might alter the plant cell cycle and induce the host DNA synthesis apparatus in a manner similar to mammalian small DNA tumour viruses (Fig. 1).

pRRB activity is not modified in the same way by all geminiviruses. The REP proteins of begomoviruses and curtoviruses do not contain the LXCXE motif (19). In spite of this fact, REP proteins such as the TGMV AL1 protein interact with maize pRRB homologues in yeast two-hybrid and *in vitro* studies (17). A conserved central domain of pRb, termed the A/B pocket, mediates binding between the LXCXE motif and pRb (24). Mutational analysis of this domain in maize pRRB1 demonstrated that the AL1 protein and SV40 large T-antigen binding can be affected by mutations that disrupt folding of the pocket. Furthermore, truncations of the pocket showed that AL1 requires a more extensive pocket region for efficient binding than does the SV40 large T antigen (19). The primary determinants of AL1 binding were mapped to a region containing two alpha helices between amino acids 119 and 180 (19). Thus, begomoviruses and possibly curtoviruses use alternative methods to inactivate pRRBs in plants when compared to mastreviruses and nanoviruses.

Interaction of REP with pRRB may determine tissue specificity in geminivirus infections (19). The KEE146 mutant of TGMV AL1 was unable to bind pRRB1 *in vitro*, replicated to low levels in plants, caused mild symptoms, and was limited to the vascular system when compared with wild-type TGMV, which is also found in mesophyl and epidermal cells. The E-N140 mutant of TGMV AL1 was able to bind pRRB, replicated to low levels in plants, caused mild symptoms, but had a distribution within plant tissues similar to the wild-type virus. Comparison of the behaviour of these mutants in plants suggested that interaction with pRRB might be a determinant of tissue specificity during infection. This experiment might explain why some geminiviruses are phloem limited, whereas others are distributed in more tissue types. This conclusion leads to the hypothesis that if a geminivirus is unable to suppress the appropriate pRRB, then it will be limited to cell types where it can suppress pRRB function. An inability to suppress pRRB function might be caused by one of at least two mechanisms. First, multiple pRRBs can be encoded by plant genomes (17). In this case, different pRRBs could control cell-cycle progression in specific tissue types, and REP proteins might inactivate only a subset of pRRBs. However, this model may not apply because pRb suppressors can interact with pRbs from a broad range of species. A second possible mechanism for tissue specificity based on pRRB suppression would be whether or not the REP proteins can function in a particular cell type. This would require REPs to be regulated in a tissue-specific manner. Another important question is: Why don't geminiviruses generally cause tumours? This observation suggests that geminiviruses are able to activate DNA synthesis but not cell division. It is possible that the inactivation of pRRB alone is an insufficient trigger for progression out of S-phase and subsequent cell division. These questions of tissue tropism and control of cell-cycle progression are interesting for the future. One way these issues might be addressed is to separate pRRB/REP interaction from the G1 to S phase transition. Such mutants have been identified in mammalian pRb (25), and similar plant mutants may allow the processes of pRRB inactivation, DNA synthesis, and cell-cycle progression to be further dissected.

3. Viral suppression of post-transcriptional gene silencing

Another way that viruses create conducive environments for infection is to suppress host defence responses. Although host cells can possess all factors required by viruses for infection, which makes them vulnerable, if they are able to recognize the invading virus, then infection can be attenuated or eliminated. On the other hand, if the virus is able to counteract the defence response, then it can continue to multiply and spread systemically throughout host cells. There are many examples of counter-defensive strategies in the animal virus literature (1, 3, 4). Recently this was demonstrated for plant viruses by their ability to suppress post-transcriptional gene silencing (PTGS).

PTGS has been the subject of many recent reviews (26–31). PTGS is associated with methylation of coding sequences and post-transcriptional degradation of RNA in the cytoplasm. PTGS is distinguished from transcriptional gene silencing, which is associated with methylation of promoter regions and inhibition of gene transcription. Here we discuss PTGS as an induced defence response during host–virus interactions. The accumulating evidence indicates that PTGS is a mechanism to protect eukaryotic cells and genomes from nucleic acid-based invaders and endogenous transposable elements. A natural role for PTGS in plants is to mediate general, adaptive defence responses to protect host cells from exogenous DNA and RNA viruses. This is supported by two major lines of evidence:

- first, PTGS is a mechanism that, once activated, is sufficient to protect plants from an invading viral pathogen (32–34); and
- secondly, plant viruses possess mechanisms to suppress PTGS (35–39).

When PTGS is suppressed, viruses can accumulate to higher levels than normal. This anti-silencing ability is the first example of plant viruses creating conducive environments for their replication and movement by interfering with a host defence mechanism.

PTGS was originally discovered due to the simultaneous silencing (co-suppression) of chalcone synthase transgenes and the homologous endogenous genes in transgenic petunia plants (40). More recently, PTGS was shown to function as a defence mechanism against viral pathogens. PTGS was first correlated with the ability to restrict virus infection in studies of pathogen-derived resistance against tobacco etch potyvirus (TEV) in tobacco (41). Studies on TEV resistance established that the RNA sequence, not the encoded protein, of a coat protein (CP) transgene construct was responsible for pathogen-derived resistance. The steady state level of the CP transgene mRNA was much reduced in plants where resistance to TEV was expressed when compared to sibling plants in which resistance was not induced.

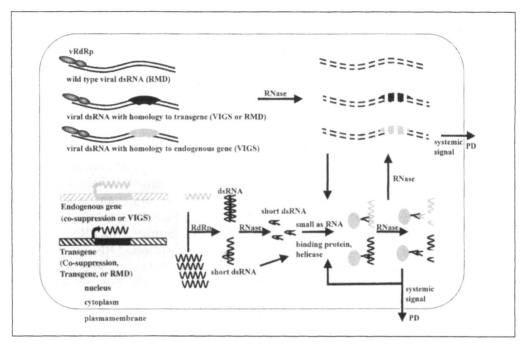

Fig. 2 A speculative model for PTGS and viral defence in plants. Source and target RNAs of PTGS are derived from nuclear endogenous genes and transgenes and extranuclear viral RNAs. Endogenous genes are targets for co-suppression and virus-induced gene silencing (VIGS). Transgenes are targets for co-suppression and transgene silencing. Silenced transgenes of viral origin can initiate co-suppression, transgene silencing, and RNA-mediated defence (RMD) against homologous viruses. Transgene silencing may require RNA-dependent RNA polymerase (RdRp) to generate double-stranded RNAs (dsRNAs) or short antisense RNAs (asRNAs) that can subsequently direct sequence specific degradation of any homologous RNA derived from nuclear or viral sources. Virus-initiated PTGS does not require host RdRp because viral RdRp (vRdRp; dark grey ovals) makes dsRNA during viral replication. These dsRNAs may be targeted by RNases to generate small RNA species. Small RNA species of viral origin can subsequently direct sequence-specific degradation of any homologous RNA derived from nuclear or viral sources. Some form of RNA or RNA/protein complex may move systemically through plasmodesmata (PD) and the vascular system to initiate PTGS at distal sites in plants. Small antisense RNA-binding proteins and helicases may help to mediate recognition of homologous RNAs and subsequent degradation by RNases (light grey ovals).

Furthermore, nuclear run-off experiments showed that the transcription of the CP transgenes was not affected, although steady state levels of transgene mRNA were significantly reduced after resistance to TEV was established. Lindbo *et al.* (41) proposed that a sequence-specific degradation process occurred in the cytoplasm that degraded both the CP transgene mRNA and the TEV RNA.

Since these studies, at least five distinct types of PTGS have been identified in plants: co-suppression, transgene silencing, RNA-mediated defence (RMD), virus-induced gene silencing (VIGS), and systemic silencing (Fig. 2). Co-suppression is the simultaneous PTGS of a transgene and a homologous endogenous gene, as discussed above for chalcone synthase. Transgene silencing is PTGS of a single- or multiple-copy transgene directed against itself. RMD occurs when a virally expressed sequence directs silencing against itself and any other viral RNA of sufficient homology. RMD includes pathogen-derived resistance, which was described above for TEV CP-mediated resistance. RMD is distinguished from VIGS, which is virally-induced PTGS of a transgene or an endogenous plant gene. Finally, systemic-induced PTGS occurs when a sequence is introduced into plant cells locally, but is able to silence expression of homologous transgenes or endogenous genes at distal sites. These types of PTGS are initiated by distinct sources of RNA and have different targets, but the common factor is that RNA species are degraded in the cytoplasm by a sequence- specific mechanism.

The specific degradation of mRNA and viral RNAs that become targeted by PTGS was confirmed by the detection of small RNA fragments of about 25 nucleotides in length. The small RNA species were identified from plant tissues in which four distinct types of PTGS were occurring: co-suppression, transgene silencing, RMD, and systemic-induced transgene silencing (42). Similar small RNA species of 21–23 nucleotides were observed *in vitro* when double-stranded RNA (dsRNA) was added to a *Drosophila* embryo lysate (43). These small RNAs are comprised of a mixture of sense and antisense fragments. The role of these small RNA species is currently not understood in detail, but it is likely that they are not simply the degradation products of PTGS. PTGS is exquisitely sequence specific, suggesting that some form of nucleic acid must direct the process. The small RNAs may be acquired by size-specific RNA binding proteins that can use the antisense sequences to scan the cytoplasm for new homologous RNA molecules and then recruit degradation machinery onto these RNAs (Fig. 2). This model is similar to those previously proposed by (43, 44). In some cases of PTGS, such as silencing of a highly expressed single-copy transgene, there must be an RNA synthesis step to generate the antisense RNA, which would require a host RNA-dependent RNA polymerase (RdRp) (Fig. 2). In the cases of RMD or VIGS, the host RdRp may not be necessary because viral RdRps generate dsRNA during virus replication (45, 46).

Evidence that RdRp activity is involved in PTGS comes from mutants of *Caenorhabditis elegans* (*ego1*; 47), *Neurospora crassa* (*qde1*; 48), and *Arabidopsis thaliana* (*sgs2*, 45; *sde1*, 46). *sgs2* and *sde1* are alleles of the same locus isolated from the Columbia and C24 *Arabidopsis* ecotypes, respectively. All of these loci encode genes with similarity to known RdRps, and they are necessary for some types of PTGS, including RNA interference in *C. elegans*, quelling in *N. crassa*, and PTGS of trans-genes in *Arabidopsis*. Plant RdRp is apparently not required for VIGS or RMD in *Arabidopsis* because *sde1* mutant plants developed photo bleaching similar to wild-type plants when infected by tobacco rattle virus containing the *Arabidopsis* phytoene desaturase gene (46). Photo bleaching was due to VIGS of the endogenous phytoene desaturase gene. The ability of *sde1* plants to support VIGS of phytoene desaturase suggests that appropriate dsRNA for initiating PTGS are provided by viral RdRps.

PTGS is initiated by RNA species that are recognized as foreign or aberrant by plant cells. RNAs that possess extensive double-stranded structures or originate from complex loci, which have potential to form double-stranded structures, are very efficient at eliciting PTGS (49–53). RNAs that are expressed at very high levels, above a certain threshold, are also targets for PTGS (54). Both of these models for initiation of PTGS are relevant to viruses. Plant viruses often accumulate to high levels in plant cells, and they are sources of dsRNAs. Plus-sense RNA viruses enter host cells as virion-sense RNA then synthesize a complementary-sense RNA as a template to generate more virion-sense RNA. Therefore, viral replication strategies naturally produce dsRNAs that could initiate PTGS.

Additional evidence that the host RdRp does not have a major role in VIGS or RMD is that *sde1* mutant plants supported accumulation of tobacco rattle virus and tobacco mosaic virus (crucifer strain) similar to wild-type plants (46). Similarly, *sgs2* mutants supported accumulation of turnip mosaic virus and turnip vein-clearing virus similar to wild-type plants (45). The exception was that accumulation of cucumber mosaic virus (CMV) was enhanced fivefold in the *sgs2* mutant when compared to accumulation in wild-type plants (45). It is interesting that CMV accumulation is suppressed but not completely inhibited in wild-type *Arabidopsis*. This observation suggests that PTGS can quantitatively reduce viral accumulation. It will be interesting to test if any viruses that are completely incapable of infecting wild-type *Arabidopsis* can now infect any of the existing silencing-suppressed mutants (46, 55). Results of these experiments could prove whether PTGS can be the sole determinant of viral host range.

Some compelling evidence that PTGS is a viral defence mechanism is that many DNA and RNA viruses encode proteins that suppress PTGS. This first became evident through the study of synergism. Synergism results from co-infection of two unrelated viruses, one of which is often a potyvirus. Co-infection leads to increased disease symptoms and accumulation of the non-potyvirus compared to single infection. The ability of potyviruses to stimulate accumulation of heterologous viruses was mapped to the P1/HC-Pro region of the potyvirus genome (56). This finding led to the hypothesis that P1/HC-Pro stimulated accumulation of heterologous viruses by suppressing a viral defence mechanism. The HC-Pro proteins of TEV and potato virus Y (both potyviruses) were subsequently shown to prevent initiation of PTGS and to revert transgene silencing of GUS or GFP transgenes to non-silenced states (35–37). The anti-gene silencing activity of HC-Pro appears to be specific for PTGS as methylation and TGS are unaffected in the presence of HC-Pro (57). Several additional viruses encode anti-PTGS genes, including 2b of CMV and P19 of tomato bushy stunt virus (TBSV) (38, 57). Similar to HC-Pro, 2b has no effect on TGS. HC-Pro, 2b, and P19 are required for systemic spread of TEV, CMV, and TBSV, respectively, but not for local movement (58–60). Therefore, the ability to suppress PTGS may be a key determinant of whether plant viruses can move systemically. Long-distance movement of RNA species and silencing signals might converge at the vascular system to create an important host check point for viral surveillance against which viruses have developed counterdefensive strategies (61, 62).

HC-Pro and 2b appear to interfere with different points in the PTGS pathway, because expression of these genes from the heterologous virus potato virus X (PVX), which does not encode its own silencing suppressor, results in distinct anti-silencing phenotypes (36). PTGS is thought to have at least three major stages: initiation, maintenance, and systemic signalling. PVX-HC reverted GFP transgene silencing both in older leaves where it was established and in newer leaves where silencing was being initiated. PVX-2b reverted GFP transgene silencing only in newly emerging leaves, but not in older leaves where silencing was established. Therefore, HC-Pro may interfere with a later stage in the silencing pathway, perhaps a maintenance stage, whereas 2b may interfere with an early stage that involves initiation. To understand how HC-Pro and 2b may be interfacing with host cells, the yeast two-hybrid system has been used to identify host proteins that interact with HC-Pro or 2b (63, 64). A protein named rgs-CaM (regulator of gene silencing-calmodulin-like protein) that interacted with HC-Pro was determined to be a host-encoded suppressor of PTGS (63). Transgenic *Nicotiana benthamiana* plants that over-expressed rgs-CaM phenocopied plants that constitutively over-expressed the HC-Pro protein. Interestingly, rgs-CaM expression was induced by HC-Pro. Plants that expressed rgs-CaM failed to initiate PTGS of a GFP transgene, and when rgs-CaM was expressed from a PVX vector it was able to revert a silenced GFP transgene to a non-silenced state. The interaction of HC-Pro with rgs-CaM and induction of its mRNA expression suggests a mechanism by which HC-Pro might interfere directly with the PTGS pathway. Moreover, the similarity of rgs-CaM to calmodulin suggests a role for calcium signalling in PTGS. Further understanding of how viral silencing suppressors function will continue to provide insight into the regulation of the PTGS pathway and the role of PTGS as a viral defence mechanism.

4. Alterations in host gene expression during virus infection

The previous two sections present specific examples of how viruses can create favourable environments for infection by altering control of the cell cycle or by suppressing PTGS. Numerous other changes in host biology have been described, but the significance of these changes on the infection process has not been elucidated. For example, several viruses disrupt host gene expression and translation through viral replication and/or specific viral proteins. The disruption of these central processes may provide additional molecular mechanisms by which plant viruses create favourable environments for infection and subsequently cause the symptoms in infected plants. In many cases, viral proteins can directly elicit or influence symptomatology of the host (65). A challenge will be to determine mechanisms by which viral proteins bring about changes in host biology that promote infection and elicit symptoms. Defining these mechanisms will require a detailed understanding of how viral proteins accumulate, interact with host proteins, and influence host signalling pathways.

The symptoms caused by cauliflower mosaic virus (CaMV) infection have been mapped to gene VI. CaMV gene VI encodes a multifunctional protein (P6) that is involved in symptomatology, host range, translational transactivation, virion assembly, and inclusion bodies (66). Constitutive expression of gene VI in transgenic plants results in symptoms similar to those observed during virus infection (66, 67). In *Arabidopsis*, the phenotypes of independent gene VI transgenic lines can range from severe chlorosis, stunting, and altered leaf morphology through mild chlorosis, stunting, and late flowering to vein chlorosis (67, 68). The severity of the phenotype is proportional to the expression level of the gene VI mRNA and accumulation of P6 protein. Among the factors that might influence the ability of P6 to induce symptoms are specific alterations in host gene expression.

Host gene expression changes in response to CaMV infection and constitutive expression of gene VI were investigated by using differential display polymerase chain reaction (DD-PCR). DD-PCR is a method to discover mRNA species that are expressed at dissimilar levels in control versus experimental mRNA samples by comparing the relative intensities of the resulting PCR products. By using this method, a portion of the *Arabidopsis* genome was surveyed, and 18 *Arabidopsis* genes were identified whose expression was clearly altered by constitutive gene VI expression and CaMV infection (69). An additional 15 genes were identified whose expression was altered in a more subtle fashion. Thus, up to 33 host genes were coordinately regulated by CaMV infection and P6 accumulation. Interestingly, many of these genes were correlated with known defence or stress responses. DD-PCR results establish correlations between gene expression changes and viral infection or expression of the viral transgene, but do not demonstrate that these genes are involved in symptomatology. Therefore, each gene identified by DD-PCR is a candidate for further analyses by techniques such as reverse genetics. These analyses will lead to the identification of the signalling pathways governing the effects on host gene expression and possibly symptomatology.

Common alterations of host gene expression occur in pea embryonic tissues in response to a variety of RNA and DNA viruses (70). Expression of the heat-shock protein (Hsp) 70 and polyubiquitin genes was transiently induced at sites of pea seed-borne mosaic virus (PSbMV) replication (71). After the PSbMV infection front had passed, the expression levels of these genes returned to normal or slightly below normal levels as compared with non-infected tissue. Conversely, expression of a number of host genes, including lipoxygenase (*lox*), was transiently repressed at sites of PSbMV replication (72). Subsequently, similar changes in Hsp70 and *lox* gene expression were demonstrated for three other viruses; pea early-browning tobravirus (PEBV), white clover mosaic potexvirus (WClMV), and beet curly top geminivirus (BCTV) (70). PEBV and WClMV are positive-sense RNA viruses whereas BCTV is an ssDNA virus. Therefore, induction or repression of certain classes of host genes is common to distinct classes of plant viruses and may be a conserved feature of viral pathogenesis. The viral triggers of these gene-expression changes may be specific proteins encoded by each virus or they may be generic properties of viral replication. To begin to address this question, PEBV was used to eliminate some viral components as potential elicitors of the gene-expression changes. PEBV is a bipartite virus with replication and movement proteins on RNA1 and capsid and vector transmission proteins on RNA2. Infection by PEBV RNA1 alone was sufficient to induce the gene-expression changes, which demonstrated that capsid protein and vector transmission proteins are not required to elicit these host responses.

The heat-inducible expression of Hsp70 and other heat-shock genes is facilitated by heat-shock transcription factors (Hsfs). Upon heat shock, Hsfs form trimers and bind to heat-shock elements (HSEs) in the promoters of heat-inducible genes. In plants, HSEs have a canonical sequence of 5'-nGAAn-3' and are usually present in tandem arrays in either orientation. The induction of genes containing HSEs in their promoters was investigated in pea tissues in response to virus infection and heat stress (73). A single Hsf (*HsfA*) was identified in pea. *HsfA* contains HSEs in its promoter as does the Hsp70 gene. When pea embryos were infected by PSbMV, induction of Hsp70 mRNA was observed, but this was not the case for *HsfA*. Transcription of both genes was induced in response to heat stress. These results suggest that in the pea the induction of Hsps by virus infection may not be regulated through HSEs, but through some other stress-responsive pathway. Identifying the viral inducers of this response and additional elements in the promoters of heat-shock genes will be very helpful in discerning how this response is provoked.

5. Induction of *R*-gene-mediated defence responses

5.1 Virus-resistance gene structure, function, and regulation

Plants contain many resistance (*R*) genes that confer resistance to different strains of viruses and other phytopathogens. Plant *R* genes are hypothesized to encode receptors that directly or indirectly recognize elicitors (ligands) produced by corresponding avirulence (*avr*) genes of invading pathogens. This initial recognition event triggers signal transduction pathways to activate defence responses that ultimately halt pathogen spread. Absence of an *R* gene or the corresponding *avr* gene results in successful colonization by the pathogen and development of disease. This race-specific resistance response is dubbed 'gene-for-gene' type of resistance (74).

The typical *R*-gene-mediated defence response includes host-cell death that is referred to as the hypersensitive response (HR). Many biochemical and physiological changes accompany *R*-gene-triggered resistance and host-cell death (75). Some of the processes associated with gene-for-gene resistance are: generation of reactive oxygen species (ROS) and nitric oxide (NO), production of antimicrobial compounds, lipid peroxidation, ion fluxes, cell wall strengthening, lignin deposition, and induction of defence genes. The locally induced HR response often precedes the induction of a non-specific general defence response throughout the plant, called systemic acquired resistance (SAR) (76). During SAR, salicylic acid (SA) levels increase throughout the plant, defence genes such as pathogenesis-related (PR) genes are expressed, and the plant becomes more resistant to further pathogen attack.

To date three *R* genes that confer gene-for-gene type of resistance to viral pathogens have been cloned (77–79). The *N* gene from tobacco confers resistance to tobacco mosaic virus (TMV) (77); *Rx1* from potato confers resistance to potato virus X (PVX) (78); *HRT* confers resistance to turnip crinkle virus (TCV) (79). The *N*- and

HRT-mediated resistance involves HR, whereas, *Rx1* mediates resistance in the absence of HR. *Rx1*-mediated resistance is active in isolated protoplasts, whereas *N*-mediated HR-type resistance is not active in protoplasts but active in callus or cell culture level. Therefore, *Rx1*-mediated resistance is referred to as extreme resistance because it arrests PVX accumulation in initially infected cells. In gene-for-gene resistance, HR at the site of pathogen ingress has been suspected to function in containment of the pathogen. However, *Rx1*-mediated resistance suggests that HR might be separate and independent from the virus-limiting defence response and, thus, the role HR plays in halting pathogen spread is unclear.

N, *Rx1*, and *HRT* contain a centrally located nucleotide-binding site (NBS) and C-terminal leucine-rich repeats (LRR) of various lengths and therefore belong to the NBS-LRR class of *R* genes (80) (Fig. 3). This primary structure is highly conserved in *R* genes that confer resistance to virus, fungal, bacterial, nematode, and insect pathogens (80). In addition, these *R* genes contain a highly conserved domain of unknown function called GLPLAL/V/I in between NBS and LRR domains. The NBS regions of *R* genes share sequence homology with the NBS region of cell-death genes, *CED4*, from *Caenorhabditis elegans* and *Apaf-1*, FLASH, CARD4, and Nod1 from humans (81, 82). These virus *R* genes differ in their structure at the amino-terminus domain. *N* contains a Toll/interleukin-1 receptor homology region (TIR), *HRT* contains a leucine zipper (LZ) at the amino terminus, and *Rx1* does not contain either domain.

5.1.1   Structure–function analysis of virus resistance genes

The presence of conserved TIR, NBS, GLPLAL/V/I, and LRR structural motifs implies their involvement in protein complexes that recognize pathogen-derived ligands and trigger signal transduction leading to defence responses. A compre-

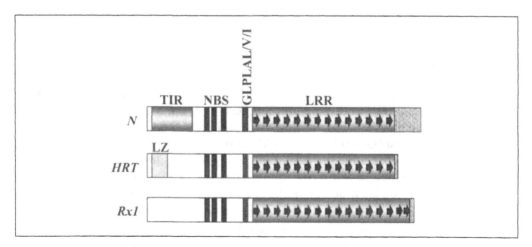

Fig. 3 Structure of different virus resistance genes. The *N* gene confers resistance to TMV, *HRT* confers resistance to TCV, and *Rx1* confers resistance to PVX. TIR, toll/interleukin-1 homology region; NBS, nucleotide-binding site; LRR, leucine-rich repeat; LZ, leucine zipper.

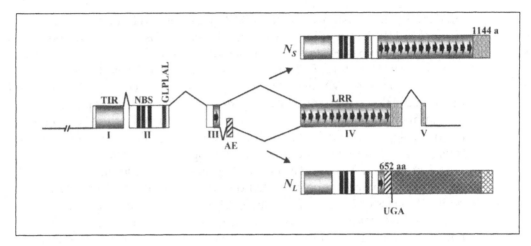

Fig. 4 The *N* gene encodes two transcripts by alternative splicing. The *N* gene contains five exons (rectangle boxes) and four introns (lines). The transcript N_S is generated by default splicing and encodes full-length N protein of 1144 amino acids. The N_L transcript is generated by alternative splicing that includes the AE in intron III and encodes a truncated protein of 652 amino acids because of the premature translation termination codon (UGA).

hensive amino acid deletion and site-directed mutagenesis analysis of the *N* gene indicates that TIR, NBS, and LRR domains are necessary for *N* function (83). The TIR domain is necessary for *Toll*- and IL-1R-activated signalling pathways that trigger defence responses in *Drosophila* and humans, respectively (84, 85). Amino acids found at positions conserved between *Toll*/IL-1R and the TIR-NBS-LRR class of *R* genes are essential for the *N* gene function (83). However, the exact role of the *N* gene TIR domain in TMV resistance signalling is unknown.

Many mutations in the NBS region of *N* result in loss of function, some mutations in the putative Mg^{2+}-binding site in the P-loop lead to a partial loss of function or dominant change of function (83). Plants bearing partial loss-of-function or dominant change-of-function alleles respond to TMV infection with HR. Despite the elicitation of HR, TMV was still able to replicate and move systemically in these plants, causing a systemic hypersensitive response (SHR). This SHR phenotype is virus-dependent. The onset of HR in these plants suggests that the mutant N protein may be able to recognize the TMV ligand directly or indirectly and initiate HR, but that cell death alone is not sufficient to inhibit TMV replication and systemic movement. A possible explanation for why these mutant *N* plants develop HR but fail to inhibit virus movement may be due to delayed occurrence of associated biochemical and physiological events such as generation of ROS, induction of salicylic acid, and defence genes, etc. In the future, these mutant forms of *N* will provide useful tools in understanding the role of localized cell death in antiviral processes during HR. Despite the requirement of NBS for disease resistance function, to date there is no report of biochemical evidence to support direct binding of nucleotide to the NBS domains of *R* genes.

5.1.2 Post-transcriptional regulation of the N gene expression

The TIR-NBS-LRR class of R genes encodes multiple transcripts (86). The N gene encodes two transcripts, N_S and N_L, via alternative splicing from a single gene *in vivo* (87) (Fig. 4). The N_S transcript encodes a full-length N protein containing the TIR, NBS, and LRR domains. The N_L transcript results from the alternative splicing of a 70 bp alternative exon (AE) within intron III of the N gene. The inclusion of AE in N_L results in a shift in the reading frame, leading to premature translational termination after the first LRR. The longer N_L message therefore encodes a putative truncated N (N^{tr}) protein. Alternative splicing of N transcripts is regulated by TMV-induced signals. The N_S transcript is more prevalent before and up to 3 hours after TMV infection, whereas the N_L transcript is more prevalent 4–8 hours after TMV infection. RT-PCR analysis and cDNA reconstruction experiments in transgenic plants suggest that an N_S-cDNA that harbours intron III, which contains the AE, and 3′-genomic regulatory sequences, is the minimum sequence required to confer complete resistance to TMV. It will be interesting to dissect how TMV signals regulate the splicing of N, to identify the *cis-* and *trans-*acting factors involved in such regulation, and to determine what roles N and N^{tr} play in N-TMV-mediated signalling.

5.2 Avirulence components of virus-resistance genes

A number of virus-encoded *avr* components induce HR in corresponding R-gene-containing plants when expressed using a viral-based expression vector or constitutive 35S promoter inside plant cells. The coat proteins of PVX and TCV elicit $Rx1$ (88) and HRT (79) responses, respectively. In the case of the N gene, the 126 kDa TMV replicase protein encodes the *avr* component (89). Within the 126 kDa protein, the C-terminal 50 kDa ($p50$) helicase portion is sufficient to induce HR (90–92). Helicase activity is not required to induce HR, which suggests features of the helicase domain, independent of its enzymatic activity, are recognized by N (92). To date there is no evidence to support direct interaction between viral-encoded *avr* products and corresponding R gene products. A direct interaction between Avr-R products has been demonstrated only in two gene-for-gene types of resistance: tomato bacterial resistance gene Pto and corresponding *avr* gene product avrPto (93, 94), and rice fungal resistance gene $Pi-ta$ and corresponding *avr* gene product AVR-Pita (95).

5.3 Virus-resistance gene signalling

Although many R genes that confer resistance to different phytopathogens have been cloned, the mechanisms by which R genes activate defence responses are poorly understood. A striking theme emerging from the isolation of various R genes is that they encode structurally similar proteins (80, 96, 97). This similarity suggests that pathogen recognition and/or defence-signalling mechanisms may be conserved in plants. Although evidence exists to support this view, some studies suggest that

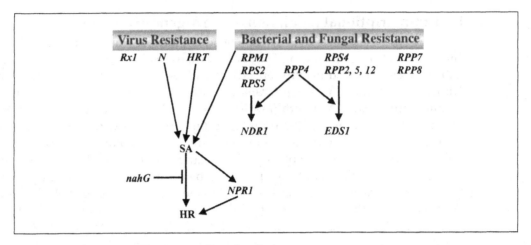

Fig. 5 Resistance-gene signalling (see text for explanation).

plants also employ specific components of defence-signalling pathways for different pathogens or different strains of the same pathogen (98, 99) (Fig. 5).

Analysis of *Arabidopsis* mutant *ndr1* and *eds1* suggests that they participate in bacterial- and fungal-resistance gene pathways (100, 101). *NDR1* is required for the action of LZ-NBS-LRR-type *R* genes (*RPS2*, *RPM1*, and *RPS5*), whereas, *EDS1* is required for the function of the TIR-NBS-LRR class of *R* genes (*RPP5*, *RPP2*, *RPP4*, *RPP12*, and *RPS4*) (102, 103). However, *RPP8*, a member of the LZ-NBS-LRR class, shows no requirement for either *NDR1* or *EDS1* (103). The requirement for either *EDS1* or *NDR1*, but not both, clearly suggests that different *R* genes employ different signal transduction pathways. *NDR1* encodes a transmembrane protein with no significant homology to any protein in GenBank (104). *EDS1* encodes a protein with similarity to eukaryotic lipase (105) and may be involved in generation of a second messenger derived from phospholipids or acyl glycerols. At the moment, the role of *EDS1* and *NDR1* in the viral-resistance gene pathway is not known.

SA plays an important role in the *N*-mediated resistance to TMV (106) and *HRT*-mediated resistance to TCV (107). During the *N*-mediated HR to TMV, SA accumulates around the HR lesions and throughout the plant (108). Furthermore, *nahG* transgenic tobacco plants convert SA into catechol, fail to accumulate SA, fail to develop SAR, and exhibit enhanced susceptibility to pathogen infection (106). In these plants, HR lesions induced by TMV infection continue to expand, leading to rolling HR phenotype (106). Whether SA is a translocated signal in *N*-mediated resistance to TMV is debatable (76).

The genetic analysis suggests that the *NPR1/NIM1/SAI1* gene encodes a key component of the SA pathway (109–111). The *Arabidopsis npr1/nim1* mutant fails to express PR genes upon treatment with chemical inducers such as SA, INA (2,6-dichloroisonicotinic acid) or benzothiazole (BTH). The *npr1/nim1* mutants exhibit enhanced disease susceptibility to virulent and avirulent *Pseudomonas syringae* and some fungal pathogens (111–114). *NPR1* encodes a novel protein containing ankyrin repeats (109, 110), which are found in many eukaryotic proteins with diverse functions mediated by protein–protein interactions. NPR1 is most similar to mammalian ankyrin 3 and IκB, inhibitors of the NF-κB transcription factor. Based on these homologies, NPR1 may act as a transcriptional regulator of PR gene expression. Furthermore, overexpression of *NPR1* confers resistance to fungal and bacterial pathogens in a dose-dependent manner (115). The role of *NPR1* in *N*- and *Rx1*-mediated virus-resistance pathway is not known. However, recent studies by Kachroo *et al.* (116) suggest that HRT-induced HR and resistance to TCV are independent of *NPR1*.

ROS, like hydrogen peroxide (H_2O_2), play an important role in the *N*-mediated resistance to TMV in tobacco (117, 118). In mammalian systems, ROS play a key role in immune responses. ROS production in neutrophils is mediated by NADPH oxidase, which is composed of gp91, p22, p47, p67, and two subunits of cytochrome b_{558} (119). In plants a gp91 homologue has been cloned from rice (120) and *Arabidopsis* (121). However, other components of the NADPH oxidase complex have not been identified in plants. Therefore, the role of plant gp91 in generating ROS during plant defence requires further investigation.

SA and NO may potentiate ROS production during plant defence response (122). An SA-binding protein (SABP) that shows high affinity to SA encodes a catalase, and is involved in breaking down H_2O_2 into O_2 and H_2O. Therefore, Chen *et al.* (118) speculated that the SA might increase ROS by inhibiting the breakdown of H_2O_2. However, transgenic tobacco plants expressing reduced levels of catalase fail to show increased levels of SA and expression of PR-1 (123, 124). Furthermore, H_2O_2 fails to induce PR-1 in *nahG* tobacco plants (125). Therefore, the exact role of ROS in *N*-mediated TMV resistance is still unclear.

During *N*-mediated resistance to TMV, two mitogen activated protein kinases (MAPKs), SIPK (SA-induced protein kinase) (126), and WIPK (wound-induced protein kinase) (127), are induced (128). The MAPK genes are also induced by the *Cf-9* fungal-resistance pathway (129), SA treatment, mechanical stress, and non-host elicitors (130). In *N*-TMV-mediated signalling, the WIPK transcript level increases and the protein undergoes post-translational phosphorylation (128). Furthermore, analysis of *nahG::NN* transgenic plants suggests that WIPK activation is independent of SA. Together, these results suggest that MAPKs play an important role in *R*-gene-mediated resistance to phytopathogens. However, the exact biological role of these MAPKs in plant defence remains to be determined.

6. Future approaches to host–virus interactions and conclusions

Traditional approaches to understand events that occur during the onset of viral infection have centred on study of viral proteins and their roles during viral replication and movement. Recently, the host side of the interaction has gained more attention and will continue to be an expanding area of study. Genetically tractable model hosts such as *Arabidopsis* allow forward genetics to be used to identify host mutants with altered infection phenotypes. Altered susceptibility mutants have and will be found for a variety of plant viruses that will enable host genes involved in viral defence and susceptibility to be identified. The limitation to forward genetics approaches will be development of robust genetic screens to identify the altered susceptibility mutants. In addition to forward genetics, *Arabidopsis* is an ideal model system for a variety of complementary approaches including genomics and reverse genetics. The field of genomics will contribute tremendously to the study of host–virus interactions. Transcriptional profiling techniques such as cDNA microarray (131) and GeneChip® oligoarrays (132) will allow studies on how the expression of every host gene varies in response to compatible and incompatible viruses. The results from these microarray experiments will open up a variety of avenues for further study. In particular, candidate genes from transcriptional profiling experiments or yeast two-hybrid screens can be knocked out using transposon or T-DNA collections available in *Arabidopsis*. The use of these complementary approaches, known collectively as functional genomics, provides the power to examine the specific role of many host genes involved in virus susceptibility and defence.

Acknowledgements

The authors thank Yu-Ming Hou for critical reading of the manuscript and Vicki B. Vance for providing a manuscript prior to publication. S.A.W. is supported by the Novartis Agricultural Discovery Institute, Inc. S.P.D.K. is supported by National Science Foundation DBI-0077510.

References

1. Ploegh, H. L. (1998) Viral strategies of immune evasion. *Science*, **280**, 248.
2. Vousden, K. H. (1995) Regulation of the cell cycle by viral oncoproteins. *Semin. Cancer Biol.*, **6**, 109.
3. Gillet, G. and Brun, G. (1996) Viral inhibition of apotosis. *Trends Microbiol.*, **4**, 312.
4. Roulston, A., Marcellus, R. C. and Branton, P. E. (1999) Viruses and apoptosis. *Annu. Rev. Microbiol.*, **53**, 577.
5. Carrington, J. C. and Whitham, S. A. (1998) Viral invasion and host defense: strategies and counter-strategies. *Curr. Opin. Plant Biol.*, **1**, 336.
6. Gutierrez, C. (2000) DNA replication and cell cycle in plants: learning from geminiviruses. *Embo J.*, **19**, 792.
7. Weinberg, R. A. (1995) The retinoblastoma protein and cell cycle control. *Cell*, **81**, 323.
8. Xie, Q., Suarez-Lopez, P. and Gutierrez, C. (1995) Identification and analysis of a retinoblastoma binding motif in the replication protein of a plant DNA virus: requirement for efficient viral DNA replication. *EMBO J.*, **14**, 4073.
9. Horvath, G. V., Pettko-Szandtner, A., Nikovics, K., Bilgin, M., Boulton, M., Davies, J. W., Gutierrez, C. and Dudits, D. (1998) Prediction of functional regions of the maize streak

virus replication-associated proteins by protein-protein interaction analysis. *Plant Mol. Biol.*, **38**, 699.

10. Liu, L., Saunders, K., Thomas, C. L., Davies, J. W. and Stanley, J. (1999) Bean yellow dwarf virus RepA, but not rep, binds to maize retinoblastoma protein, and the virus tolerates mutations in the consensus binding motif. *Virology*, **256**, 270.

11. Wanitchakorn, R., Hafner, G. J., Harding, R. M. and Dale, J. L. (2000) Functional analysis of proteins encoded by banana bunchy top virus DNA-4 to -6. *J. Gen. Virol.*, **81**, 299.

12. Sano, Y., Wada, M., Hashimoto, Y., Matsumoto, T. and Kojima, M. (1998) Sequences of ten circular ssDNA components associated with the milk vetch dwarf virus genome. *J. Gen. Virol.*, **79**, 3111.

13. Aronson, M. N., Meyer, A. D., Gyorgyey, J., Katul, L., Vetten, H. J., Gronenborn, B. and Timchenko, T. (2000) Clink, a nanovirus-encoded protein, binds both pRB and SKP1. *J. Virol.*, **74**, 2967.

14. Xie, Q., Sanz-Burgos, A. P., Hannon, G. J. and Gutierrez, C. (1996) Plant cells contain a novel member of the retinoblastoma family of growth regulatory proteins. *EMBO J.*, **15**, 4900.

15. Huntley, R., Healy, S., Freeman, D., Lavender, P., de Jager, S., Greenwood, J., Makker, J., Walker, E., Jackman, M., Xie, Q., Bannister, A. J., Kouzarides, T., Gutierrez, C., Doonan, J. H. and Murray, J. A. (1998) The maize retinoblastoma protein homologue ZmRb-1 is regulated during leaf development and displays conserved interactions with G1/S regulators and plant cyclin D (CycD) proteins. *Plant Mol. Biol.*, **37**, 155.

16. Grafi, G., Burnett, R. J., Helentjaris, T., Larkins, B. A., DeCaprio, J. A., Sellers, W. R. and Kaelin Jr, W. G. (1996) A maize cDNA encoding a member of the retinoblastoma protein family: involvement in endoreduplication. *Proc. Natl Acad. Sci. USA*, **93**, 8962.

17. Ach, R. A., Durfee, T., Miller, A. B., Taranto, P., Hanley-Bowdoin, L., Zambryski, P. C. and Gruissem, W. (1997) RRB1 and RRB2 encode maize retinoblastoma-related proteins that interact with a plant D-type cyclin and geminivirus replication protein. *Mol. Cell. Biol.*, **17**, 5077.

18. Nakagami, H., Sekine, M., Murakami, H. and Shinmyo, A. (1999) Tobacco retinoblastoma-related protein phosphorylated by a distinct cyclin-dependent kinase complex with Cdc2/cyclin D in vitro. *Plant J.*, **18**, 243.

19. Kong, L. J. Orozco, B. M., Roe, J. L., Nagar, S., Ou, S., Feiler, H. S., Durfee, T., Miller, A. B., Gruissem, W., Robertson, D. and Hanley-Bowdoin, L. (2000) A geminivirus replication protein interacts with the retinoblastoma protein through a novel domain to determine symptoms and tissue specificity of infection in plants. *EMBO J.*, **19**, 3485.

20. Soni, R., Carmichael, J. P., Shah, Z. H. and Murray, J. A. (1995) A family of cyclin D homologs from plants differentially controlled by growth regulators and containing the conserved retinoblastoma protein interaction motif. *Plant Cell*, **7**, 85.

21. Ramirez-Parra, E., Xie, Q., Boniotti, M. B. and Gutierrez, C. (1999) The cloning of plant E2F, a retinoblastoma-binding protein, reveals unique and conserved features with animal G(1)/S regulators. *Nucleic Acids Res.*, **27**, 3527.

22. Sekine, M., Ito, M., Uemukai, K., Maeda, Y., Nakagami, H. and Shinmyo, A. (1999) Isolation and characterization of the E2F-like gene in plants. *FEBS Lett.*, **460**, 117.

23. Nagar, S., Pedersen, T. J., Carrick, K. M., Hanley-Bowdoin, L. and Robertson, D. (1995) A geminivirus induces expression of a host DNA synthesis protein in terminally differentiated plant cells. *Plant Cell*, **7**, 705.

24. Lee, J. O., Russo, A. A. and Pavletich, N. P. (1998) Structure of the retinoblastoma tumour-suppressor pocket domain bound to a peptide from HPV E7. *Nature*, **391**, 859.

25. Dick, F. A., Sailhamer, E. and Dyson, N. J. (2000) Mutagenesis of the pRB pocket reveals that cell cycle arrest functions are separable from binding to viral oncoproteins. *Mol. Cell. Biol.*, **20**, 3715.

26. Kooter, J. M., Matzke, M. A. and Meyer, P. (1999) Listening to the silent genes: transgene silencing, gene regulation and pathogen control. *Trends Plant Sci.*, **4**, 340.

27. Baulcombe, D. C. (1999) Fast forward genetics based on virus-induced gene silencing. *Curr. Opin. Plant Biol.*, **2**, 109.

28. Waterhouse, P. M., Smith, N. A. and Wang, M. B. (1999) Virus resistance and gene silencing: killing the messenger. *Trends Plant Sci.*, **4**, 452.

29. Vaucheret, H., Beclin, C., Elmayan, T., Feuerbach, F., Godon, C., Morel, J. B., Mourrain, P., Palauqui, J.C. and Vernhettes, S. (1998) Transgene-induced gene silencing in plants. *Plant J.*, **16**, 651.

30. Wassenegger, M. and Pelissier, T. (1998) A model for RNA-mediated gene silencing in higher plants. *Plant Mol. Biol.*, **37**, 349.

31. Grant, S. R. (1999) Dissecting the mechanisms of posttranscriptional gene silencing: divide and conquer. *Cell*, **96**, 303.

32. Ratcliff, F., Harrison, B. D. and Baulcombe, D. C. (1997) A similarity between viral defense and gene silencing in plants. *Science*, **276**, 1558.

33. Al-Kaff, N. S., Covey, S. N., Kreike, M. M., Page, A. M., Pinder, R. and Dale, P. J. (1998) Transcriptional and posttranscriptional plant gene silencing in response to a pathogen. *Science*, **279**, 2113.

34. Ratcliff, F. G., MacFarlane, S. A. and Baulcombe, D. C. (1999) Gene silencing without DNA. RNA-mediated cross-protection between viruses. *Plant Cell*, **11**, 1207.

35. Kasschau, K. D. and Carrington, J. C. (1998) A counter-defensive strategy of plant viruses: Suppression of posttranscriptional gene silencing. *Cell*, **95**, 461.

36. Brigneti, G., Vionnet, O., Li, W.-X., Ji, L.-H., Ding, S.-W. and Baulcombe, D. C. (1998) Viral pathogenicity determinants are suppressors of transgene silencing in *Nicotiana benthamiana*. *EMBO J.*, **17**, 6739.

37. Anandalakshmi, R., Pruss, G. J., Ge, X., Marathe, R., Mallory, A. C., Smith, T. H. and Vance, V. B. (1998) A viral suppressor of gene silencing in plants. *Proc. Natl Acad. Sci. USA*, **95**, 13079.

38. Voinnet, O., Pinto, Y. M. and Baulcombe, D. C. (1999) Suppression of gene silencing: a general strategy used by diverse DNA and RNA viruses of plants. *Proc. Natl Acad. Sci. USA*, **96**, 14147.

39. Beclin, C., Berthome, R., Palauqui, J. C., Tepfer, M. and Vaucheret, H. (1998) Infection of tobacco or *Arabidopsis* plants by CMV counteracts systemic post-transcriptional silencing of nonviral (trans)genes. *Virology*, **252**, 313.

40. Napoli, C., Lemieux, C. and Jorgensen, R. (1990) Introduction of a chimeric chalcone synthase gene into petunia results in reversible co-suppression of homologous genes *in trans*. *Plant Cell*, **2**, 279.

41. Lindbo, J. A., Silva-Rosales, L., Proebsting, W. M. and Dougherty, W. G. (1993) Induction of a highly specific antiviral state in transgenic plants: implications for regulation of gene expression and virus resistance. *Plant Cell*, **5**, 1749.

42. Hamilton, A. J. and Baulcombe, D. C. (1999) A species of small antisense RNA in post-transcriptional gene silencing in plants. *Science*, **286**, 950.

43. Zamore, P. D., Tuschl, T., Sharp, P. A. and Bartel, D. P. (2000) RNAi: double-stranded RNA directs the ATP-dependent cleavage of mRNA at 21 to 23 nucleotide intervals. *Cell*, **101**, 25.

44. Metzlaff, M., O'Dell, M., Cluster, P. D. and Flavell, R. B. (1997) RNA-mediated RNA degradation and chalcone synthase A silencing in petunia. *Cell*, **88**, 845.

45. Mourrain, P., Beclin, C., Elmayan, T., Feuerbach, F., Godon, C., Morel, J. B., Jouette, D., Lacombe, A. M., Nikic, S., Picault, N., Remoue, K., Sanial, M., Vo, T. A. and Vaucheret, H. (2000) *Arabidopsis* SGS2 and SGS3 genes are required for posttranscriptional gene silencing and natural virus resistance. *Cell*, **101**, 533.

46. Dalmay, T., Hamilton, A., Rudd, S., Angell, S. and Baulcombe, D. C. (2000) An RNA-dependent RNA polymerase gene in *Arabidopsis* is required for posttranscriptional gene silencing mediated by a transgene but not by a virus. *Cell*, **101**, 543.

47. Smardon, A., Spoerke, J. M., Stacey, S. C., Klein, M. E., Mackin, N. and Maine, E. M. (2000) EGO-1 is related to RNA-directed RNA polymerase and functions in germ-line development and RNA interference in *C. elegans. Curr. Biol.*, **10**, 169.

48. Cogoni, C. and Macino, G. (1999) Gene silencing in *Neurospora crassa* requires a protein homologous to RNA-dependent RNA polymerase. *Nature*, **399**, 166.

49. Stam, M., Viterbo, A., Mol, J. N. and Kooter, J. M. (1998) Position-dependent methylation and transcriptional silencing of transgenes in inverted T-DNA repeats: implications for posttranscriptional silencing of homologous host genes in plants. *Mol. Cell. Biol.*, **18**, 6165.

50. Waterhouse, P. M., Graham, M. W. and Wang, M. B. (1998) Virus resistance and gene silencing in plants can be induced by simultaneous expression of sense and antisense RNA. *Proc. Natl Acad. Sci. USA*, **95**, 13959.

51. Fire, A., Xu, S., Montgomery, M. K., Kostos, S. A., Driver, S. E. and Mello, C. C. (1998) Potent and specific genetic interference by double-stranded RNA in *Caenorhabditis elegans. Nature*, **391**, 806.

52. Chuang, C. F. and Meyerowitz, E. M. (2000) Specific and heritable genetic interference by double-stranded RNA in *Arabidopsis thaliana. Proc. Natl Acad. Sci. USA*, **97**, 4985.

53. Smith, N. A., Singh, S. P., Wang, M. B., Stoutjesdijk, P. A., Green, A. G. and Waterhouse, P. M. (2000) Total silencing by intron-spliced hairpin RNAs. *Nature*, **407**, 319.

54. Goodwin, J., Chapman, K., Swaney, S., Parks, T. D., Wernsman, E. A. and Dougherty, W. G. (1996) Genetic and biochemical dissection of transgenic RNA-mediated virus resistance. *Plant Cell*, **8**, 95.

55. Elmayan, T., Balzergue, S., Beon, F., Bourdon, V., Daubremet, J., Guenet, Y., Mourrain, P., Palauqui, J. C., Vernhettes, S., Vialle, T., Wostrikoff, K. and Vaucheret, H. (1998) *Arabidopsis* mutants impaired in cosuppression. *Plant Cell*, **10**, 1747.

56. Vance, V. B., Berger, P. H., Carrington, J. C., Hunt, A. G. and Shi, X. M. (1995) 5′ Proximal potyviral sequences mediate potato virus X/potyviral synergistic disease in transgenic tobacco. *Virology*, **206**, 583.

57. Marathe, R., Smith, T. H., Anandalakshmi, R., Bowman, L. H., Fagard, M., Mourrain, P., Vaucheret, H. and Vance, V. B. (2000) Plant viral suppressors of post-transcriptional silencing do not suppress transcriptional silencing. *Plant J.*, **22**, 51.

58. Scholthof, H. B., Scholthof, K.-B. G., Kikkert, M. and Jackson, A. O. (1995) Tomato bushy stunt virus spread is regulated by two nested genes that function in cell-to-cell movement and host-dependent systemic invasion. *Virology*, **213**, 425.

59. Kasschau, K. D., Cronin, S. and Carrington, J. C. (1997) Genome amplification and long-distance movement functions associated with the central domain of tobacco etch potyvirus helper component-proteinase. *Virology*, **228**, 251.

60. Ding, S.-W., Li, W.-X. and Symons, R. H. (1995) A novel naturally occuring hybrid gene encoded by a plant RNA virus facilitates long distance virus movement. *EMBO J.*, **14**, 5762.

61. Crawford, K. M. and Zambryski, P. C. (1999) Plasmodesmata signaling: many roles, sophisticated statutes. *Curr. Opin. Plant Biol.*, **2**, 382.
62. Lucas, W. J. and Wolf, S. (1999) Connections between virus movement, macromolecular signaling and assimilate allocation. *Curr. Opin. Plant Biol.*, **2**, 192.
63. Anandalakshmi, R., Marathe, R., Ge, X., Herr, J. M., Mau, C., Mallory, A., Pruss, G., Bowman, L. and Vance, V. B. (2000) A calmodulin-related protein suppresses post-transcriptional gene silencing in plants. *Science*, **290**, 142.
64. Ham, B. K., Lee, T. H., You, J. S., Nam, Y. W., Kim, J. K. and Paek, K. H. (1999) Isolation of a putative tobacco host factor interacting with cucumber mosaic virus-encoded 2b protein by yeast two-hybrid screening. *Mol. Cells*, **9**, 548.
65. Rao, A. L. N. (1999) Molecular basis of symptomatology. In Mandahar, C. L. (ed.), *Molecular Biology of Plant Viruses*, pp. 201. Kluwer Academic Publishers, Norwell, MA, USA.
66. Rothnie, H. M., Chapdelaine, Y. and Hohn, T. (1994) Pararetroviruses and retroviruses: a comparative review of viral structure and gene expression strategies. *Adv. Virus Res.*, **44**, 1.
67. Cecchini, E., Gong, Z., Geri, C., Covey, S. N. and Milner, J. J. (1997) Transgenic *Arabidopsis* lines expressing gene VI from cauliflower mosaic virus variants exhibit a range of symptom-like phenotypes and accumulate inclusion bodies. *Mol. Plant–Microbe Interact.*, **10**, 1094.
68. Zijlstra, C., Scharer-Hernandez, N., Gal, S. and Hohn, T. (1996) *Arabidopsis thaliana* expressing the cauliflower mosaic virus ORF VI transgene has a late flowering phenotype. *Virus Genes*, **13**, 5.
69. Geri, C., Cecchini, E., Giannakou, M. E., Covey, S. N. and Milner, J. J. (1999) Altered patterns of gene expression in *Arabidopsis* elicited by cauliflower mosaic virus (CaMV) infection and by a CaMV gene VI transgene. *Mol. Plant–Microbe Interact.*, **12**, 377.
70. Escaler, M., Aranda, M. A., Thomas, C. L. and Maule, A. J. (2000) Pea embryonic tissues show common responses to the replication of a wide range of viruses. *Virology*, **267**, 318.
71. Aranda, M. A., Escaler, M., Wang, D. and Maule, A. J. (1996) Induction of HSP70 and polyubiquitin expression associated with plant virus replication. *Proc. Natl Acad. Sci. USA*, **93**, 15289.
72. Wang, D. and Maule, A. J. (1995) Inhibition of host gene expression associated with plant virus replication. *Science*, **267**, 229.
73. Aranda, M. A., Escaler, M., Thomas, C. L. and Maule, A. J. (1999) A heat shock transcription factor in pea is differentially controlled by heat and virus replication. *Plant J.*, **20**, 153.
74. Flor, H. (1971) Current status of the gene-for-gene concept. *Annu. Rev. Phytopathol.*, **9**, 275.
75. Lamb, C. J. (1994) Plant disease resistance genes in signal perception and transduction. *Cell*, **76**, 419.
76. Ryals, J. A., Neuenschwander, U. H., Willits, M. G., Molina, A., Steiner, H.-Y. and Hunt, M. D. (1996) Systemic acquired resistance. *Plant Cell*, **8**, 1809.
77. Whitham, S., Dinesh-Kumar, S. P., Choi, D., Hehl, R., Corr, C. and Baker, B. (1994) The product of the tobacco mosaic virus resistance gene N: similarity to toll and the interleukin-1 receptor. *Cell*, **78**, 1101.
78. Bendahmane, A., Kanyuka, K. and Baulcombe, D. C. (1999) The Rx gene from potato controls separate virus resistance and cell death responses. *Plant Cell*, **11**, 781.
79. Cooley, M. B., Pathirana, S., Wu, H.-J., Kachroo, P. and Klessig, D. F. (2000) Members of the *Arabidopsis* HRT/RPP8 family of resistance genes confer resistance to both viral and oomycete pathogens. *Plant Cell*, **12**, 663.

80. Baker, B., Zambryski, P., Staskawicz, B. and Dinesh-Kumar, S. P. (1997) Signaling in plant-microbe interactions. *Science*, **276**, 726.
81. van der Biezen, E. A. and Jones, J. D. (1998) The NB-ARC domain: a novel signalling motif shared by plant resistance gene products and regulators of cell death in animals. *Curr. Biol.*, **8**, R226.
82. Arvind, L., Dixit, V. M. and Koonin, E. V. (1999) The domains of death: evolution of the apoptosis machinery. *Trends Biochem. Sci.*, **24**, 47.
83. Dinesh-Kumar, S. P., Tham, W.-H. and Baker, B. (2000) The structure-function analysis of the tobacco mosaic resistance gene, N. *Proc. Natl Acad. Sci. USA*, 97, 14789.
84. Heguy, A., Baldari, C. T., Macchia, G., Telford, J. L. and Melli, M. (1992) Amino acids conserved in interleukin-1 receptors (IL-1Rs) and the *Drosophila* Toll protein are essential for IL-1R signal transduction. *J. Biol. Chem.*, **267**, 2605.
85. Schneider, D. S., Hudson, K. L., Lin, T. Y. and Anderson, K. V. (1991) Dominant and recessive mutations define functional domains of Toll, a transmembrane protein required for dorsal-ventral polarity in the *Drosophila* embryo. *Genes Dev.*, **5**, 797.
86. Ellis, J., Dodds, P. and Pryor, T. (2000) Structure, function and evolution of plant disease resistance genes. *Curr. Opin. Plant Biol.*, **3**, 278.
87. Dinesh-Kumar, S.p. and Baker, B. (2000) Alternatively spliced N resistance gene transcripts: their possible role in tobacco mosaic virus resistance. *Proc. Natl Acad. Sci. USA*, **97**, 1908.
88. Bendahmane, A., Kohn, B. A., Dedi, C. and Baulcombe, D. C. (1995) The coat protein of potato virus X is a strain-specific elicitor of Rx1-mediated virus resistance in potato. *Plant J.*, **8**, 933.
89. Padgett, H. S. and Beachy, R. N. (1993) Analysis of a tobacco mosaic virus strain capable of overcoming N gene-mediated resistance. *Plant Cell*, **5**, 577.
90. Padgett, H. S., Watanabe, Y. and Beachy, R. N. (1997) Identification of the TMV replicase sequence that activates the *N* gene-mediated hypersensitive response. *Mol. Plant–Microbe Interact.*, **10**, 709.
91. Abbink, T. E. M., Tjernberg, P. A., Bol, J. F. and Linthorst, J. M. (1998) Tobacco mosaic virus helicase domain induces necrosis in N gene-carrying tobacco in the absence of virus replication. *Mol. Plant–Microbe Interact.*, **11**, 1242.
92. Erickson, F. L., Holzberg, S., Calderon-Urrea, A., Handley, V., Axtell, M., Corr, C. and Baker, B. (1999) The helicase domain of the TMV replicase proteins induces the *N*-mediated defense response in tobacco. *Plant J.*, **18**, 67.
93. Scofield, S., Tobias, C., Rathjen, J., Chang, J., Lavelle, T., Michelmore, R. and Staskawicz, B. (1996) Molecular basis of gene-for-gene recognition in bacterial speck disease of tomato. *Science*, **274**, 2063.
94. Tang, X., Frederick, R. D., Zhou, J., Halterman, D. A., Jia, Y. and Martin, G. B. (1996) Initiation of plant disease resistance by physical interaction of avrPto and Pto kinase. *Science*, **274**, 2060.
95. Jia, Y., McAdams, S. A., Bryan, G. T., Hershey, H. P. and Valent, B. (2000) Direct interaction of resistance gene and avirulence gene products confers rice blast resistance. *EMBO J.*, **19**, 4004.
96. Dangl, J. L. (1995) Piece de resistance: novel classes of plant disease resistance genes. *Cell*, **80**, 363.
97. Staskawicz, B. J., Ausubel, F. M., Baker, B. J., Ellis, J. G. and Jones, J. D. (1995) Molecular genetics of plant disease resistance. *Science*, **268**, 661.

98. Boyes, D. C., McDowell, J. M. and Dangl, J. L. (1996) Plant pathology: many roads lead to resistance. *Curr Biol*, **6**, 634.

99. Innes, R. W. (1998) Genetic dissection of *R* gene signal transduction pathways. *Curr. Opin. Plant Biol.*, **1**, 229.

100. Century, K. S., Holub, E. B. and Staskawicz, B. J. (1995) NDR1, a locus of *Arabidopsis thaliana* that is required for disease resistance to both a bacterial and fungal pathogen. *Proc. Natl Acad. Sci. USA*, **92**, 6597.

101. Parker, J., Holub, E., Frost, L., Falk, A., Gunn, N. and Daniels, M. (1996) Characterization of *eds1*, a mutation in *Arabidopsis* suppressing resistance to *Peronospora parasitica* specified by several different *RPP* genes. *Plant Cell*, **8**, 2033.

102. Aarts, N., Metz, M., Holub, E., Staskawicz, B. J., Daniels, M. J. and Parker, J. E. (1998) Different requirements for EDS1 and NDR1 by disease resistance genes define at least two R gene-mediated signaling pathways in *Arabidopsis*. *Proc. Natl Acad. Sci. USA*, **95**, 10306.

103. McDowell, J. M., Cuzick, A., Can, C., Beynon, J., Dangl, J. L. and Holub, E. B. (2000) Downy mildew (*Peronospora parasitica*) resistance genes in *Arabidopsis* vary in functional requirements for NDR1, EDS1, NPR1 and salicylic acid accumulation. *Plant J.*, **22**, 523.

104. Century, K. S., Shapiro, A. D., Repetti, P. P., Dahlbeck, D., Holub, E. and Staskawicz, B. J. (1997) NDR1, a pathogen-induced component required for *Arabidopsis* disease resistance. *Science*, **278**, 1963.

105. Falk, A., Feys, B. J., Frost, L. N., Jones, J. D. G., Daniels, M. J. and Parker, J. E. (1999) EDS1, an essential component of R gene-mediated disease resistance in *Arabidopsis* has homology to eukaryotic lipases. Proc. Natl Acad. Sci. USA, **96**, 3292.

106. Delaney, T., Uknes, S., Vernooij, B., Friedrich, L., Weymann, K., Negrotto, D., Gaffney, T., Güt-Rella, M., Kessmann, H., Ward, E. and Ryals, J. (1994) A central role of salicylic acid in plant disease resistance. *Science*, **266**, 1247.

107. Dempsey, D. A., Pathirana, M. S., Wobbe, K. K. and Klessig, D. F. (1997) Identification of an *Arabidopsis* locus required for resistance to turnip crinkle virus. *Plant J.*, **11**, 301.

108. Malamy, J., Carr, J. P., Klessig, D. F. and Raskin, I. (1990) Salicylic acid: a likely endogenous signal in the resistance response of tobacco to viral infection. *Science*, **250**, 1002.

109. Cao, H., Glazebrook, J., Clarke, J. D., Volko, S. and Dong, X. (1997) The *Arabidopsis* NPR1 gene that controls systemic acquired resistance encodes a novel protein containing ankyrin repeats. *Cell*, **88**, 57.

110. Ryals, J., Weymann, K., Lawton, K., Friedrich, L., Ellis, D., Steiner, H. Y., Johnson, J., Delaney, T. P., Jesse, T., Vos, P. and Uknes, S. (1997) The *Arabidopsis* NIM1 protein shows homology to the mammalian transcription factor inhibitor I kappa B. *Plant Cell*, **9**.

111. Shah, J., Tsui, F. and Klessig, D. (1997) Characterization of a salicylic acid-insensitive mutant (sai1) of *Arabidopsis thaliana*, identified in a selective screen utilizing the SA-inducible expression of *tms2* gene. *Mol. Plant–Microbe Interact.*, **10**, 69.

112. Cao, H., Bowling, S. A., Gordon, A. S. and Dong, X. (1994) Characterization of an *Arabidopsis* mutant that is nonresponsive to inducers of systemic acquired resistance. *Plant Cell*, **6**, 1583.

113. Delaney, T., Friedrich, L. and Ryals, J. (1995) *Arabidopsis* signal transduction mutant defective in chemically and biologically induced disease resistance. *Proc. Natl Acad. Sci. USA*, **92**, 6602.

114. Glazebrook, J., Zook, M., Mert, F., Kagan, I., Rogers, E. E., Crute, I. R., Holub, E. B., Hammerschmidt, R. and Ausubel, F. M. (1997) Phytoalexin-deficient mutants of *Arabidopsis* reveal that PAD4 encodes a regulatory factor and that four PAD genes contribute to downy mildew resistance. *Genetics*, **146**, 381.

115. Cao, H., Li, X. and Dong, X. (1998) Generation of broad-spectrum disease resistance by overexpression of an essential regulatory gene in systemic acquired resistance. *Proc. Natl Acad. Sci. USA*, **95**, 6531.

116. Kachroo, P., Yoshioka, K., Shah, J., Dooner, H. K. and Klessig, D. F. (2000) Resistance to turnip crinkle virus in *Arabidopsis* is regulated by two host genes and is salicylic acid dependent but NPR1, ethylene, and jasmonate independent. *Plant Cell*, **12**, 677.

117. Doke, N. and Ohashi, Y. (1988) Involvement of an O_2-generating system in the induction of necrotic lesions on tobacco leaves infected with tobacco mosaic virus. *Physiol. Mol. Plant Pathol.*, **32**, 163.

118. Chen, Z., Silva, H. and Klessig, D. F. (1993) Active oxygen species in the induction of plant systemic acquired resistance by salicylic acid. *Science*, **262**, 1883.

119. Segal, A. W. and Abo, A. (1993) The biochemical basis of the NADPH oxidase of phagocytes. *Trends Biochem. Sci.*, **18**, 43.

120. Groom, Q. J., Torres, M. A., Fordham-Skelton, A. P., Hammond-Kosack, K. E., Robinson, N. J. and Jones, J. D. G. (1996) rbohA, a rice homologue of the mammalian gp91phox respiratory burst oxidase gene. *Plant J.*, **10**, 515.

121. Torres, M. A., Onouchi, H., Hamada, S., Machida, C., Hammond-Kosack, K. E. and Jones, J. D. (1998) Six *Arabidopsis thaliana* homologues of the human respiratory burst oxidase (gp91phox). *Plant J*, **14**, 365.

122. McDowell, J. M. and Dangl, J. L. (2000) Signal transduction in the plant immune response. *Trends Biochem. Sci.*, **25**, 79.

123. Chamnongpol, S., Willekens, H., Langebartels, C., Van Montagu, M., Inze, D. and Van Camp, W. (1996) Transgenic tobacco with a reduced catalase activity develops necrotic lesions and induces pathogenesis related expression under high light. *Plant J.*, **10**, 491.

124. Takahashi, H., Chen, Z., Du, H., Liu, Y. and Klessig, D. F. (1997) Development of necrosis and activation of disease resistance in transgenic tobacco plants with severely reduced catalase levels. *Plant J.*, **11**, 993.

125. Neuenschwander, U., Vernooij, B., Friedrich, L., Uknes, S., Kessmann, H. and Ryals, J. (1995) Is hydrogen peroxide a second messenger of salicylic acid in systemic acquired resistance? *Plant J.*, **8**, 227.

126. Zhang, S. and Klessig, D. F. (1997) Salicylic acid activates a 48-kDa MAP kinase in tobacco. *Plant Cell*, **9**, 809.

127. Seo, S., Okamoto, M., Seto, H., Ishizuka, K., Sano, H. and Ohashi, Y. (1995) Tobacco MAP kinase: a possible mediator in wound signal transduction pathways. *Science*, **270**, 1988.

128. Zhang, S. and Klessig, D. F. (1998) Resistance gene N-mediated de novo synthesis and activation of a tobacco mitogen-activated protein kinase by tobacco mosaic virus infection. *Proc. Natl Acad. Sci. USA*, **95**, 7433.

129. Romeis, T., Piedras, P., Zhang, S., Klessig, D. F., Hirt, H. and Jones, J. D. G. (1999) Rapid, Avr9- and Cf9-dependent, activation of MAP kinase in tobacco cell cultures and leaves: convergence in resistance gene, elicitor, wound and salicylate responses. *Plant Cell*, **11**, 273.

130. Scheel, D. (1998) Resistance response physiology and signal transduction. *Curr. Opin. Plant Biol.*, **1**, 305.

131. Schena, M., Shalon, D., Davis, R. W. and Brown, P. O. (1995) Quantitative monitoring of gene expression patterns with a complementary DNA microarray. *Science*, **270**, 467.

132. Lockhart, D., Dong, H., Byrne, M. C., Follettie, M. T., Gallo, M. V., Chee, M. S., Mittmann, M., Wang, C., Kobayashi, M., Horton, H. and Brown, E. L. (1996) Expression monitoring by hybridization to high-density oligonucleotide arrays. *Nat. Biotechnol.*, **14**, 1675.

11 | LCO signalling in the interaction between rhizobia and legumes

ROSSANA MIRABELLA, HENK FRANSSEN AND TON BISSELING

1. Introduction

Bacteria belonging to the genera *Rhizobium*, *Bradyrhizobium*, *Azorhizobium*, *Mesorhizobium*, and *Synorhizobium*, collectively referred to as rhizobia, are able to invade the roots of their legume host plant. There they trigger the formation of a new organ, the root nodule in which nitrogen fixation occurs (1–6). In a nutshell, the formation of root nodules starts with the exchange of signals between the root and the bacteria, which is followed by the colonization of the root surface. Subsequently, the root hairs deform and curl and the bacteria invade the plant by newly formed infection threads (Plate 1). These threads grow towards the basis of the root hair and then they traverse cortical cells. Concomitantly, cortical cells are mitotically activated, by which a nodule primordium is formed. Infection threads grow towards the primordium and there the bacteria, surrounded by a plant membrane, are released into the cytoplasm of the host cell. Subsequently, the primordium develops into a nodule and the bacteria start to reduce nitrogen into ammonia that can be utilized by the plant. During each step of nodule formation, nodule-specific plant genes, the so-called nodulin genes, are activated. For example, *Enod12* is activated in the epidermis and *Enod40* in the root pericycle and both nodulin genes are also expressed in the nodule primordium.

During nodule development, the host and the rhizobia probably exchange several signals; however, two classes of signal molecules play a crucial role in the initiation of this symbiotic interaction (Fig. 1). The first are compounds excreted by the roots of the host, in general flavonoids; they induce the transcription of bacterial modulation genes (*nod* genes). One of these *nod* genes, *nodD*, is constitutively active and upon recognition of a certain flavonoid becomes a transcriptional activator of the other *nod* genes. The proteins encoded by these genes are involved in the biosynthesis or secretion of the second class of signal molecules, specific lipo-chito-oligosaccharides (LCOs), or so-called Nod factors (7, 8), which are perceived by the plant.

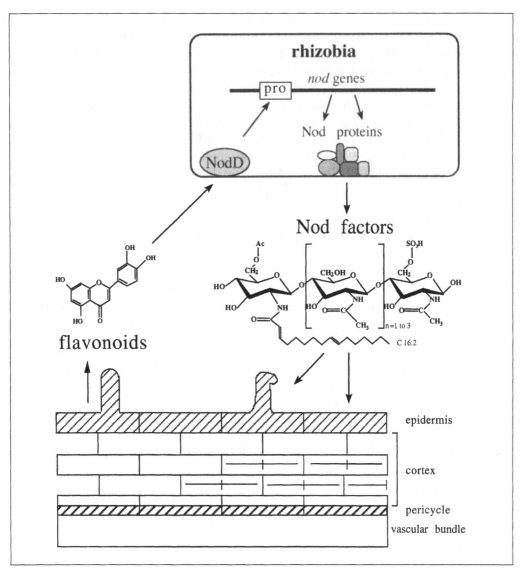

Fig. 1 Early events in the interaction between legumes and rhizobia (modified from 36). NodD, constitutively formed, is activated by specific flavonoids (an activator of *Sinorhizobium meliloti* NodD is luteolin) excreted by the plant and becomes a transcriptional activator of the other *nod* genes. The Nod proteins are involved in the biosynthesis and the secretion of Nod factor. In this figure the major *S. meliloti* Nod factor is depicted. Nod factors induce various responses in three root tissues, e.g. root-hair deformation and curling (epidermis), nodulin gene expression (pericycle), and cell division (cortex).

In Fig. 1, the structure of the major *Sinorhizobium meliloti* Nod factor is shown. The structure of Nod factors produced by most rhizobia has now been elucidated (9–12), showing that they have a generic LCO structure. In general, they consist of a β-1,4-linked *N*-acetyl-D-glucosamine backbone with four or five residues, although some

Nod factors have a backbone of three or six sugar units. The non-reducing terminal sugar moiety is substituted at the C2 position with a fatty acid that, in general, has a length of 16 or 18 carbons. Depending on the rhizobial species, additional substitutions are present on the terminal sugar residues. The major *S. meliloti* Nod factor (Fig. 1) contains a sulphuryl group at its reducing sugar residue. This substitution is essential for the biological activity on *Medicago* roots. Other rhizobia produce molecules with different substitutions at this position. For example, *Mesorhizobium loti* Nod factors can have an acetylated-fucose group. *Rhizobium leguminosarum* biovar *viciae* strains make, in general, molecules without a substitution, but if they have a *nodX* gene they produce Nod factors with an acetyl substitution at the reducing terminal sugar residue. For the interaction with certain pea lines this acetyl group is essential for infection thread growth (see Section 3.2). The major *S. meliloti* Nod factor (Fig. 1) has an acetyl substitution at its non-reducing terminal sugar; this substitution is important for the induction of the infection process, but is not essential for most other responses (see Section 3.2). Other rhizobial species can have different substitutions at this position (11, 12). The acyl chain of the major *S. meliloti* Nod factor has α,β unsaturated bonds. Nod factors with such α,β unsaturated acyl chains are especially made by rhizobia that interact with legumes belonging to the galegoid phylum (including *Medicago*, *Pisum*, and *Vicia*; 13). These specific acyl groups are especially important in the induction of the infection process (see Section 3.2).

Nod factors are involved in the induction of various responses in the root, such as deformation and curling of root hairs, infection thread formation, mitotic activation of cortical cells, and induction of gene expression in the epidermis, cortex (nodule primordia), and pericycle. However, whether the cortical and pericycle cells perceive Nod factors or whether the Nod factor responses are induced by secondary signals, e.g generated in the epidermis, is unclear. For these reasons, this chapter focuses on the mechanism of Nod factor perception and signalling active in the root epidermis. In the first part the morphological changes induced in the root hairs are summarized. Then, the biochemical and genetic studies concerning perception and transduction are reviewed and finally how future research could provide insight in the mechanisms by which the Nod factor signal can induce the morphological changes in root hairs is discussed.

2. Morphological changes induced in root hairs

2.1 Root-hair growth

The root epidermis is the plant tissue that is in direct contact with Nod-factor-secreting rhizobia, when they colonize the root. It also forms a cell layer that the rhizobia have to traverse to reach the nodule primordia, formed in the cortex. In most legumes the bacteria induce the formation of a sophisticated infection structure, the infection thread (Plate 1). This is a tube-like structure, the formation of which is strictly controlled by the host plant, and it exemplifies how rhizobia recruit and modify the cell biological machinery of the host to create an apparently novel structure. In order to appreciate the morphological responses induced by rhizobial Nod factors, some background information about root-hair growth is provided.

Root hairs elongate by tip growth, a process shared with pollen tubes, and fungal hyphae (14, 15). A common feature of tip growing cells is the presence of a high Ca^{2+} gradient at the apex. The deposition of cell wall and membrane material by Golgi vesicles is confined to the tip and this determines the localized growth. To maintain the delivery of vesicles at this site of the cell, growing root hairs have a polar organization of the cytoplasm, with an apical region at the tip containing a large number of Golgi vesicles; this region is devoid of large organelles and therefore is sometimes called an 'empty zone' (16–20).

The actin cytoskeleton has been shown to be essential for the maintenance of the polar cytoarchitecture in growing root hairs and in other tip growing plant cells such as pollen tubes (14, 21–23). The organization of the actin cytoskeleton in legume root hairs has been studied in most detail in *Vicia sativa* (21). Growing root hairs have thick axial actin bundles along the length of the hair. In the subapical region of the hair, the bundles flare out into thinner bundles. These subapical fine bundles have a net-axial orientation and are called FB-actin (fine bundles; 21). The apex of growing root hairs appears to be devoid of actin (21). In root hairs that are terminating growth, FB-actin has almost disappeared and the region devoid of actin at the tip is not present; in full grown root hairs the FB-actin is not present and the thick actin bundles loop through the tip (21). Cytochalasin D, which affects actin polymerization, causes a disappearance of the FB-actin and of the vesicle-rich region and as a result root hairs stop growing. This suggests that the FB-actin is essential for vesicle delivery at the tip and therefore tip growth of root hairs (21).

The role of the microtubular cytoskeleton during root-hair development is less clear, but, in contrast to the actin, it is not essential for tip growth, as is the case in other tip-growing plant cells such as pollen tubes (24–27). In root hairs, microtubules are predominantly located in the cortical cytoplasm next to the plasma membrane with a longitudinal or helical orientation (28–32), and with a random orientation in the root-hair tip. In addition, bundles of endoplasmatic microtubules connecting the nucleus to the root-hair tip are present in *Vicia hirsuta* (33) and growing *Medicago* root hairs (34, 35), whereas in full-grown root hairs they are no longer present (B. Sieber and A. M. C. Emons, personal communication). Therefore, the endoplasmatic microtubules may function in nuclear positioning and movement. Recent work by Bibikova (24) suggests that the CMTs (cortical microtubules) could determine the direction of root-hair growth. When *Arabidopsis* was grown in the presence of drugs (taxol and oryzalin, respectively) stabilizing or depolymerizing the CMTs, a loss of directionality in root-hair growth was observed, causing a waving of the root hairs as they elongated. However, the rate of growth was not affected. Thus CMTs are not required for tip growth but are involved in controlling the direction of growth (24).

Since the actin and microtubular skeleton are important for root-hair tip growth and for the direction of growth, respectively, it is very probable that both skeletons play an important role in the rhizobial Nod-factor-induced morphological responses.

2.2 Root-hair deformation

Purified Nod factors are sufficient to induce root-hair deformation at a concentration as low as 10^{-12} M (9, 10). In *Vicia* especially, the root hairs that are temiinating growth deform (36), whereas in *Medicago truncatula* as well as in *Lotus japonicus* younger growing root hairs do deform (35; G. van der Krogt and T. Bisseling, personal communication). Within a susceptible zone, the vast majority of root hairs respond to Nod-factor application with a swelling of the root-hair tip, followed by a new outgrowth from the swelling. In some species, e.g. *M. truncatula*, the growth direction of this new outgrowth is markedly different from the original growth direction of the responding hair by which the deformation has the appearance of a branch. The swelling is the result of isotropic growth, whereas the outgrowth/branch is the result of a reinitiation of polar growth (21, 36, 37). In fact, the outgrowth has all the typical features of tip-growing cells, e.g. calcium gradient at the tip (37), polar cytoarchitecture (21), actin organization typical of growing root hairs (21).

Since deformation/branching is caused by a modification of the tip-growth process, it is probable that Nod factors induce changes in the configuration of the actin skeleton. The first reports on the effect of Nod factors on the actin cytoskeleton are from Allen and colleagues (38, 39). They reported a fragmentation of the actin cytoskeleton in *Medicago sativa* root hairs within 15 minutes of Nod-factor application. Later, Cárdenas (40) showed a similar fragmentation in *Phaseolus vulgaris* root hairs (40). In *Vicia sativa* root hairs, no actin fragmentation was observed after Nod factor application (21, 41, 42). De Ruijter (41) reported an increase of the number of fine actin bundles in the subapical region of most root hairs within 3–15 minutes after Nod-factor application. This increase occurs in root hairs at all stages of development and also in those hairs that do not deform. At a later stage when the outgrowth is formed, the architecture of the actin cytoskeleton is the same as in growing root hairs and the formation of the outgrowth can be inhibited by cytochalasin D (21), showing that the actin skeleton is essential for root-hair deformation.

Although the observed changes in the actin configuration markedly vary in the different studies, it has become clear that the actin skeleton is one of the first targets of Nod-factor signalling in root hairs. Nod factors induce these rearrangements in the actin configuration within minutes, but the molecular mechanisms controlling these changes are unknown.

2.3 Root-hair curling and infection

Unlike root-hair deformation and branching, root-hair curling only occurs in a relatively small number of hairs upon inoculation with rhizobia. Nod factors are essential for the induction of curling but in most legume species curling is only induced when the bacteria are present (43). During curling a so-called shepherd's crook is formed (Plate 1), which is the result of a redirection of root-hair tip growth by more then 360° (19).

Curling is probably caused by a gradual and constant reorientation of tip growth (44, 45). During curling the bacteria become entrapped within the curl and there the plant cell wall is modified in a very local manner (46–48). At this site the plasma membrane invaginates and the host deposits new material around the site of infection. In this way a tube-like structure, the infection thread, is formed by which the bacteria enter the root.

Studies on cytoskeletal changes during curling and infection have focused on the microtubular skeleton. Since microtubules play an important role in controlling the direction of root-hair growth, it seems probable that this part of the cytoskeleton may play a pivotal role in the curling process. The microtubular cytoskeleton in root hairs markedly rearranges during curling, initiation of infection, and infection thread growth. Electron microscopic studies showed that microtubules are present at the site of infection thread initiation (19, 47, 49). Recently, Timmers (34) studied in more detail the rearrangements in *Medicago* root hairs during infection thread formation by immunocytochemistry. At the start of infection the nucleus migrates to the root-hair tip and the microtubules concentrate in the region between the nucleus and the hair tip. When curling is initiated, these endoplasmatic microtubules move to the area within the curl where the infection will initiate. This results in an asymmetric organization of the skeleton. When thread formation is initiated (24–48 hours post-inoculation), the microtubules are predominantly located within the curl at the site of infection thread initiation and they no longer connect the nucleus with the root-hair tip. Subsequently, the infection thread grows into the root hair and it is always preceded by the nucleus. At this stage the infection thread tip is surrounded by a dense network of microtubules connected with the nucleus; the complete body of the thread is covered with a longitudinal array of microtubules (34, 50). The described rearrangements in the microtubule cytoskeleton are in agreement with the idea that microtubules play a key role in controlling the direction of root-hair growth. However, the molecular mechanism by which Nod factors induce these changes still remains to be discovered.

3. Nod-factor perception

Some of the changes induced by Nod factor in the plant cytoskeleton occur very rapidly. Therefore it is very likely that the cytoskeleton is the target of Nod-factor-activated signalling. For this reason, the following paragraphs describe the mechanism involved in the perception and transduction of the Nod-factor signal.

When roots are treated with purified Nod factors (or inoculated with rhizobia), responses are induced in the majority of the epidermal cells within a susceptible zone. This is most apparent when the activation of early nodulin promoters is visualized by a reporter gene. For example, *Enod12-Gus* (β-glucuronidase; see Section 4.1) is activated in all epidermal cells, with or without a hair, in direct contact with Nod factors, in the zone where root hairs are formed and grow (51). The induction of responses in root epidermal cells appears to require a direct contact with Nod factors (51) and some of the responses occur within seconds (52, 53). Therefore, it seems probable that these cells respond in a cell-autonomous manner and it is probable that all epidermal cells (within a susceptible zone) have an LCO perception mechanism. Whether, in addition to intact LCOs, cleavage products could also have a signalling function can not be excluded. Legumes produce several enzymes that can degrade Nod factors (6, 11). However, it has not been demonstrated that Nod-factor cleavage products can induce epidermal responses. By contrast, it is possible that certain cleavage products might play a role in the induction of some responses in the inner layers of the roots. For example, chitin fragments are sufficient to induce cortical cell divisions (54).

The following paragraphs review the knowledge about the subcellular location of the perception process as well as the candidate molecules that could have a function in LCO perception.

3.1 Site of Nod-factor perception

The amphiphilic nature of Nod factors with their hydrophobic lipid tail and more hydrophilic sugar backbone suggests that Nod factors have a high affinity for membranes. Indeed, *in vitro* studies show that Nod factors rapidly insert into membranes, but are unable to flip-flop between membrane leaflets (55). This indicates that Nod factors are perceived by a receptor located in the plasma membrane. However, the behaviour of fluorescent Nod factors on roots show that perception might also occur at the cell wall and it can not be completely excluded that Nod factors are perceived in the cytoplasm. Using fluorescence correlation microscopy, it was shown that biologically active Nod factors, which were tagged with a fluorescent group, do accumulate in the cell wall of epidermal cells at a concentration 50-fold higher than in the medium that was applied to the roots (56). Further, it was shown that Nod factors are present at low quantities in the plasma membrane and a low level of fluorescence was even detected in the cytoplasm. In addition, another study indicates that Nod factor could be internalized (57), but it remains to be demonstrated that intact Nod factors are present in the cytoplasm.

The fluorescence correlation microscopy studies furthermore revealed that Nod factors become highly immobilized in the cell wall. This accumulation of Nod factors in the cell wall as well as its immobilization also occurs in a non-legume (55). Nevertheless, these properties might be of importance in the curling of root hairs. To obtain a proper shepherd's crook, it is probable that a certain dominant microcolony of rhizobia induces a gradual and constant reorientation of tip growth by which a 360° turn is created. It seems probable that the bacteria within such a colony have to provide information about their position to the hair. It is hypothesized that the bacteria, within a microcolony, locally secrete Nod factor, which becomes more or less immobilized in the plant cell wall at the site of the colony and this could provide the positional information to the hair (45, 55).

3.2 Multiple receptors

Studies with purified Nod factors and mutant rhizobia provided insight into the Nod factor structure requirements for various epidermal responses. Such studies have especially been done in *Medicago*, *Pisum*, and *Vicia*. These showed that responses like root-hair deformation, *Enod12* activation and ion fluxes (see Section 4.1), have similar Nod-factor structural requirements. In *Medicago* these responses depend highly on the presence of the sulphate group at the reducing terminal-sugar residue, whereas in *Pisum* and *Vicia* at this position no substitution (for the H atom) is required. Substitutions at the non-reducing terminal sugar are not essential for these responses and also there are no specific demands concerning the structure of the lipid moiety except that it has to be longer than 10 carbon atoms (55). By contrast, for the initiation and growth of the infection thread in the epidermis, the structure of the fatty acyl group and the presence of the O-acetyl group at the non-reducing end are important. Rhizobial mutants that produce Nod factors that do not carry the appropriate unsaturated fatty acid and the O-acetyl group at the non-reducing terminal sugar (*nodEL* mutant) have almost completely lost the ability to initiate the infection process (58–60). These data support the hypothesis that two receptors are operational in the epidermal cells (60): an 'entry' receptor involved in the rhizobia infection with high stringency for the Nod factor structure and a 'signalling' receptor involved in the other epidermal responses with a lower stringency.

The characterization of the *sym2* allele of the Afghanistan pea also showed that infection depends highly on the Nod factor structure. This *sym2* allele is responsible for the inability of European *R. leguminosarum* biovar *viciae* strains to nodulate Afghanistan pea. When pea plants containing the Afghanistan *sym2* allele are inoculated with European *R. leguminosarum* biovar *viciae*, the bacteria induce all responses except the infection process which is affected; the number of infections is markedly reduced and the few infection threads that are formed in the root hairs are aborted (61). However, certain *R. leguminosarum* biovar *viciae* strains carrying *nodX* can form nodules on pea lines containing the *sym2* allele (62). NodX is an acetyl transferase responsible for the addition of an O-acetyl group at the reducing sugar residue (63), where *S. meliloti* Nod factors have a sulphuryl group. Therefore, the O-acetyl group is essential for the infection of (Afghanistan) *sym2* peas but not for the induction of all the other responses. Therefore, *sym2* is probably involved in the recognition of the NodX-modified Nod factors and it may act as an 'entry' receptor.

3.3 Nod-factor-binding proteins

By a biochemical approach, two Nod-factor-binding sites, NFBS1 and NFBS2, have been identified in *Medicago* (64, 65). NFBS1 has an affinity of 86 nM for *S. meliloti* Nod factors. However, it does not discriminate between factors with or without the sulphate substitution, although the sulphate group is essential for most of the *S. meliloti* Nod-factor-induced response during *Medicago* nodulation. Furthermore, a similar binding site has been identified in a non-legume. NFBS1 has been identified in a 1000 g fraction of *Medicago* roots and therefore might be located in the cell wall (64). This in combination with its lack of specificity and its occurrence in non-legumes make it a good candidate to be involved in the cell-wall binding of Nod factors (55; see Section 3.1).

NFSB2 has been identified in *Medicago* cell suspension and might be located in the plasma membrane (65). It has a higher affinity (4 nM) for Nod factors than NFBS1, but like this binding site it also does not discriminate between sulphated and non-sulphated Nod factors.

Due to the lack of specificity, it is unlikely that these NFBSs by themselves can function as Nod-factor-signalling or entry receptor. However, both NFBSs only efficiently bind chitin fragments with a covalently attached acyl, showing a very different specificity from that of chitin fragment receptors so far reported (66). Therefore, these sites might be part of a Nod-factor-binding complex, in combination with other components that confer specificity to the complex.

Lectins have been shown to be a determinant of host specificity (67). White clover plant expressing a pea lectin have an extended host range since they can form nodules with the microsymbiont of pea, *R. leguminosarum* biovar *viciae*. Similarly, *Lotus corniculatus* plants, normally nodulated by *M. loti*, obtained the ability to interact with *Bradyrhizobium japonicum*, that form nodules on soybean, when they express a soybean lectin (68). Although these studies show that lectins play a role in the nodulation process, it was never demonstrated that they could bind Nod factors. By contrast, a lectin from the roots of *Dolichos biflorus* was shown to bind Nod factors, albeit with a rather low affinity. This lectin is a member of the eukaryotic ATPase superfamily, with an apyrase specificity that is activated upon Nod-factor binding. Moreover, it is an extracellular protein located at the tip of the root hairs and an antibody against the protein block nodulation (69, 70). Although these studies strongly suggest that this lectin most likely has a role in nodulation, its precise role in the Nod-factor signalling remains to be demonstrated.

4. Nod-factor signalling

4.1 Biochemical approach

Nod factors induce rapid changes in ion concentrations in the root epidermis (71). The earliest response to Nod factor is a calcium influx, induced within seconds after Nod-factor addition (53). This Ca^{2+} influx is followed by a Cl^- efflux and together they cause the observed plasma membrane depolarization, which is induced about 1 minute after Nod-factor addition (53, 72–74). Upon depolarization of the membrane, a K^+ efflux is induced by which the membrane can repolarize (53).

The Ca^{2+} influx is sufficient to cause a membrane depolarization since the Ca^{2+} ionophore A23187 was able to induce Ca^{2+} and Cl^- fluxes leading to membrane depolarization. Moreover, the Ca^{2+} influx appears also to be essential for Nod-factor-induced membrane depolarization since nifedipine, a Ca^{2+} channel antagonist, as well as EGTA, which chelates Ca^{2+} ions, inhibited this response in Nod-factor-treated plants (53, 75).

The Ca^{2+} influx causes an increase of the cytoplasmatic Ca^{2+} concentration (75–78) but whether this increase is only due to the Ca^{2+} influx or also to Ca^{2+} release from internal stores is not clear. Furthermore, about 6–10 minutes after Nod-factor application, Ca^{2+} spiking has been observed in the perinuclear region of root hairs of *Medicago, Vicia, Pisum*, and *Lotus* (78–81).

Further insight in Nod-factor signalling came from pharmacological approaches where it was aimed to mimic or block certain Nod-factor responses with drugs interfering with a specific signalling step. An assay based on the transcriptional activation of an early nodulin gene (e.g. *Enod12*) was shown to be very useful. *Medicago* plants expressing a fusion between the *M. truncatula Enod12* promoter (*MtEnod12*) and the reporter gene *Gus* were used for these studies. The *Enod12* promoter is induced in the root epidermal cells 2–3 hours after Nod-factor application (51, 82, 83). It was tested whether trimeric G-proteins could be involved in Nod-factor signalling by studying the effect of mastoparan. Mastoparan is a cationic tetradecapeptide that acts as an activator of G-proteins in the animal system (84). Strikingly, mastoparan induced the expression of *MtEnod12-Gus* in the root epidermis in a similar spatial and temporal way as Nod factors. Further, the pertussis toxin, an inhibitor of G-protein activity, blocked the Nod-factor-induced *MtEnod12-Gus* expression (51).

Further evidence for the involvement of a G-protein-coupled receptor in Nod-factor signalling came from studies by den Hartog *et al.* (85). They showed that mastoparan was able to induce deformation of *Vicia sativa* root hairs. Since trimeric G-proteins are known to activate phospholipid signalling in animal cells, it was hypothesized that phospholipids were involved in Nod-factor signalling and biochemical studies showed that this is indeed the case. *Vicia* plants were grown in the presence of radioactive phosphate, to label phospholipids and it was shown that Nod factor as well as mastoparan induced an increase of the concentration of phosphatidic acid (PA) and diacylglycerol pyrophosphate (DGPP). Changes in the concentration of other phospholipids, like phosphatidylinositol 4,5-bisphosphate (PIP_2), were not observed. The concentration of PA and DGPP increases in a dose- and time-dependent manner, reaching a maximum 10–15 minutes after Nod-factor/mastoparan application (85). PA is a potential signal molecule in plants (86–92) and it is converted to DGPP by PA kinase. The increase of PA concentration was shown to be caused by the activation of two plant enzymes, namely, phospholipase C (PLC) and phospholipase D (PLD). PLD activity leads to PA formation directly by hydrolysis of structural phospholipids as phosphatidylcholine. By contrast, PLC hydrolyses PIP_2 into inositol 1,4,5-triphosphate (IP_3) and diacylglycerol (DAG). DAG is then quickly phosphorylated to PA. By blocking the activation of PLC (neomycin) or PLD (primary alcohols), it was demonstrated that the activation of both phospholipases is essential to induce root-hair deformation. PLC activation was also shown to be essential for *Enod12* induction (51). PLC activation most probably leads to IP_3 formation, but the methods used by den Hartog (85) were probably not sensitive enough to detect this increase. Whether IP_3 and/or PA are involved in the induction of *Enod12* is not known.

4.2 Genetic dissection of the Nod-factor signalling

The comparison of the *Rhizobium*–legume interaction and the mychorrhiza–plant association has been of direct importance for the genetic dissection of Nod-factor signalling in legume nodulation. Before these genetic studies are reviewed, some background information is provided on endomycorrhizal symbiosis.

About two decades ago a genetic analysis showed that nodulation and mycorrhizae formation in pea share some common steps (93). This was a surprising discovery since at first glance these two endosymbiotic interactions seem to have little in common. Such genetic analyses have now been done in several legumes and provided insight into the relationship of these two endosymbiotic interactions. The vast majority of higher plants are able to form an endosymbiotic association with fungi belonging to the order *Glomales*, which results in the formation of arbuscular mycorrhizas (AMs; 94–97). These fungi enter inner cortical cells of the root where they differentiate into highly branched structures, the so-called arbuscles. Thus the induced morphological responses in the root appear to be very different from those induced by rhizobia. Furthermore, the two interactions are extremes in terms of host specificity: whereas the rhizobial nodulation is highly specific and restricted to legumes (with one exception, namely *Parasponia*), in AM formation there is very little host specificity. Despite these differences, genetic studies showed that several common steps are involved in establishing these symbioses. Furthermore, some nodulin genes are induced during both the interactions, e.g. *Enod12* (96).

Several plant mutants defective in nodulation (Nod$^-$) have been isolated in legumes such as pea (98–100), *M. truncatula* (101–104), *L. japonicus* (105–108), and *M. sativa* (79). In all these host species it was shown that some mutants are blocked in nodulation as well as AM formation. The mutants of pea and *M. truncatula* have been studied in greatest detail with respect to Nod-factor responses and their ability to form AM. Therefore, the data obtained with these host species are summarized here. It has been tested whether rhizobia (or purified Nod factors) could induce Ca^{2+} spiking, activation of early nodulin gene expressions (e.g. *Enod11/12*), root-hair deformation/branching, and infection thread formation, respectively, in pea and *Medicago* mutants. Recently, in *M. truncatula* five loci, *dmi-1*, *dmi-2*, *dmi-3*, *hcl*, and *nsp-1*, were analysed. Mutants in these loci are Nod$^-$ and are blocked at a very early stage of the symbiotic interaction (35, 58). These mutants could be divided into a few groups. The mutants in the loci *dmi-1*, *dmi-2*, and *dmi-3* showed root-hair tip swelling upon Nod-factor treatment, but were blocked in branching (Fig. 2). Moreover, the early nodulin genes *MtEnod11*, root-hair curling, infection as well as cortical cell divisions could not be induced. Unlike the *dmi* mutants, the root hairs of *hcl* and *nsp-1* mutants were able to branch but curling and initiation of infection thread formation were blocked (35, 58). These data suggest that HCL and NSP act downstream of the *dmi* genes in the Nod-factor-induced signalling pathway (Fig. 2). The *dmi* mutants could be further classified by examining their Ca^{2+} response to Nod factor: *dmi-1* and *dmi-2* mutants were blocked for Ca^{2+} spiking, whereas *dmi-3* mutants showed Ca^{2+} spiking indistinguishable from wild-type plants (81). This indicates that the *dmi-1* and *dmi-2* genes act upstream of *dmi-3* in the Nod-factor-signalling pathway (81; Fig. 2).

Several pea mutants have also been identified that are affected in early steps of the symbiotic interaction (these all are all named *sym* mutants; 98–100, 109). Like the

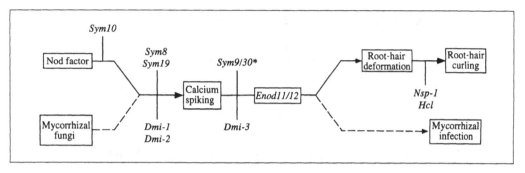

Fig. 2 Genetic dissection of Nod-factor and mycorrhizal signalling in the root epidermis (modified from 80). Mutations in *Sym8*, *Sym19*, *Dmi-1*, and *Dmi-2* loci blocks all the Nod-factor-induced responses in the epidermis as well as mycorrhizal infection; mutations in *Sym9* and *Dmi-3* inhibits *Enod12* induction, root-hair deformation and curling, and mychorrizal infection but not Ca^{2+} spiking. Since mutation in *Sym10* blocks Ca^{2+} spiking but not mycorrhizal infection, *Sym10* acts upstream of the common steps to rhizobia and mychorrizal infection; on the other site, mutations in *Nsp-1* and *Hcl* block root-hair curling but not deformation, Ca^{2+} spiking, or mycorrhizal infection. The induction of Ca^{2+} spiking during mycorrhizal infection has not been tested. *Preliminary data (80) suggest that *Sym9* and *Sym30* belong to the same complementation group.

Medicago dmi-1 and *dmi-2* mutants, *sym8*, *sym10*, and *sym19* mutants were blocked in Nod-factor-induced Ca^{2+} spiking, *Enod12* induction, and root-hair deformation, whereas *Sym9*, like the *Medicago Dmi-3* is positioned in between Ca^{2+} spiking and *Enod12* induction (80, 110; Fig. 2).

When these pea and *Medicago* mutants were analysed for their ability to form AM, all the *dmi* mutants as well as the pea *sym8*, *sym19*, and *sym9* mutants were also unable to interact with endomycorrhizal fungi (Myc⁻), whereas the *Medicago hcl* and *nsp* mutants were not affected (Myc⁺). This observation indicates that the *Dmi* and the orthologous pea *Sym* genes control common steps in the signal transduction cascade induced by rhizobial and mycorrhizal signals leading (among others) to *Enod12* activation. The signalling pathway obviously branches upstream of *Enod11/12* induction, since the *hcl* and *nsp* mutants are Myc⁺ (Fig. 2). The pea *Sym10* gene has a unique property. It is obviously involved in a very early stage of the interaction since: Ca^{2+} spiking is blocked in the *sym10* mutant and, in contrast with the other genes required for Ca^{2+} spiking, it is not essential for AM formation. Since the ultimate morphological responses induced by mycorrhizal fungi and rhizobia are different, it seems probable that the mycorrhizal signal molecule differs in some way from the rhizobial Nod factors. This assumption would imply that the perception of Nod factors and maybe early steps of signalling are not shared by the two interactions. Therefore, *Sym10* could be a Nod factor receptor or could be involved in a very early stage of the Nod-factor signalling.

The cloning of the above-described genes will ultimately provide insight into the molecular mechanisms controlling signalling events in both interactions. Since several genes are also involved in AM formation, it is probable that these genes are widespread in the plant kingdom and are not restricted to legumes. Such genes will therefore also be important tools to study the phylogenetic origin of nodulation.

5. Concluding remarks

The second part of this review summarizes the biochemical and genetic studies on the mechanisms of Nod-factor perception (Sections 3.1–3.3) and transduction (Sections 4.1, 4.2) in root hairs. It is obvious that these studies have revealed many candidate pieces of the Nod-factor perception/signalling jigsaw, but in most cases their relevance and precise role needs to be carefully evaluated.

5.1 Cloning of legume genes involved in Nod-factor perception or transduction

It is clear that the genetically identified genes are essential for certain steps in the Nod-factor-induced signalling and cloning of these genes will be of pivotal importance to advance our knowledge of Nod-factor perception and transduction.

Up until recently, legume mutants had especially been made in agronomically important crop plants, such as pea, soybean, and clover. Unfortunately, these legumes have large and/or complex genomes, which has impeded a map-based cloning with the result that none of these genes has been cloned. To overcome these problems, two legume model systems have now been developed: *Medicago truncatula* and *Lotus japonicus* (111, 112). Both model legumes are diploid, have a rather small genome (four to five times bigger than *Arabidopsis*) and worldwide genome projects on these legumes have and will provide the tools to clone and analyse the genes of interest. However, it still might be important to clone certain genes of non-model legumes, especially when these genes play a key role in the process and have not been identified by genetic approaches in the model legumes. An example is the pea *Sym10* gene, which is a good candidate for the Nod-factor receptor (80). Fortunately pea is highly microsyntenic with *M. truncatula* and therefore it is very probable that *M. truncatula* can be used as an intergenomic cloning vehicle for the cloning of pea genes of interest (113).

Many mutants in *L. japonicus* and *M. truncatula* have been identified, that are specifically affected in the infection by rhizobia. Moreover, Nod-factor signalling in *M. truncatula* has now been genetically dissected (see Section 4.2). The achievements of the genome projects on these model legumes will now make it possible to clone the identified loci. Recently, the first *L. japonicus* gene required for infection has been cloned. This is the *Nin* gene, which encodes a putative transcription factor (106). It is to be expected that most *M. truncatula* genes indicated in Fig. 2 will be cloned within a couple of years and this will provide tools to test hypotheses as depicted in Fig. 2. It is probable that this proposed signalling cascade is far more complex and this is illustrated by the preliminary data of the group of Kiss (see 114). They showed that a *M. sativa* gene with a similar function as *Dmi-1 or Dmi-2*, and which is probably an orthologue of one of these genes, encodes a putative receptor kinase containing an extracellular leucine-rich repeat domain. However, the scheme depicted in Fig. 2 suggests that this gene is active downstream of Nod-factor perception. Therefore, a further analysis of this gene might falsify the proposed hypothesis or alternatively, this gene could have a very intriguing function in the signalling cascade.

5.2 Integration of biochemical and genetic studies

The cell biological, biochemical, and genetic studies on Nod-factor signalling each provide insight in Nod-factor perception mechanisms. Fortunately, most studies are now focused on the model legumes and many of the data described in this review were obtained with *Medicago* species. This facilitates the integration of data obtained with the various approaches.

The analysis of Ca^{2+} spiking in the various *Medicago* and pea mutants is a first step towards an integration of biochemical and genetic approaches. This study made it possible to classify the identified legume mutants but it also provided a strong indication that Ca^{2+} spiking is a relevant step in Nod-factor signalling. Further biochemical and electrophysiological studies on the available mutants will be essential to integrate the signalling pathways obtained with the two different approaches. Moreover, this can also reveal the relevance of certain proposed steps of the signalling cascade process. For example, the set of *Medicago* mutants will make it possible to assess whether, indeed, trimeric G-proteins play a role in signalling. This is especially important since not all subunits of trimeric G-proteins have been identified in plants (115). Furthermore, trimeric G-proteins are almost exclusively activated by serpentine receptors (116), which are also scarce in plants.

5.3 From signal to form

The cell biological analysis of root-hair curling and infection thread formation indicate that these processes are regulated by mechanisms that are derived from those controlling 'normal' root-hair growth. Furthermore, it is clear that both the actin and microtubule skeleton play a key role in these processes. The configuration of the actin skeleton rapidly (within 5 minutes) changes after Nod-factor addition. This shows that this part of the skeleton is one of the first targets of Nod-factor signalling and, therefore, it will be important to correlate the elements involved in Nod-factor signalling, e.g. ion fluxes, G-protein activation and phosphoinositides, with the actin cytoskeleton rearrangements. Although the mechanism by which Nod factors induce these changes in the actin cytoskeleton is not known, it is likely to involve the activation of several actin-binding proteins (ABPs). These proteins, including monomer-sequestering, end-capping, cross-linking, severing, and side-binding proteins, regulate the rate of actin polymerizing and the configuration of the skeleton (F-actin; 117, 118). A few ABPs have been identified in plants and some of these can be regulated by phosphoinositides [e.g. actin-depolymerizing factor (ADF)/cofilin and profilin] or by Ca^{2+} (ADF/cofilin and villin; 119, 120). Since both intracellular Ca^{2+} and phosphoinositide concentrations rapidly change after Nod-factor addition, it is possible that ABP, like ADF and profilin, are involved in the Nod-factor-induced changes of the actin cytoskeleton configuration.

Both root-hair deformation and curling have in common that tip growth is put under control of Nod-factor-secreting rhizobia and, moreover, during root-hair curling, the direction of tip growth is also strictly controlled. Curling is accompanied by marked changes in the configuration of the microtubule skeleton. Since it has been shown that this part of the cytoskeleton is especially involved in controlling the direction of growth, it seems probable that the microtubules are a key player in the curling process. It will be a major challenge to unravel the mechanism controlling curling (and the initiation of infection) since bacteria secreting Nod factor are essential and purified Nod factors are not sufficient to induce these processes. Furthermore, these processes are only induced in a minority of the root hairs by which (classical) biochemical studies are almost impossible. Fortunately, cell biological approaches are being developed by which important components of signal transduction cascades can be monitored in living cells. Examples are chameleon construct by which free Ca^{2+} concentrations can be monitored in living cells (121) or PIP_2-binding domain (pleckstrin homology domain) that allow localization studies in living cells (122). This, in combination with the fact that both actin and microtubule skeleton can be visualized *in vivo* with green fluorescent protein (GFP; 123–125)-based fusions, will make it possible to study fascinating processes such as curling and infection thread initiation. Such a cell biological approach should be integrated with molecular genetic studies aiming to clone genes that are specifically involved in curling and infection, like *Hcl*, *Nsp*, and *Sym2*. In years to come, the symbioses between the achievements of the genetic analysis and the effort put into establishing the cell biological tools, will improve our knowledge on Nod-factor signalling.

References

1. Cohn, J., Day, B. and Stacey, G. (1998) Legume nodule organogenesis. *Trends Plant Sci.*, **3**, 105.
2. Bladergroen, M. R. and Spaink, H. P. (1998) Genes and signal molecules involved in the rhizobia-leguminoseae symbiosis. *Curr. Opin. Plant Biol.*, **1**, 353.
3. Hadri, A.-E., Spaink, H. P., Bisseling, T. and Brewin, N. J. (1998) Diversity of root nodulation and rhizobial infection processes. In Spaink, H. P., Kondorosi, A. and Hooykaas, P. J. J. (eds), *Rhizobiaceae*. Kluwer Academic, Dordrecht, **1**, 347.
4. Long, S. R. (1996) *Rhizobium* symbiosis: nod factors in perspective. *Plant Cell*, **8**, 1885.
5. Mylona, P., Pawlowski, K. and Bisseling, T. (1995) Symbiotic nitrogen fixation. *Plant Cell*, **7**, 869.
6. Schultze, M. and Kondorosi, A. (1998) Regulation of symbiotic root nodule development. *Annu. Rev. Genet.*, **32**, 33.
7. Schlaman, H. R. M., Phillips, D. A. and Kondorosi, E. (1998) Genetic organization and transcriptional regulation of rhizobial nodulation genes. In Spaink, H. P., Kondorosi, A. and Hooykaas, P. J. J. (eds), *Rhizobiaceae*. Kluwer Academic, Dordrecht, **1**, 361.
8. Downie, J. A. (1998) Functions of rhizobial nodulation genes. In Spaink, H. P., Kondorosi, A. and Hooykaas, P. J. J. (eds), *Rhizobiaceae*. Kluwer Academic, Dordrecht, **1**, 387.
9. Lerouge, P., Roche, P., Faucher, C., Maillet, F., Truchet, G., Prome, J. C. and Dénarié, J. (1990) Symbiotic host-specificity of *Rhizobium meliloti is* determined by a sulphated and acylated glucosamine oligosaccharide signal. *Nature*, **344**, 781.
10. Spaink, H. P., Sheeley, D. M., van Brussel, A. A., Glushka, J., York, W. S., Tak, T., Geiger, O., Kennedy, E. P., Reinhold, V. N. and Lugtenberg, B. J. (1991) A novel highly

unsaturated fatty acid moiety of lipo-oligosaccharide signals determines host specificity of *Rhizobium. Nature*, **354**, 125.

11. Perret, X., Staehelin, C. and Broughton, W. J. (2000) Molecular basis of symbiotic promiscuity. *Microbiol. Mol. Biol. Rev.*, **64**, 180.

12. Dénarié, J., Debelle, F. and Prome, J. C. (1996) *Rhizobium* lipochitooligosaccharide nodulation factors: signalling molecules mediating recognition and morphogenesis. *Annu. Rev. Biochem.*, **65**, 503.

13. Yang, G. P., Debelle, F., Savagnac, A., Ferro, M., Schiltz, O., Maillet, F., Prome, D., Treilhou, M., Vialas, C., Lindstrom, K., Dénarié, J. and Prome, J. C. (1999) Structure of the *Mesorhizobium huakuii* and *Rhizobium galegae* Nod factors: a cluster of phylogenetically related legumes are nodulated by rhizobia producing Nod factors with alpha, beta-unsaturated N-acyl substitutions. *Mol. Microbiol.*, **34**, 227.

14. Cai, G., Moscatelli, A. and Cresti, M. (1997) Cytoskeleton organization and pollen tube growth. *Trends Plant Sci.*, **2**, 86.

15. Kropf, D. L. (1997) Induction of polarity in fucoid zygotes. *Plant Cell*, **9**, 1011.

16. Sieber, B. and Emons, A. M. C. (2000) Cytoarchitecture and pattern of cytoplasmatic streaming in root hair of *Medicago truncatula* during development and deformation by nodulation factor. *Protoplasma*, **214**, 118.

17. Miller, D. D., de Ruijter, N. C. A. and Emons A. M. C. (1997) From signal to form: aspects of the cytoskeleton-plasma membrane-cell wall continuum in root hair tips. *J. Exp. Bot.*, **48**, 1881.

18. Miller, D. D., Leferink-ten Klooster, H. B. and Emons, A. M. C. (2000) Lipochito-oligosaccharide nodulation factors stimulate cytoplasmatic polarity with longitudinal endoplasmatic reticulum and vesicles at the tip in vetch root hairs. *Mol. Plant–Microbe Interact.*, **13**, 1385.

19. Ridge, R. W. (1992) A model of legume root hair growth and *Rhizobium* infection. *Symbiosis*, **14**, 359.

20. Galway, M. E. (2000) Root hair ultrastructure and tip growth. In Ridge, R. W. and Emons, A. M. C. (eds), *Root Hair: Cell and Molecular Biology*. Springer, Berlin, **1**, 1.

21. Miller, D. D., de Ruijter, N. C. A., Bisseling, T. and Emons, A. M. C. (1999) The role of actin in root hair morphogenesis: studies with lipochitooligosaccharide as a growth stimulator and cytochalasin as an actin perturbing drug. *Plant J.*, **17**, 141.

22. Esseling, J., de Ruijter, N. C. A. and Emons, A. M. C. (2000) The root hair actin cytoskeleton as backbone, highway. Morphogenic instrument and target for signalling. In Ridge, R. W. and Emons, A. M. C. (eds), *Root Hair: Cell and Molecular Biology*. Springer, Berlin, **1**, 29.

23. Kropf, D. L., Bisgrove, S. R. and Hable, W. E. (1998) Cytoskeletal control of polar growth in plant cells. *Curr. Opin. Cell Biol.*, **10**, 117.

24. Bibikova, T. N., Blancaflor, E. B. and Gilroy, S. (1999) Microtubules regulate tip growth and orientation in root hairs of *Arabidopsis thaliana. Plant J.*, **17**, 657.

25. Emons, A. M. C., Wolters-Arts, J. A., Traas, J. and Derksen, J. (1990) The effects of colchicine on microtubules and microfibrils in root hairs. *Acta Bot. Neerl.*, **39**, 19.

26. Sievers, A. and Schnepf, E. (1981) Morphogenesis and polarity in tubular cells with tip growth. In Kiermayer, O. (ed.), *Cytomorphogenesis in Plants*. Springer-Verlag, New York, 265.

27. That, T. C., Rossier, C., Barja, F., Turian, G. and Roos, U. P. (1988) Induction of multiple germ tubes in *Neurospora crassa* by antitubulin agents. *Eur. J Cell Biol.*, **46**, 68.

28. Traas, J. A., Braat, P., Emons, A. M., Meekes, H. and Derksen, J. (1985) Microtubules in root hairs. *J. Cell Sci.*, **76**, 303.

29. Lloyd, C. W. and Wells, B. (1985) Microtubules are at the tips of root hairs and form helical patterns corresponding to inner wall fibrils. *J. Cell Sci.*, **75**, 225.

30. Emons, A. M. C. (1989) Helicoidal microfibril deposition in a tip growing cell and microtubule alignment during tip morphogenesis: a dry cleaving and freeze-substitution study. *Can. J. Bot.*, **67**, 2401.

31. Ketelaar, T. and Emons, A. M. C. (2000) The role of microtubules in root hair growth and cellulose microfibril deposition. In Ridge, R. W. and Emons, A. M. C. (eds), *Root Hair: Cell and Molecular Biology*. Springer, Berlin, **1**, 17.

32. Ridge, R. W. (1996) Root hairs: cell biology and development. In Waisel, Y., Eshel, A., Kalkafi, U. (eds), *Plant the Hidden Half*. Marcel Dekker Inc., New York, **1**, 127.

33. Lloyd, C. W., Pearce, K. J., Rawlins, D. J, Ridge, R. W. and Shaw, P. J. (1987) Endo-plasmatic microtubules connect the advancing nucleus to the tip of legume root hairs, but F-actin is involved in basipetal migration. *Cell Motil. Cytoskel.*, **8**, 27.

34. Timmers, A. C., Auriac, M. C. and Truchet, G. (1999) Refined analysis of early symbiotic steps of the *Rhizobium-Medicago* interaction in relationship with microtubular cyto-skeleton rearrangements. *Development*, **126**, 3617.

35. Catoira, R., Galera, C., de Billy, F., Penmetsa, R. V., Joumet, E. P., Maillet, F., Rosenberg, C., Cook, D., Gough, C. and Dénarié, J. (2000) Four genes of *Medicago truncatula* con-trolling components of a Nod-factor transduction pathway. *Plant Cell*, **12**, 1647.

36. Heidstra, R., Geurts, R., Franssen, H., Spaink, H., van Kammen, A. and Bisseling, T. (1994) Root hair deformation activity of nodulation factors and their fate on *Viciae Sativa*. *Plant Physiol.*, **105**, 787.

37. de Ruijter, N. C. A., Rook, M. B., Bisseling, T. and Emons, A. M. C. (1998) Lipochito-oligosaccharides re-initiate root hair tip growth in *Vicia sativa* with high calcium and spectrin-like antigen at the tip. *Plant J.*, **13**, 341.

38. Allen, N. S., Bennett, M. N., Cox, D. N., Shipley, A., Ehrhardt, D. W. and Long, S. R. (1994) Effect of Nod factor on alfalfa root hair Ca^{2+} and H^+ currents and on cytoskeletal behavior. In Daniels, M. J., Downie, J. A., Osboum, A. E. (eds), *Advances in Molecular Genetics of Plant–Microbe Interactions*. Kluwer Academic Publishers, Dordrecht, The Netherlands, **1**, 107–113.

39. Allen, N. S. and Bennett, M. N. (1996) Electro-optical imaging F-actin and endoplasmatic reticulum in living and fixed plant cells. *Scanning Microsc. Suppl.*, **10**, 177.

40. Cárdenas, L., Vidali, L., Dominguez, J., Perez, H., Sanchez, F., Hepler, P. K. and Quinto, C. (1998) Rearrangement of actin microfilaments in plant root hairs responding to *Rhizobium etli* nodulation signals. *Plant Physiol.*, **116**, 871.

41. de Ruijter, N. C. A., Bisseling, T. and Emons, A. M. C. (1999) *Rhizobium* Nod factors induce an increase in sub-apical fine bundles of actin filaments in *Vicia sativa* root hairs within minutes. *Mol. Plant–Microbe Interact.*, **12**, 829.

42. Emons, A. M. C. and de Ruijter, N. C. A. (2000) Actin: a target of signal transduction in root hairs. In Staiger, C. J. (ed.), *Actin: a Dynamic Framewotk of Multiple Plant Cell Functions*. Kluwer Academic, The Netherlands, **1**, 373.

43. Relic, B., Talmont, F., Kopcinska, J., Golinowski, W., Prome, J. C. and Broughton, W. J. (1993) Biological activity of *Rhizobium sp. NGR234* Nod-factors on *Macroptilium atropurpureum*. *Mol. Plant–Microbe Interact.*, **6**, 764.

44. van Batenburg, F. H. D., Jonker, R. and Kijne, J. W. (1986) *Rhizobium* induces marked root hair curling by redirection of tip growth: a computer simulation. *Physiol. Plant*, **66**, 476.

45. Emons, A. M. C. and Mulder, B. (2000) Nodulation factor triggers an increase of fine bundles of subapical actin filaments in *Vicia* root hair: implication for root hair curling

around bacteria. In De Wit, P. J. G. M., Bisseling, T. and Stiekema, W. J. (eds), *Biology of Plant–Microbe Interactions*. International Society of Molecular Plant–Microbe Interaction, St Paul, MN, USA, **2**, 272.

46. van Spronsen, P. C., Bakhuizen, R., van Brussel, A. A. and Kijne, J. W. (1994) Cell wall degradation during infection thread formation by the root nodule bacterium *Rhizobium leguminosarum* is a two-step process. *Eur. J. Cell Biol.*, **64**, 88.

47. Ridge, R. W., Rolfe, B. G. (1985) *Rhizobium* sp. degradation of legume root hair cell wall at the site of infection thread origin. *Appl. Environ. Microbiol.*, **50**, 717.

48. Bauer, W. D. (1981) Infection of legumes by rhizobia. *Annu. Rev. Plant Physiol.*, **32**, 407.

49. Bakhuizen, R. (1988) The plant cytoskeleton in the *Rhizobium*-legume symbiosis. PhD thesis, Leiden University, The Netherlands.

50. Timmers, A. C. J. (2000) Infection of root hairs by rhizobia: infection thread development with emphasis on the microtubular cytoskeleton. In Ridge, R. W. and Emons, A. M. C. (eds), *Root Hair: Cell and Molecular Biology*. Springer, Berlin, **1**, 223.

51. Pingret, J. L., Journet, E. P. and Barker, D. G. (1998) *Rhizobium* Nod factor signaling. Evidence for a G protein-mediated transduction mechanism. *Plant Cell*, **10**, 659.

52. Felle, H. H., Kondorosi, E. and Schultze, M. (1996) Rapid alkalinization in alfalfa root hairs in response to rhizobial lipochitooligosaccharide signals. *Plant J.*, **13**, 455.

53. Felle, H. H., Kondorosi, E., Kondorosi, A. and Schultze, M. (1998) The role of ion fluxes in Nod factor signaling in *Medicago sativa*. *Plant J.*, **13**, 455.

54. Schlaman, H. R., Gisel, A. A., Quaedvlieg, N. E., Bloemberg, G. V., Lugtenberg, B. J., Kijne, J. W., Potrykus, I., Spaink, H. P. and Sautter, C. (1997) Chitin oligosaccharides can induce cortical cell division in roots of *Vicia sativa* when delivered by ballistic microtargeting. *Development*, **124**, 4887.

55. Goedhart, J., Rohrig, H., Hink, M. A., van Hoek, A., Visser, A. J., Bisseling, T. and Gadella, T. W., Jr. (1999) Nod factors integrate spontaneously in biomembranes and transfer rapidly between membranes and to root hairs, but transbilayer flip-flop does not occur. *Biochemistry*, **38**, 10898.

56. Goedhart, J., Hink, M. A., Visser, A. J., Bisseling, T. and Gadella, T. W., Jr. (2000) *In vivo* fluorescence correlation microscopy (FCM) reveals accumulation and immobilization of Nod factors in root hair cell walls. *Plant J.*, **21**, 109.

57. Philip-Hollingsworth, S., Dazzo, F. B. and Hollingsworth, R. I. (1997) Structural requirements of *Rhizobium* chitolipooligosaccharides for uptake and bioactivity in legume roots as revealed by synthetic analogs and fluorescent probes. *J. Lipid Res.*, **38**, 1229.

58. Catoira, R., Timmers, A. C. J., Maillet, F., Galera, C., Penmetsa, R. V., Cook, D., Dénarié, J. and Gough, C. (2001) The HCL gene of *Medicago truncatula* controls *Rhizobium*-induced root hair curling. *Development*, **128**, 1507.

59. Walker, S. A. and Downie, J. A. (2000) Entry of *Rhizobium leguminosarum* bv. *viciae* into root hairs requires minimal Nod factor specificity, but subsequent infection thread growth requires *nodO* or *nodE*. *Mol. Plant–Microbe Interact.*, **13**, 754.

60. Ardourel, M., Demont, N., Debelle, F., Maillet, F., de Billy, F., Promé, J. C., Dénarié, J. and Truchet, G. (1994) *Rhizobium meliloti* lipooligosaccharide nodulation factors: different structural requirements for bacterial entry into target root hair cells and induction of plant symbiotic developmental responses. *Plant Cell*, **6**, 1357.

61. Geurts, R., Heidstra, R., Hadri, A.-E., Downie, A., Franssen, H., van Kammen, A. and Bisseling, T. (1997) *Sym2* of *Pisum sativum* is involved in a Nod factor perception mechanism that controls the infection process in the epidermis. *Plant Physiol.*, **115**, 351.

62. Davis, E. O., Evans, I. J. and Johnston, A. W. (1988) Identification of *nodX*, a gene that allows *Rhizobium leguminosarum* biovar *viciae* strain TOM to nodulate Afghanistan peas. *Mol. Gen. Genet.*, **212**, 531.

63. Firmin, J. L., Wilson, K. E., Carlson, R. W., Davies, A. E. and Downie, J. A. (1993) Resistance to nodulation of cv. Afghanistan peas is overcome by *nodX*, which mediates an O-acetylation of the *Rhizobium leguminosarum* lipo-oligosaccharide nodulation factor. *Mol. Microbiol.*, **10**, 351.

64. Bono, J. J., Riond, J., Nicolaou, K. C., Bockovich, N. J., Estevez, V. A., Cullimore, J. V. and Ranjeva, R. (1995) Characterization of a binding site for chemically synthesized lipo-oligosaccharidic NodRm factors in particulate fractions prepared from roots. *Plant J.*, **7**, 253.

65. Gressent, F., Drouillard, S., Mantegazza, N., Samain, E., Geremia, R. A., Canut, H., Niebel, A., Driguez, H., Ranjeva, R., Cullimore, J. and Bono, J. J. (1999) Ligand specificity of a high-affinity binding site for lipochitooligosaccharidic Nod factors in *Medicago* cell suspension cultures. *Proc. Natl Acad. Sci. USA*, **96**,4704.

66. Stacey, G. and Shibuya, N. (1997) Chitin recognition in rice and legumes. *Plant Soil*, **194**, 161.

67. Diaz, C. L., Spaink, H. P. and Kijne, J. W. (2000) Heterologous rhizobial lipochitin oligosaccharides and chitin oligomers induce cortical cell divisions in red clover roots, transformed with the pea lectin gene. *Mol. Plant–Microbe Interact.*, **13**, 268.

68. van Rhijn, P., Goldberg, R. B. and Hirsch, A. M. (1998) *Lotus corniculatus* nodulation specificity is changed by the presence of a soybean lectin gene. *Plant Cell*, **10**, 1233.

69. Thomas, C., Sun, Y., Naus, K., Lloyd, A. and Roux, S. (1999) Apyrase functions in plant phosphate nutrition and mobilizes phosphate from extracellular ATP. *Plant Physiol.*, **119**, 543.

70. Day, R. B., McAlvin, C. B., Loh, J. T., Denny, R. L., Wood, T. C., Young, N. D. and Stacey, G. (2000) Differential expression of two soybean apyrases, one of which is an early nodulin. *Mol. Plant–Microbe Interact.*, **13**,1053.

71. Cárdenas, L., Holdaway-Clarke, T. L., Sanchez, F., Quinto, C., Feijó, J. A., Kunkel, J. G. and Hepler, P. K. (2000) Ion changes in legume root hairs responding to Nod factors. *Plant Physiol.*, **123**, 443.

72. Ehrhardt, D. W., Atkinson, E. M. and Long, S. R. (1992) Depolarization of alfalfa root hair membrane potential by *Rhizobium meliloti* Nod factors. *Science*, **256**, 998.

73. Kurkdjian, A. C. (1995) Role of the differentiation of root epidermal cells in Nod factor from *Rhizobium meliloti*-induced depolarization of *Medicago sativa*. *Plant Physiol.*, **107**, 783.

74. Felle, H. H., Kondorosi, E., Kondorosi, A. and Schultze, M. (1995) Nod signal-induced plasma membrane potential changes in alfalfa root hairs are differently sensitive to structural modifications of the lipochitooligosaccharide. *Plant J.*, **7**, 939.

75. Felle, H. H., Kondorosi, E., Kondorosi, A. and Schultze, M. (1999) Nod factors modulate the concentration of cytosolic free calcium differently in growing and non-growing root hairs of *Medicago sativa* L. *Planta*, **209**, 207.

76. Gehring, C. A., Irving, H. R., Kabbara, A. A., Parish R. W., Boukli, N. M. and Broughton, W. J. (1997) Rapid, plateau-like increases in intracellular free calcium are associated with Nod-factor-induced root hair deformation. *Mol. Plant–Microbe Interact.*, **7**, 791.

77. Felle, H. H., Kondorosi, E., Kondorosi, A. and Schultze, M. (1999) Elevation of the cytosolic free [Ca $^{2+}$] is indispensable for the transduction of the Nod factor signal in alfalfa. *Plant Physiol.*, **121**, 273.

78. Cárdenas, L., Feijo, J. A., Kunkel, J. G., Sanchez, F., Holdaway-Clarke, T., Hepler, P. K. and Quinto, C. (1999) *Rhizobium* Nod factors induce increases in intracellular free calcium and extracellular calcium influxes in bean root hairs. *Plant J.*, **19**, 347.

79. Ehrhardt, D. W., Wais, R. and Long, S. R. (1996) Calcium spiking in plant root hairs responding to *Rhizobium* nodulation signals. *Cell*, **85**, 673.

80. Walker, S. A., Viprey, V. and Downie, J. A. (2000) Dissection of nodulation signaling using pea mutants defective for calcium spiking induced by Nod factors and chitin oligomers. *Proc. Natl Acad. Sci. USA*, **979**, 13413.

81. Wais, R. J., Galera, C., Oldroyd, G., Catoira, R., Penmetsa, R. V., Cook, D., Gough, C., Dénarié, J. and Long, S. R. (2000) Genetic analysis of calcium spiking responses in nodulation mutants of *Medicago truncatula*. *Proc. Natl Acad. Sci. USA*, **97**,13407.

82. Pichon, M., Journet, E. P., Dedieu, A., de Billy, F., Truchet, G. and Barker, D. G. (1992) *Rhizobium meliloti* elicits transient expression of the early nodulin gene *ENOD12* in the differentiating root epidermis of transgenic alfalfa. *Plant Cell*, **4**, 1199.

83. Chabaud, M., Larsonneau, C., Marmouget, C. and Huguet, T. (1996) Transformation of barrel medis (*Medicago truncatula Gaertn.*) by *Agrobacterium tumefaciens* and regeneration via somatic embryogenesis of transgenic plants with the *MtEnod12* nodulin promoter fused to the *gus* reporter gene. *Plant Cell Rep.*, **15**, 305.

84. Ross, E. M. and Higashijima, T. (1994) Regulation of G-protein activation by mastoparans and other cationic peptides. *Methods Enzymol.*, **237**, 26.

85. den Hartog, M., Musgrave, A. and Munnik, T. (2001) Nod factor-induced phosphatidic acid and diacylglycerol pyrophosphate formation: a role for phospholipase C and D in root-hair deformation. *Plant J*, **25, 1.**

86. Munnik, T., Arisz, S. A., De Vrije, T. and Musgrave, A. (1995) G-protein activation stimulates phospholipase D signaling in plants. *Plant Cell*, **7**, 2197.

87. Munnik, T., Van Himbergen, J. A. J., Ter Riet, B., Braun, F. J., Irvine, R. F., Van den Hende, H. and Musgrave, T. (1998) Detailed analysis of the turnover of poliphosphoinositides and phosphatidic acid upon activation of phospholipases C and D in *Chlamydomonas* cells treated with non-permeabilizing concentrations of mastoparan. *Planta*, **207**, 133.

88. Munnik, T., Meijer, H. J., Ter Riet, B., Hirt, H., Frank, W., Bartels, D. and Musgrave, A. (2000) Hyperosmotic stress stimulates phospholipase D activity and elevates the levels of phosphatidic acid and diacylglycerol pyrophosphate. *Plant J.*, **22**, 147.

89. Ritchie, S. and Gilroy, S. (1998) Abscisic acid signal transduction in the barley aleurone is mediated by phospholipase D activity. *Proc. Natl Acad. Sci. USA*, **95**, 2697.

90. Ritchie, S. and Gilroy, S. (2000) Abscisic acid stimulation of phospholipase D in the barley aleurone is G-protein-mediated and localized to the plasma membrane. *Plant Physiol.*, **124**, 693.

91. Jacob, T., Ritchie, S., Assmann, S. M. and Gilroy, S. (1 999) Abscisic acid signal transduction in guard cells is mediated by phospholipase D activity. *Proc. Natl Acad Sci USA*, **96**, 12192.

92. Munnik, T., Irvine, R. F. and Musgrave, A. (1998) Phospholipid signalling in plants. *Biochim. Biophys. Acta*, **1389**, 222.

93. Duc, G., Trouvelot, A., Gianinazzi-Pearson, V. and Gianinazzi, S. (1989) First report of non-mychorrizal plant mutants (Myc−) obtained in pea *(Pisum sativum)* and Fababean *(Vicia faba L.)*. *Plant Sci.*, **60**, 215.

94. Harrison, M. J. (1997) The arbuscular mycorrhizal symbiosis: an underground association. *Trends Plant Sci.*, **2**, 54B.

95. Harrison, M. J. (1998) Development of the arbuscular mycorrhizal symbiosis. *Curr. Opin. Plant Biol.*, **1**, 360.

96. Albrecht, C., Geurts, R., Lapeyrie, F. and Bisseling, T. (1998) Endomycorrhizae and rhizobial Nod factors both require *sym8* to induce the expression of the early nodulin genes *PsENOD5* and *PsENOD12A*. *Plant J.*, **15**, 605.

97. Albrecht, C., Geurts, R. and Bisseling, T. (1999) Legume nodulation and mycorrhizae formation; two extremes in host specificity meet. *EMBO J.*, **18**, 281.

98. Kneen, B. E., Weeden, N. F. and LaRue, T. A. (1994) Non nodulating mutants of *Pisum sativum (L.) cv. Sparkle. J. Heredity*, **85**, 129.

99. Sagan, M., Huguet, T. and Duc, G. (1994) Phenotypic characterization and classification of nodulation mutants of pea *(Pisum sativum). Plant Sci.*, **100**, 59.

100. Duc, G. and Messager, A. (1989) Mutagenesis of pea *(Pisum sativum L)* and the isolation of mutants for nodulation and nitrogen fixation. *Plant Sci.*, **60**, 207.

101. Sagan, M., Morandi, D., Tarenghi, E. and Duc, G. (1995) Selection of nodulation and mycorrhizal mutants in the model plant *Medicago truncatula (Gaertn)* after γ-ray mutagenesis. *Plant Sci.*, **111**, 63.

102. Sagan, M., de Larambergue, H. and Morandi, D. (1998) Genetic analysis of symbiosis mutants in *Medicago truncatula*. In Elmerich, C., Kondorosi, A., Newton, W. E. (eds), *Biological Nitrogen Fixation for the 21st Century*. Kluwer Academic Publisher, Dordrecht, The Netherlands, **1**, 317.

103. Penmetsa, R. V. and Cook, D. R. (1997) A legume ethylene-insensitive mutant hyper-infected by its rhizobial symbiont. *Science*, **275**, 527.

104. Penmetsa, R. V. and Cook, D. R. (2000) Production and characterization of diverse developmental mutants of *Medicago truncatula*. *Plant Physiol.*, **123**, 1387.

105. Schauser, L., Handberg, K., Sandal, N., Stiller, J., Thykjaer, T., Pajuelo, E., Nielsen, A. and Stougaard, J. (1998) Symbiotic mutants deficient in nodule establishment identified after T-DNA transformation of *Lotus japonicus*. *Mol. Gen. Genet.*, **259**, 414.

106. Schauser, L., Roussis, A., Stiller, J. and Stougaard, J. (1999) A plant regulator controlling development of symbiotic root nodules. *Nature*, **402**, 191.

107. Wopereis, J., Pajuelo, E., Dazzo, F. B., Jiang, Q., Gresshoff, P. M., De Bruijn, F. J., Stougaard, J. and Szczyglowski, K. (2000) Short root mutant *of Lotus japonicus* with a dramatically altered symbiotic phenotype. *Plant J.*, **23**, 97.

108. Szczyglowski, K., Shaw, R. S., Wopereis, J., Copeland, S., Hamburger, D., Kasiborski, B., Dazzo, F. B. and de Bruijn, F. J. (1998) Nodule organogenesis and symbiotic mutants of the model legume *Lotus Japonicus*. *Mol. Plant–Microbe Interact.*, **11**, 684.

109. Markwei, C. M. and La Rue, T. A. (1992) Phenotypic characterization of *Sym8* and *Sym9*, two genes conditioning non nodulation in *Pisum sativum 'sparkle'. Can. J. Microbiol.*, **38**, 548.

110. Schneider, A., Walker, S. A., Poyser, S., Sagan, M., Ellis, T. H. and Downie, J. A. (1999) Genetic mapping and functional analysis of a nodulation-defective mutant *(sym19)* of pea *(Pisum sativum L.)*. *Mol. Gen. Genet.*, **262**, 1.

111. Cook, D. R. (1999) *Medicago truncatula* – a model in the making! *Curr. Opin. Plant Biol.*, **2**, 301.

112. Jiang, Q. and Gresshoff, P. M. (1997) Classical and molecular genetics of the model legume *Lotus japonicus*. *Mol. Plant–Microbe Interact.*, **10**, 59.

113. Gualtieri, G., Kulikova, O., Limpens, E., Kim, D.-J., Bisseling, T. and Geurts, R. (2001) The potential of *Medicago truncatula* as an intergenomic cloning vehicle for pea genes. Submitted

114. Cullimore, J. V., Ranjeva, R. and Bono, J.-J. (2001) Perception of lipochitooligosaccharidic Nod factor in legumes. *Trends Plant Sci.*, **6**, 24.
115. Millner, P. A. and Causier, B. E. (1995) G-protein coupled receptors in plant cells. *J. Exp. Bot.*, **47**, 983.
116. Dohlman, H. G., Thomer, J., Caron, M. G. and Lefkowitz, R. J. (1991) Model systems for the study of seven-transmembrane-segment receptors. *Annu. Rev. Biochem.*, **60**, 653.
117. Schmidt, A. and Hall, M. N. (1998) Signalling to the actin cytoskeleton. *Annu. Rev. Cell Dev. Biol.*, **14**, 305.
118. Ayscough, K. R. (1998) *In vivo* functions of actin-binding proteins. *Curr. Opin. Cell Biol.*, **10**, 102.
119. Staiger, C. J., Gibbon, B. C., Kovar, D. R. and Zonia, L. E. (1997) Profilin and actin-depolymerizing factor: modulators of actin organization in plants. *Trends Plant Sci.*, **2**, 275.
120. Klahre, U., Friederich, E., Kost, B., Louvard, D. and Chua, N.-H. (2000) Villin-like actin-binding proteins are expressed ubiquitously in *Arabidopsis*. *Plant Physiol.*, **122**, 35.
121. Allen, G. J., Kwak, J. M., Chu, S. P., Llopis, J., Tsien, R. Y., Harper, J. F. and Schroeder, J. I. (1999) Cameleon calcium indicator reports cytoplasmic calcium dynamics in *Arabidopsis* guard cells. *Plant J.*, **19**, 735.
122. Kost, B., Lemichez, E., Spielhofer, P., Hong, Y., Tolias, K., Carpenter, C. and Chua, N. H. (1999) Rac homologues and compartmentalized phosphatidylinositol 4,5-bisphosphate act in a common pathway to regulate polar pollen tube growth. *J. Cell Biol.*, **145**, 317.
123. Ludin, B. and Matus, A. (1998) GFP illuminates the cytoskeleton. *Trends Cell Biol.*, **8**, 72.
124. Kost, B., Spielhofer, P. and Chua, N. H. (1998) A GFP-mouse talin fusion protein labels plant actin filaments *in vivo* and visualizes the actin cytoskeleton in growing pollen tubes. *Plant J.*, **16**, 393.
125. Marc, J., Granger, C. L., Brincat, J., Fisher, D. D., Kao, T., McCubbin, A. G. and Cyr, R. J. (1998) A GFP-MAP4 reporter gene for visualizing cortical microtubule rearrangements in living epidermal cells. *Plant Cell*, **10**, 1927.

12 | Rhizospheric signals and early molecular events in the ectomycorrhizal symbiosis

F. MARTIN, S. DUPLESSIS, F. A. DITENGOU, H. LAGRANGE, C. VOIBLET AND F. LAPEYRIE

1. Ectomycorrhiza development: a multistep process

The rhizosphere, the region of soil surrounding and including the plant, is of crucial importance for plant nutrition and fitness (1, 2). The rhizosphere hosts a large and diverse community of prokaryotic and eukaryotic microbes that compete and interact with each other and with plant roots. Within the rhizospheric microbial community, mycorrhizal mutualistic associations are almost ubiquitous, and their effects on the ability of plants to grow productively in suboptimal environments are profound (3). In forest ecosystems of temperate and boreal regions, ectomycorrhizal fungi are found on most absorbing short roots of tree species where they affect not only mineral and nutrient absorbtion (3, 4), but also adaptation to adverse soil chemical conditions (5) and susceptibility to diseases (6). Ectomycorrhizal fungi mainly belong to the Ascomycotina and Basidiomycotina and the switch between saprobic and mycorrhizal lifestyles probably happened convergently and perhaps many times during evolution of these fungal lineages (7). The first mycorrhizal associations must have been derived from earlier types of plant–fungus interactions, such as endophytic fungi in the bryophite-like precursors of vascular plants. The fungus partner likely colonized land long before plants and would have been a saprobic soil fungus, or a symbiotic fungus such as *Geosiphon* that associated with cyanobacteria (8).

The ectomycorrhiza is characterized by the presence of three structural components:

- a mantle of aggregated fungal hyphae which ensheaths the lateral root;
- a labyrinthine inward growth of hyphae between the epidermal and cortical cells; and
- an outwardly growing system of hyphal elements which form essential connections with both the soil, fruit bodies of the fungi forming the ectomycorrhizas (Plate 3) and other surrounding plants (4, 9, 10).

The extramatrical hyphae, the ectomycorrhizal mantle, and the intraradicular hyphal network are active metabolic entities that provide essential nutrient resources (e.g. phosphate, amino acids) to the host plant. These nutrient contributions are reciprocated by the provision of a stable carbohydrate-rich niche in the roots for the fungal partner, making the relationship a mutualistic symbiosis (11). The ecological performance of ectomycorrhizal fungi is a complex phenotype affected by many different traits and by environmental factors. Without any doubt, anatomical features (e.g. extension of the extramatrical hyphae) resulting from the symbiosis development is of paramount importance to the metabolic (and ecophysiological) fitness of the mature mycorrhiza (12). Understanding the complexity of the interactions between ectomycorrhizal symbionts and how this mutualistic association adapts and responds to changes in the biological, chemical, and physical properties of the rhizosphere remains a significant challenge for plant and microbial biologists. Identification of the primary determinants controlling the symbiosis development and its metabolic activity (e.g. P and N acquisition) will open the door to understanding the ecological fitness of the ectomycorrhizal symbiosis.

One can imagine that distinct but partially overlapping sets of fungal and plant factors are required and produced at each stage of the symbiotic infection. For most ectomycorrhizal associations, the stages involve:

- survival of hyphae in the rhizosphere;
- primary attachment of hyphal tips to host roots;
- invasion of root tissues;
- avoidance of and/or resistance to host immune defences;
- transfer/acquisition of photoassimilates; and
- evacuation of hyphae from colonized senescent roots to either a new root or an environmental reservoir (11, 13).

Ectomycorrhiza development involves striking alterations in root and hyphae morphology (e.g. increased hyphal branching, root tip swelling) (9, 10). The developmental processes in the plant involve architectural changes at the tissular and organ levels (Plate 3) (e.g. enhanced formation of root tips, root-hair suppression) as well as cellular differentiation that includes cell-wall (14) and cytoskeleton reorganization. In the mycobiont, the development of ectomycorrhiza involves the differentiation of structurally specialized fungal tissues with hyphae aggregation and dramatic alterations of hyphae morphogenesis (14). Mycobionts, belonging to a wide range of ascomycetous and basidiomycetous species, induce similar symbiotic structures in widely diverse host plant species (7), suggesting that ectomycorrhizal symbioses have likely evolved by exploiting some common core genetic programmes in their hosts, and a hypothetical set of symbiotic factors has been proposed and supported by various genetic and molecular studies (11, 13). Some of these primary functions likely include sensing the appropriate plant rhizospheric signals, cell-wall reorganization and membrane synthesis to build up the symbiotic interface, cytoskeleton rearrangements, and topological redistribution of nutrient transporters (11, 13). These mechanisms are likely to be similar if not identical between the numerous ectomycorrhiza morphotypes and between mutualistic and some parasitic symbioses. Identification of these primary genetic determinants (i.e. symbiosis-regulated genes) is a daunting task. It involves the typical gene-to-phenotype approach (i.e. identification of traits through characterization of gene expression and subsequent gene inactivation). However, it should be kept in mind that this powerful approach may be restricted by the fact that symbiosis development and activity may be partly determined by complex epistatic interactions among different genes showing subtle quantitative variation (6).

For optimal development of the symbiosis, symbionts need to coordinate complex developmental processes and, at the same time, sense and respond to novel physiological factors and environmental cues (11, 13). In both plant cells and fungal hyphae, responses should involve five main events:

- an initial stimulus,
- a subsequent generation and release of rhizospheric signals,
- secondary stimulus and perception,
- generation and transduction of intracellular signals, and
- subsequent changes in downstream biochemical process.

An understanding of the molecular communication that underlies the temporal and spatial control of genes involved in symbiosis development is now within reach, as more sophisticated techniques of functional genomics are applied to mycorrhizal interactions. The development of the technology to sequence expressed genes on a large scale and to analyse this DNA (bioinformatics, cDNA arrays) will have enormous impact on the way we think about the biology of mycorrhizal associations. These approaches have been used with success in the *Eucalyptus globulus–Pisolithus tinctorius* symbiosis, and a number of promising candidates for morphogenetic proteins that may be involved in ectomycorrhiza development have been identified (15). In addition, major roles have recently emerged for indolic compounds and flavonoids in regulating the early steps of mycorrhiza development and recent advances in these fields are discussed in this review.

2. Rhizospheric signals are involved in the symbiosis developmental sequence

The rhizospheric phase, or preinfection stage, commences before the host plant and its compatible fungal associate recognize each other as potential partners on a cellular basis. Exchange of signals between the partners is clearly a key step in a series of interaction events, leading to contact at the host surface and subsequent development of the microbial structures in the host roots (16). These signals coordinate and organize the responses of partner cells and, on some level, control their development (13, 17, 18). Determining the mechanisms that control the information flux between mycorrhizal fungi and root is presently a major challenge because fungal spore germination, chemoattraction of the mycelium by the root, attachment to the host, root penetration, and development of fungal multicellular structures in the root are dependent on precisely tuned mycobiont- and host-derived signals. To date, auxins, cytokinins, flavonoids, and alkaloids have been shown to play a role in the symbiosis initiation and development.

Host-released metabolites, such as the flavonol, rutin, and the cytokinin, zeatin, strikingly modified the hyphal morphology (19). Rutin stimulated *Pisolithus* growth when present in the growth medium at very low level, whereas zeatin modified hyphae branch angle (19). These rhizospheric molecules are therefore able to induce morphological changes similar to those observed during actual ectomycorrhizal development.

Auxins are involved in a number of developmental processes in plant, such as root elongation, lateral root development, meristem maintenance, and senescence (17). Indole-3-acetic acid is a naturally occurring auxin which is produced by many bacteria (e.g. *Sinorhizobium meliloti, Pseudomonas savastanoi, Erwinia herbicola*) (20) and fungi, including ectomycorrhizal fungi (17, 18). Many studies indicate that changes in auxin balance are a prerequisite for mycorrhiza organogenesis (17, 18, 21–26). Nevertheless, the role of indole-3-acetic acid or/and auxin-stimulated ethylene in the different stages of the ectomycorrhizal development is still unclear and has been challenged (6, 27). Ectomycorrhizal fungi and their exudates stimulate root proliferation (28). Thus, the ectomycorrhizal mycelium enhances uptake of soil nutrients by the host plant and multiplies the infection sites. The growth promotion of tree seedlings observed after mycorrhizal inoculation has been related to the biosynthesis and secretion of indole-3-acetic acid by symbiotic hyphae (22, 25, 29). The presence of plant-derived tryptophan in the rhizosphere could be sufficient for ectomycorrhizal fungi to enhance the biosynthesis of fungal indole-3-acetic acid (22). Pine inoculated with mutant *of Hebeloma cylindrosporum* strains overproducing indole-3-acetic acid produced an increased number of ectomycorrhizal roots (25), which presented a strikingly altered morphology (i.e. multiseriate Hartig net) (26), confirming that some morphogenetic steps controlling the mycorrhiza development are regulated by fungal indole-3-acetic acid. Whether tryptophan or/and other components of the plant exudates induces an indole-3-acetic acid amplification loop in the rhizospheric mycelium, the indole-3-acetic acid synthesis must be tightly controlled or compensated by other factors since above a certain level, exogenously supplied indole-3-acetic acid inhibits root development. Inhibitors of polar auxin transport, such as 2,3,5-triiodobenzoic acid, restrict the stimulation of lateral root formation and the colonization of the tap-root cortex of conifer seedlings by ectomycorrhizal fungi (23). The fungal alkaloid, hypaphorine, a betaine of tryptophan, is the major indolic compound isolated from the ectomycorrhizal fungus *Pisolithus* (30). It is produced in large amount by this fungus during mycorrhiza development (31) and upon triggering by root exudates (31, H. Lagrange and F. Lapeyrie, unpublished results). Hypaphorine acts as an auxin antagonist (32) and it affects root hairs of *Eucalyptus* seedlings by reducing their elongation rate, whilst it has no activity on root elongation and development (33). At inhibitory concentrations (100 μM), hypaphorine also induced a transitory root hair swelling (Fig. 1). At higher hypaphorine concentrations (500 μM and above), root hair elongation stops 15 minutes after hypaphorine application. However, the root-hair initiation from trichoblasts is not affected by hypaphorine, confirming that initiation and elongation are distinct phenomena as previously reported using various mutants affected on root-hair development (34). There is emerging evidence that actin microfilament, which extends as long cables in untreated eucalypt root hairs, are markedly induced to form thicker bundles (F. A. Ditengou *et al.*, unpublished results) following the application of hypaphorine. Whether these drastic cytoskeleton changes are related to interactions with calcium channels, cofilin/actin-depolymerizing proteins, and/or auxin-signalling pathways is currently under investigation.

Thus, during ectomycorrhizal development, the absence of root hairs might be due partly to inhibitors of root-hair elongation, such as hypaphorine, transferred by colonizing hyphae to differentiating epidermal cells. Mycorrhizal colonization of *Eucalyptus* seedlings is initiated in the root-cap region and then propagates by an acropetal extension of root and fungal tissue (Fig. 2) (35, 36). Only new epidermal cells formed after fungal invasion of the apex are involved in this infection process, and there is no backward-formation of mycorrhiza along previously differentiated root tissues. Epidermal cells of eucalypt, those bearing root hairs in uninfected root

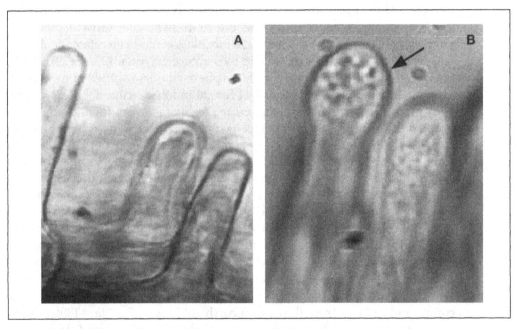

Fig. 1 The morphological effect of hypaphorine, secreted by the ectomycorrhizal fungus *Pisolithus*, on root hairs of *Eucalyptus globulus*. (A) Root hair treated with water appears normal without deformation. (B) In response to hypaphorine (100 μM), root hairs show a swelling (arrows) after 45 minutes of incubation. (After reference 33.)

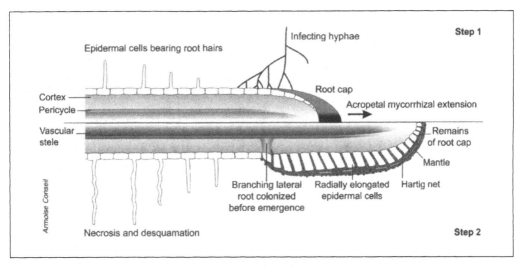

Fig. 2 Dynamic of the colonization of the *Eucalyptus* root by the ectomycorrhizal fungus *Pisolithus* during symbiosis development. Step 1, colonization by the infecting hyphae is initiated in the root-cap region and then propagates by an acropetal extension of root and fungal tissues. Step 2, epidermal cells, which initiate root hairs in non-mycorrhizal root zones, elongate radially as a specific response to the hyphae. The intracellular hyphal network, so-called Hartig net, develops only between these elongated epidermal cells. (Drawn after data from references 33, 35, and 36.)

portions, elongate radially as a specialized response to the fungus (35, 36). The intercellular network of hyphae (i.e., the Hartig net) develops only between these elongated epidermal cells which therefore remain a part of the active ectomycorrhiza. In such circumstances, root hairs will not emerge on colonized root surface, suggesting that their development has been inhibited very early either mechanically or chemically (e.g. through the action of hypaphorine), by the fungus (33).

In addition to its rapid effect on the cytoskeleton of host root hairs, hypaphorine has additional long-term effects on the expression of auxin-regulated genes. The transcript level of the hypaphorine- and auxin-responsive glutathione-*S*-transferase, *EgHypar*, in eucalypt roots was up-regulated after inoculation with *Pisolithus* and the addition of *Pisolithus*-released hypaphorine (37). Group III glutathione-*S*-transferases, to which belongs EGHYPAR, are hormone responsive and may have a function in the binding and transport of auxins (38). This suggests that the rhizospheric hypaphorine might control the auxin balance in the colonized *Eucalyptus globulus* roots (33). Although hypaphorine secretion appears to be restricted to *Pisolithus* species (30), other ectomycorrhizal fungi (e.g. *Paxillus involutus*) are able to secrete unknown auxin antagonists (A. Jambois and F. Lapeyrie, unpublished results).

This brief review has attempted to shed light on recent advances in how the ectomycorrhizal symbionts respond to rhizospheric signals, including hormones, and how these various signals interact in order to execute the appropriate developmental responses (Fig. 3). The multiplicity of identified signals (auxins, alkaloids, cytokinins, flavonols) has revealed that the situation is more complicated that first

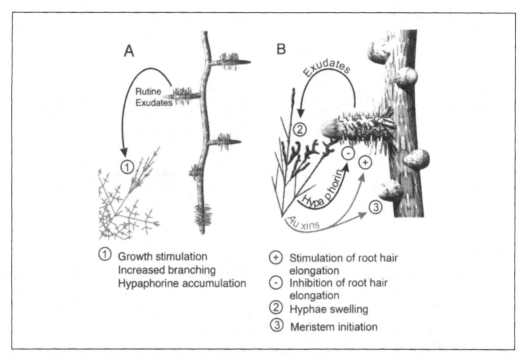

Fig. 3 The cross-talk involving rhizospheric signals during the formation of the *Eucalyptus/Pisolithus* ectomycorrhiza. (A) Root exudates stimulate hyphae growth and hypaphorine accumulation in hyphae, and alter hyphae branching. Increased hyphal growth and branching could be mimicked by incubation in the presence of rutin or zeatin. (B) Root exudates induce the hyphal tip swelling, whereas hypaphorine, released by the fungus, inhibits root-hair elongation and fungal auxins stimulate the proliferation of lateral roots.

anticipated (39). These signals can act in a synergistic (rutin/zeatin) or antagonistic (indole-3-acetic acid /hypaphorine) manner. Signal perception may culminate in the induction of downstream target genes (e.g. *EgHypar*) whose expression likely underpins physiological and/or development responses (short root proliferation, alteration of hyphal branching). Although many pieces of the puzzle remain to be elucidated, it seems inescapable that the cross-talk between rhizospheric metabolites, the hormonal balance, and signalling networks involving Ca^{2+} play an important role in coordinating the execution of the appropriate responses and it will be fascinating to learn more about the multiple facets of this communication network. In this regard, advances in the elucidation of symbiosis signalling pathways can be accelerated by the identification of more target genes by cDNA array technology.

3. Detection and analysis of gene expression during mycorrhiza development

It is increasingly clear that developmental pathways leading to the ectomycorrhizal symbiosis can be considered as modular, and that developmental transitions are accompanied by global changes in the expression of specific complements of genes under the control of rhizospheric and intracellular signals (see above) (6, 11). Goals of primary importance are achieving a complete understanding of the mechanisms induced in both symbionts at each stage of the infection and of the physiology of the symbiosis. Proteomics approaches based on two-dimensional (2D) gel electrophoresis has been used to characterize up- and down-regulated proteins in *Pisolithus/ Eucalyptus globulus* (40, 41), *Amanita muscaria/Picea abies* (42), *Paxillus involutus/Betula pendula* (43), and *Suillus bovinus/Pinus sylvestris* (44) associations. These investigations showed that ectomycorrhizal development leads to an alteration of protein patterns in both symbionts. However, this approach is limited because only a restricted set of proteins can be visualized on 2D gels. Whilst such studies have been fruitful in the past, their potential use in developing a complete and accurate understanding of the symbiotic factors during the course of the mycorrhiza development is limited.

Forward genetics is currently driving research into how plants respond to pathogen attack: the process involves the isolation of mutants in which a particular resistance response has been perturbed, and the subsequent characterization of the corresponding genes. A lack of genetic tools (i.e. ectomycorrhizal fungi are not yet amenable to gene inactivation) has prevented the application of this approach to ectomycorrhiza, and has led to the development of alternative molecular techniques for the identification of symbiosis-regulated genes. Subtractive hybridization and differential display were developed to identify plant and fungal genes that are induced upon symbiosis development in ectomycorrhizal associations involving *Pisolithus/Eucalyptus globulus* (45), *Laccaria bicolor/Pinus resinosa* (46, 47), and *Amanita muscaria/Picea abies* (48). These investigations confirmed that ectomycorrhiza development is accompanied by striking changes in gene expression at the transcriptional level and allowed the identification of a dozen symbiosis-regulated genes.

Most of the genome sequencing efforts so far have targeted model (*Saccharomyces cerevisiae, Neurospora crassa, Aspergillus nidulans*) or pathogenic (*Candida albicans, Pneumocystis carinii, Magnaporthea grisea, Ustilago maydis*) fungi (e.g. http://gene. genetics.uga.edu/). Fortunately, sequencing efforts increasingly focus on ectomycorrhizal fungi (*Amanita muscaria, Hebeloma cylindrosporum, Pisolithus tinctorius, Tuber borchii*) (15, 49, 50; P. Bonfante, U. Nehls, H. Sentenac, personal communication). As the number of expressed sequence tags increases, comparisons across genera, species, ecotypes, and strains of symbiotic fungi will become possible through 'digital Northern'. With multiple sequencing programmes dealing with pathogenic and mutualistic fungi, we will have in the near future an unparalleled opportunity to ask which genetic features are responsible for traits involved in pathogenesis and symbiosis. A few of the many possible breakthroughs will be in:

- characterization of common transduction networks,
- identification of new surface proteins that play critical roles in plant–fungus interactions, and
- new insights into unique metabolic routes critical for mycorrhiza functioning.

Although modulation of the symbiont interaction involves differential mRNA stability, post-translational modifications, regulation of transport mechanisms, and protein degradation, etc., one of the key mechanisms in fungal gene regulation takes place at the transcriptional level. Consequently, detection of even subtle gene expression modulations provides a comprehensive framework for studying events that affect cellular metabolism and regulation on a genomic scale. Quantitative analysis of the transcriptome has become possible through 'hybridization signature' methods, which allow large-scale measurement of gene expression (51–53). cDNA array analyses are currently providing efficient means of acquiring large amounts of biological information for identifying processes involved in plant–microbe interactions (54–56). In terms of understanding the function of genes involved in the symbiotic interaction, knowing when, where, and to what extent a gene is expressed is central to understanding the activity and biological roles of its encoded protein. In addition, changes in the multigene patterns of expression accompanying the symbiosis formation can provide clues about regulatory mechanisms and broader cellular functions and biochemical pathways taking place in mycorrhizas.

3.1   Analysis of the transcriptome of the *Eucalyptus/Pisolithus* ectomycorrhiza

Not all sequencing projects have to be massive to be significant. Modest projects can be powerful (and not labour or cost intensive) if the appropriate species or tissue is selected (57). To begin to identify cellular functions expressed in the symbiosis, an expressed sequence tags/cDNA array programme has been developed (15, 49, 50) on *Pisolithus tinctorius* and *Eucalyptus globulus* symbiosis, an ectomycorrhizal association widely found in Australasia and introduced with eucalyptus plantations worldwide. This investigation provided insights in the global gene activities involved in ectomycorrhiza development and function. About 80 symbiosis-regulated genes (17%) were identified by differential screening of 480 arrayed cDNAs between free-living partners and symbiotic tissues (15). A similar proportion of *in planta*-induced genes has been found in *Uromyces fabae* during haustoria formation (54). By subtractive suppressive hybridization, 200 symbiosis-regulated genes were identified by screening about 4000 clones (50). The screened clones represent about 10% of the number of clones estimated to comprehensively survey the *P. tinctorius* and *E. globulus* genomes for mycorrhiza-regulated genes. Even this modest collection of genes begins to provide an indication of symbiosis environment as perceived by the symbionts. The number of symbiosis-regulated genes displaying similarity to genes involved in cell-wall and membrane synthesis, stress-defence response, protein degradation (in plant cells), and protein synthesis (in hyphae) suggests a highly dynamic environment in which symbionts are sending and receiving signals, exposed to high levels of stress conditions and remodelling tissues. A striking result of these study is the fact that all genes investigated are common to the non-symbiotic and symbiotic stages. At the developmental stage studied, symbiosis development does not induce the expression of ectomycorrhiza-specific genes, but a marked change in the gene expression in the partners.

3.2 Genes involved in signalling, adaptation, and defence reactions

The necessity to adapt to a rapidly changing environment is of great importance to the colonizing mycelium. Within hours, the ectomycorrhizal hyphae must transit the rhizosphere, colonize the root surface, and grow in the acidic apoplastic space of the host cells. Each step of this life cycle requires a variety of genes whose products play a role in adaptation: coping with changes in pH, enhanced fluxes of nutrients, the presence of radical oxygen species and plant toxics, ... Sensing this novel environment and coordination of fungal and plant developments and metabolisms require signalling networks. Several genes with a putative function in signal transduction are highly expressed (e.g. *receptor of protein kinase C*) and/or up-regulated (e.g. the beta-transducin, *periodic tryptophan protein* 2) in symbiotic tissues (15); they represent about 13% of the cloned genes in *E. globulus/Pisolithus* ectomycorrhiza. These include genes with homology to calmodulin, *ras*, heterotrimeric guanine nucleotide-binding proteins, and protein kinases and phosphatases. Transcripts coding for fungal metallothioneins are abundant in ectomycorrhiza and the expression of the thioredoxin reductase and glutathione-*S*-transferase (*EgHypar*) genes are upregulated (2.5-fold) in the symbiotic mycelium (15). This suggests that root-colonizing hyphae experience oxidative/heavy metal stresses. This may arise through exposure to radical oxygen species produced by the host which responds to the presence of the fungus by eliciting general defence reactions (58–60) as revealed by the striking up-regulation of the elicitor-induced isoflavone 7-*O*-methyltransferase and stress-inducible proteins (15).

Surprisingly, little is known about the plant genetic requirements for symbiotic interactions. For example, we do not know which plant genes are required for the accommodation of mycorrhizal fungi and whether some are common to various mutualistic and pathogenic plant–microbe interactions. cDNA array analyses (15) have shown that several of the plant-regulated genes showed a decreased expression in ectomycorrhizal tissues (e.g. ATP synthase, cytochrome *c* oxidase, Zn-binding protein), confirming previous results that the symbiosis formation induces a down-regulation of protein synthesis in colonized roots (40, 41). Degradation of these proteins likely involved the ubiquitin/proteasome pathway, explaining the increased expression of plant ubiquitin-conjugating enzymes and RING domain-containing proteins. The existence, therefore, of drastic protein degradation in the colonized roots serves to add to the growing number of studies that indicate that the boundary between symbionts and pathogens are not as distinct as once believed. Protein degradation may also result from reprogramming of root cells as a result of changes in the hormonal balance (e.g. changes in auxin levels).

3.3   Genes coding for structural proteins

Within mycorhiza-regulated genes, a large set of sequences code for several members of the multigene 32 kDa symbiosis-regulated acidic polypeptides (61) and hydrophobin (62, 63, 64) families. cDNA array analysis showed that the expression of transcripts coding these proteins is strikingly up-regulated (up to fivefold) during the mycorrhiza formation. The expression of these genes, observed as being modulated using arrays, have also been tested in Northern blot and their up-regulation was confirmed. In addition, comparing the up-regulation of transcript levels of 32 kDa symbiosis-regulated acidic polypeptides and the increased concentration of 32 kDa symbiosis-regulated acidic polypeptides in ectomycorrhiza (40, 41, 61) showed that there was a good correlation between changes in protein synthesis and transcript levels. The expression of the various hydrophobin paralogues is increased six- to eightfold in the symbiotic tissues. *Eucalyptus/Pisolithus* ectomycorrhizas are often found in air pockets in soil in the wild. Such mycorrhizas formed in air are invariably non-wettable and water-repellent (65). The most likely explanation for this lies in the observed deposition of hydrophobins. Thus, it appears that *Pisolithus* hyphae preferentially expresses some sets of structural proteins in its wall during the formation of the symbiosis, confirming that for fungi interacting with plants, the cell wall and its surface are major players in the host–symbiont interface (9, 66). The concerted induction of these structural proteins implies a tight control of expression among genes with potentially related function and opens the door for comparative studies using the conserved elements in the regulatory regions of these genes.

3.4   Genes involved in metabolism

Our initial screen did not identify symbiosis-regulated genes involved in nutrient scavenging (hexose- and amino acid transporters, enzymes of the biosynthetic pathway such as trehalose synthetase). This bias is due partly to the symbiosis stage investigated (i.e. mantle formation). Screens carried out at later developmental stages when the Hartig net and symbiotic interfaces are developed will certainly identify these metabolic genes. Hexose transporters of the symbionts in the *Amanita muscaria/ Picea abies* and *Paxillus involutus/Betula pendula* symbioses are highly regulated (67–69). Similarly, *Amanita muscaria* phenylalanine ammonium lyase gene is likely regulated in ectomycorrhiza through changes in nitrogen and sugar levels (70). There is, indeed, much to be learned about the growth physiology of the hyphae within host environments and remodelling of root metabolism feeding the mycobiont (71, 72). These findings will in turn directly reflect upon the nature of the host compartments (e.g. apoplastic space) or tissues (senescent cells) in which the mycobiont resides.

3.5   Genes of unknown function

The last class of ectomycorrhizal genes identified by tag sequencing and differential screenings are those that are predicted to encode proteins with no similarities to database sequences or that are similar to hypothetical proteins of unknown function. This intriguing class of genes (up to 50%) has been a major class identified by all previous sequencing programmes and differential screens in plant–microbe interactions. This feature points to our still-fledging knowledge of the roles that many gene products play in the growth and infection process of pathogenic / symbiotic microbes and points to the necessity for studies designed to elucidate the roles of these many genes.

The initial cDNA array screens for *E. globulus* and *P. tinctorius* symbiosis-regulated genes was not comprehensive by any means, and thus it is expected that dozens, perhaps hundreds, more ectomycorrhiza-regulated genes involved in symbiont recognition, signalling networks, nutrient acquisition and metabolite biosynthesis await identification. Current work is directed toward completing analysis of the identified symbiosis-regulated genes, determining the role and function of selected mycorrhiza-regulated gene products and the contribution of each to the symbiosis development. This will be supplemented with knowledge of the effects of specific rhizospheric and intracellular signals, such as indole-3-acetic acid, hypaphorine, zeatin, and rutin, on their expression.

4. Conclusions

There have been substantial advances in recent years in our understanding of developmental mechanisms leading to the formation of the ectomycorrhizas, only some of which have been highlighted here. The present glimpse into the complex array of signalling molecules and ectomycorrhiza gene expression will provide the basis for a more precise molecular dissection of the complex genetic networks that control symbiosis development and function. The identified symbiosis-regulated genes might be especially interesting targets for ongoing gene disruption technology. Further studies are now needed to delineate the functions of both the known and the novel genes that are differentially expressed during ectomycorrhiza development. A difficult challenge will be to understand how mycorrhiza-regulated genes exert their developmental effects at the cellular and supracellular levels. In this regard, one anticipates that increased attention will be given to the role of the cytoskeleton and the interfacial matrix in developmental processes.

It will be far harder to define genes that influence symbiont fitness. Any gene which provides the mycobiont with a growth advantage could easily influence how beneficial a particular strain is within a given host (12). Therefore, dissecting the molecular mechanisms of symbiosis requires both the identification of the functions of individual genes as well as knowledge of how genes interact to form complex traits such as those expressed in a mutualistic symbiosis. It is difficult to predict the total number of genes involved in symbiosis and, hence, the scope for each mycorrhizal association to be unique. There are more than 5000 different ecto-mycorrhizal associations (7) and each and every ectomycorrhizal type may express a specific set of genes. Despite this, a variety of morphological and molecular param-eters can be used to classify ectomycorrhizal symbioses into discrete molecular classes for further investigation. Symbiotic molecular phenotypes may in the future be correlated to ecological phenotypes. It will be rewarding to compare plant and fungal gene expression profiles among different ectomycorrhizal associations, mycorrhizal symbioses, and other plant–fungus interactions. Genomic approaches, such as transcript profilings based on cDNA microarrays, have the potential to provide 10–100 times more than conventional means and thus expression profiles may identify potential overlaps in the genetic make-up of plant–fungus interactions.

Acknowledgements

This paper is based on a contribution delivered by F. M. at the Keystone Symposium 'Signals and Signal Perception in Biotic Interactions in Plants' (February 22–27, 2000). S. D., H. L., and C. V. were supported by Doctoral Scholarships from the Ministère de l'Education Nationale, de la Recherche et de la Technologie. F. A. D. was supported by a fellowhip from the Government of Gabon. Investigations carried out in our laboratory were supported by grants from the Groupement de Recherches et d'Etude des Génomes and the INRA (Action Transversale Microbiologie Fondamentale).

References

1. Chanway, C. P., Turkington, R. and Holl F. B. (1991) Ecological implications of specificity between plants and rhizosphere micro-organisms. *Adv. Ecol. Res.*, **21**, 121.
2. Arshad, M. and Frankenberger, W. T. (1998) Plant growth-regulating substances in the rhizosphere: microbial production and functions. *Adv. Agron.*, **62**, 45.
3. Read, D. J. (1991) Mycorrhizas in ecosystems. *Experientia*, **47**, 376.
4. Simard, S. W., Perry, D. A., Jones, M. D., Myrold, D. D., Durall, D. M. and Molina, R. (1997) Net transfer of carbon between ectomycorrhizal tree species in the field. *Nature*, **388**, 579.
5. Meharg, A. A. and Cairney, J. W. G. (2000) Co-evolution of mycorrhizal symbionts and their hosts to metal-contaminated environments. *Adv. Ecol. Res.*, **30**, 69.
6. Smith, S. E. and Read, D. J. (1997) *Mycorrhizal Symbiosis*, 2nd edn. Academic Press, London.
7. Hibbett, D. S., Gilbert, L. B. and Donoghue, M. J. (2000) Evolutionary instability of ectomycorrhizal symbioses in basidiomycetes. *Nature*, **407**, 506.
8. Selosse, M. A. and Le Tacon, F. (1998) The land flora: a phototroph-fungus partnership? *Trends Ecol. Evol.*, **13**, 15.
9. Hardham, A. R. and Mitchell, H. J. (1998) Use of molecular cytology to study the structure and biology of phytopathogenic and mycorrhizal fungi. *Fungal Genet. Biol.*, **24**, 252.
10. Kottke, I. and Oberwinkler, F. (1987) The cellular structure of the Hartig net: coenocytic and transfer cell-like organization. *Nord. J. Bot.*, **7**, 85.
11. Martin, F. and Tagu, D. (1999) Developmental biology of a plant-fungus symbiosis: the ectomycorrhiza. In Varma, A. K. and Hock, B. (eds), *Mycorrhiza: Structure, Function, Molecular Biology and Biotechnology*, 2nd edn, p. 51. Springer-Verlag, Berlin and Heidelberg.
12. Burgess, T., Dell, B. and Malajczuk, N. (1994) Variation in mycorrhizal development and growth stimulation by 20 *Pisolithus* isolates inoculated on to *Eucalyptus grandis* W. Hill ex Maiden. *New Phytol.*, **127**, 731.

13. Martin, F., Lapeyrie, F. and Tagu, D. (1997) Altered gene expression during ecto-mycorrhiza development. In Lemke, P. and Caroll, G. (eds), *The Mycota.* Vol. VI. *Plant Relationships*, p. 223. Springer-Verlag, Berlin.

14. Tagu, D. and Martin, F. (1996) Molecular analysis of cell wall proteins expressed during the early steps of ectomycorrhiza development. *New Phytol.*, **133**, 73.

15. Voiblet, C., Duplessis, S., Encelot, N. and Martin, F. (2000) Identification of symbiosis-regulated genes in *Eucalyptus globulus-Pisolithus tinctorius* ectomycorrhiza by differential hybridization of arrayed cDNAs. *Plant J.*, **25**, 181.

16. Horan, D. P. and Chilvers, G. A. (1990) Chemotropism; the key to ectomycorrhizal formation? *New Phytol.*, **116**, 297.

17. Gogala, N. (1991) Regulation of mycorrhizal infection by hormonal factors produced by hosts and fungi. *Experientia*, **47**, 331.

18. Beyrle, H. (1995) The role of phytohormones in the function and biology of mycorrhizas. In Varma, A. K. and Hock, B. (eds), *Mycorrhiza: Structure, Molecular Biology and Function*, p. 365. Springer, Berlin, Heidelberg, New York.

19. Lagrange, H., Jay-Allemand, C. and Lapeyrie, F. (2001) Rutin, the phenolglycoside from eucalyptus root exudates, stimulates *Pisolithus* hyphal growth at picomolar concentrations. *New Phytol.*, **149**, 349.

20. Hamill, J. D. (1993) Alterations in auxin and cytokinin metabolism of higher plants due to expression of specific genes from pathogenic bacteria: a review. *Austr. J. Plant Physiol.*, **20**, 405.

21. Rupp, L. A. and Mudge, K. W. (1985) Ethephon and auxin induce mycorrhiza-like changes in the morphology of root organ cultures of Mugo pine. *Physiol. Plant.*, **64**, 316.

22. Rupp, L. A., Mudge, K. W. and Negm F. B. (1989) Involvement of ethylene in ectomycorrhiza formation and dichotomous branching of roots of mugo pine seedlings. *Can. J. Bot.*, **67**, 477.

23. Karabaghli-Degron, C., Sotta, B., Bonnet, M., Gay, G. and Le Tacon F. (1998) The auxin transport inhibitor 2,3,5-triiodobenzoic acid (TIBA) inhibits the stimulation of *in vitro* lateral root formation and the colonization of the tap-root cortex of Norway spruce (*Picea abies*) seedlings by the ectomycorrhizal fungus *Laccaria bicolor*. *New Phytol.*, **140**, 723.

24. Kaska, D. D., Myllylä, R. and Cooper, J. B. (1999) Auxin transport inhibitors act through ethylene to regulate dichotomous branching of lateral root meristems in pine. *New Phytol.*, **142**, 49.

25. Gay, G., Normand, L., Marmeisse, R., Sotta, B. and Debaud, J. C. (1994) Auxin over-producer mutants of *Hebeloma cylindrosporum* Romagnési have increased mycorrhizal activity. *New Phytol.*, **128**, 645.

26. Gea, L., Normand, L., Vian, B. and Gay, G. (1994) Structural aspects of ectomycorrhiza of *Pinus pinaster* (Ait.) Sol. formed by an IAA-overproducer mutant of *Hebeloma cylindrosporum* Romagnési. *New Phytol.*, **128**, 659.

27. Hampp, R., Ecke, M., Schaeffer, C., Wallenda, T., Wingler, A., Kottke, I. and Sundberg, B. (1996) Axenic mycorrhization of wild type and transgenic hybrid aspen expressing T-DNA indoleacetic acid-biosynthetic genes. *Trees*, **11**, 59.

28. Carnero Diaz, E., Martin, F. and Tagu, D. (1996) Eucalypt α-tubulin: cDNA cloning and increased level of transcripts in ectomycorrhizal root system. *Plant Mol. Biol.*, **31**, 905.

29. Frankenberger, W. T. and Poth, M. (1987) Biosynthesis of indole-3-acetic acid by the pine ectomycorrhizal fungus *Pisolithus tinctorius*. *Appl. Environ. Microbiol.*, **53**, 2908.

30. Béguiristain, T., Côté, R., Rubini, P., Jay-Allemand, C. and Lapeyrie, F. (1995) Hypaphorine accumulation in hyphae of the ectomycorrhizal fungus *Pisolithus tinctorius*. *Phytochemistry*, **40**, 1089.

31. Béguiristain, T. and Lapeyrie, F. (1997). Host plant stimulates hypaphorine accumulation in *Pisolithus tinctorius* hyphae during ectomycorrhizal infection while excreted fungal hypaphorine controls root hair development. *New Phytol.*, **136**, 525.

32. Ditengou, F. A. and Lapeyrie, F. (2000) Hypaphorine from the ectomycorrhizal fungus *Pisolithus tinctorius* counteracts activities of indole-3-acetic acid and ethylene but not synthetic auxins in eucalypt seedlings. *Mol. Plant–Microbe Interact.*, **13**, 151.

33. Ditengou, F. A., Béguiristain, T. and Lapeyrie, F. (2000) Root hair elongation is inhibited by hypaphorine, the indole alkaloid from the ectomycorrhizal fungus *Pisolithus tinctorius*, and restored by indole-3-acetic acid. *Planta*, **211**, 722.

34. Baskin, T. I., Betzner, A. S., Hoggart, R., Cork, A. and Williamson, R. E. (1992) Root morphology mutants in *Arabidopsis thaliana*. *Austr. J. Plant Physiol.*, **19**, 427.

35. Horan, D. P., Chilvers, G. A. and Lapeyrie, F. (1988) Time sequence of the infection process in eucalypt ectomycorrhizas. *New Phytol.*, **109**, 451.

36. Chilvers, G. A. (1968) Low power electron microscopy of the root cap region of eucalypt mycorrhizas. *New Phytol.*, **67**, 663.

37. Nehls, U., Beguiristain, T., Ditengou, F. A., Lapeyrie, F. and Martin, F. (1998) The expression of a symbiosis-regulated gene in eucalypt roots is regulated by auxins and hypaphorine, the tryptophan betaine of the ectomycorrhizal basidiomycete *Pisolithus tinctorius*. *Planta* , **207**, 296.

38. Edwards, R., Dixon, D. P. and Walbot, V. (2000) Plant glutathione *S*-transferases: enzymes with multiple functions in sickness and in health. *Trends Plant Sci.*, **5**, 193.

39. Slankis, V. (1950) Effect of napthaleneacetic acid on dichotomous branching of isolated roots of Pinus sylvestris. *Physiol. Plant.*, **3**, 40.

40. Hilbert, J. L., Costa, G. and Martin, F. (1991) Regulation of gene expression in ectomycorrhizas. Early ectomycorrhizins and polypeptide cleansing in eucalypt ectomycorrhizas. *Plant Physiol.*, **97**, 977.

41. Burgess, T., Laurent, P., Dell, B., Malajczuk, N. and Martin, F. (1995) Effect of the fungal isolate infectivity on the biosynthesis of symbiosis-related polypeptides in differentiating eucalypt ectomycorrhiza. *Planta*, **195**, 408.

42. Guttenberger, M. and Hampp, R. (1992) Ectomycorrhizins – symbiosis-specific or artifactual polypeptides from ectomycorrhizas? *Planta*, **188**, 129.

43. Simoneau, P., Viemont, J. D., Moreau, J. C. and Strullu, D. G. (1993) Symbiosis-related polypeptides associated with the early stages of ectomycorrhiza organogenesis in birch (*Betula pendula* Roth). *New Phytol.*, **124**, 495.

44. Tarkka, M., Niini, S. S. and Raudaskoski, M. (1998) Developmentally regulated proteins during differentiation of root system and ectomycorrhiza in Scots pine (*Pinus sylvestris*) with *Suillus bovinus*. *Physiol. Plant.*, **104**, 449.

45. Tagu, D., Python, M., Crétin, C. and Martin, F. (1993) Cloning symbiosis-related cDNAs from eucalypt ectomycorrhizas by PCR-assisted differential screening. *New Phytol.*, **125**, 339.

46. Kim, S. J., Zheng, J., Hiremath, S. T. and Podila, G. K. (1998) Cloning and characterization of a symbiosis-related gene from an ectomycorrhizal fungus *Laccaria bicolor*. *Gene*, **222**, 203.

47. Kim, S. J., Bernreuther, D., Thumm, M. and Podila, G. K. (1999) *LB-AUT7*, a novel symbiosis-regulated gene from an ectomycorrhizal fungus, *Laccaria bicolor*, is functionally related to vesicular transport and autophagocytosis. *J. Bacteriol.*, **181**, 1963.

48. Nehls, U., Mikolajewski, S., Ecke, M. and Hampp, R. (1999) Identification and expression-analysis of two fungal cDNAs regulated by ectomycorrhiza and fruit body formation. *New Phytol.*, **144**, 195.

49. Tagu, D. and Martin, F. (1995) Expressed sequence tags of randomly selected cDNA clones from *Eucalyptus globulus-Pisolithus tinctorius* ectomycorrhiza. *Mol. Plant–Microbe Interact.*, **8**, 781.

50. Voiblet, C. and Martin, F. (2000) Identifying symbiosis-regulated genes in *Eucalyptus globulus-Pisolithus tinctorius* ectomycorrhiza using suppression subtractive hybridization and cDNA arrays. In de Wit, P. G. M., Bisseling, T. and Stiekema, W. (eds), *Biology of Plant–Microbe Interactions*, p. 208. International Society for Molecular Plant–Microbe Interactions, St Paul, MN.

51. Schena, M., Shalon, D., Davis, R. W. and Brown, P. O. (1995) Quantitative monitoring of gene expression patterns with a complementary DNA microarray. *Science*, **270**, 467.

52. Wodicka, L., Dong, H. L., Mittmann, M., Ho, M. H. and Lockhart, D. J. (1997) Genome-wide expression monitoring in *Saccharomyces cerevisiae*. *Nature Biotech.*, **15**, 1359.

53. Desprez, T., Amselem, J., Caboche, M. and Höfte, H. (1998) Differential gene expression in *Arabidopsis* monitored using cDNA arrays. *Plant J.*, **14**, 643.

54. Hahn, M. and Mendgen, K. (1997) Characterization of *in planta* induced rust genes isolated from a haustorium-specific cDNA library. *Mol. Plant–Microbe Interact.*, **10**, 427.

55. Györgyey, J., Vaubert, D., Jiménez-Zurdo, J. I., Charon, C., Troussard, L., Kondorosi, A. and Kondorosi, E. (2000) Analysis of *Medicago truncatula* nodule expressed sequence tags. *Mol. Plant–Microbe Interact.*, **13**, 62.

56. Schenk, P. M., Kazan, K., Wilson, I., Anderson, J. P., Richmond, T., Somerville S. C., Manners, J. M. (2000) Coordinated plant defense responses in *Arabidopsis* revealed by microarray analysis *Proc. Natl Acad Sci USA* **97**, 11655.

57. Browse, J. and Coruzzi, G. (2000) Physiology and metabolism. Two old grannies catch fire in the new millennium. Editorial overview. *Curr. Opin. Plant Biol.*, **3**, 179.

58. Feugey, L., Strullu, D. G., Poupard, P. and Simoneau, P. (1999) Induced defence responses limit Hartig net formation in ectomycorhizal birch roots. *New Phytol.*, **144**, 541.

59. Salzer, P., Hebe, G., Reith, A., Zitterell-Haid, B., Stransky, H., Gaschler, K. and Hager, A. (1995) Rapid reactions of spruce cells to elicitors released from the ectomycorrhizal fungus *Hebeloma crustuliniforme*, and inactivation of these elicitors by extracellular spruce cell enzymes. *Planta*, **198**, 118.

60. Schwacke, R. and Hager, A. (1992) Fungal elicitors induce a transient release of active oxygen species from cultured spruce cells that is dependent on Ca^{2+} and protein-kinase activity. *Planta*, **187**, 136.

61. Laurent, P., Voiblet, C., Tagu, D., De Carvalho, D., Nehls, U., De Bellis, R., Balestrini, R., Bauw, G., Bonfante, P. and Martin, F. (1999) A novel class of ectomycorrhiza-regulated cell wall polypeptides in *Pisolithus tinctorius*. *Mol. Plant–Microbe Interact.*, **12**, 862.

62. Tagu, D., Nasse, B. and Martin, F. (1996) Cloning and characterization of hydrophobins –encoding cDNAs from the ectomycorrhizal basidiomycete *Pisolithus tinctorius*. *Gene*, **168**, 93.

63. Wessels, J. G. H. (1993) Wall growth, protein excretion and morphogenesis in fungi. *New Phytol.*, **123**, 397.

64. Kershaw, M. J. and Talbot, N. J. (1998) Hydrophobins and repellents: proteins with fundamental roles in fungal morphogenesis. *Fungal Genet. Biol.*, **23**, 18.

65. Vesk, P. A., Ashford, A. E., Markovina, A. L. and Allaway, W. G. (2000) Apoplasmic barriers and their significance in the exodermis and sheath of *Eucalyptus pilularis-Pisolithus tinctorius* ectomycorrhizas. *New Phytol.*, **145**, 333.

66. Martin, F., Laurent, P., De Carvalho, D., Voiblet, C., Balestrini, R., Bonfante, P. and Tagu, D. (1999) Cell wall proteins of the ectomycorrhizal basidiomycete *Pisolithus tinctorius*: identification, function, and expression in symbiosis. *Fungal Genet. Biol.*, **27**, 161.
67. Nehls, U., Wiese, J., Guttenberger, M. and Hampp, R. (1998) Carbon allocation in ectomycorrhizas: identification and expression analysis of an *Amanita muscaria* monosaccharide transporter. *Mol. Plant–Microbe Interact.*, **11**, 167.
68. Nehls, U., Wiese, J. and Hampp, R. (2000) Cloning of a *Picea abies* monosaccharide transporter gene and expression – analysis in plant tissues and ectomycorrhizas. *Trees*, **14**, 334.
69. Wright, D. P., Scholes, J. D., Read, D. J. and Rolfe, S. A. (2000) Changes in carbon allocation and expression of carbon transporter genes in *Betula pendula* Roth. colonized by the ectomycorrhizal fungus *Paxillus involutus* (Batsch) Fr. *Plant Cell Environ.*, **23**, 39.
70. Nehls, U., Ecke, M. and Hampp, R. (1999) Sugar- and nitrogen-dependent regulation of an *Amanita muscaria* phenylalanine ammonium lyase gene. *J. Bacteriol.*, **181**, 1931.
71. Blaudez, D., Chalot, M., Dizengremel, P. and Botton, B. (1998) Structure and function of the ectomycorrhizal association between *Paxillus involutus* and *Betula pendula*. II. Metabolic changes during mycorrhiza formation. *New Phytol.*, **138**, 543.
72. Cairney, J. W. G. and Burke, R. M. (1996) Physiological heterogeneity within fungal mycelia: an important concept for a functional understanding of the ectomycorrhizal symbiosis. *New Phytol.*, **134**, 685.

13 | Signalling in plant–insect interactions: signal transduction in direct and indirect plant defence

MARCEL DICKE AND REMCO M. P. VAN POECKE

1. Introduction

Insects make up the largest group of organisms that comprises more than 800 000 species (new species are still discovered at high frequency) and thus ca. 60% of all species on earth are insects (1). Also in numbers of individuals insects are abundant and they occur in virtually all habitats on earth (1). About half of all insect species are herbivores and the majority of these are specialists that feed on one or a few related plant species (1). The main feeding modes of insects are chewing-biting, cell-sucking, and sap-feeding. Herbivorous insects may live on or inside the plant tissue. For instance, insects may feed externally on leaves or stem, they may live in the stem or in a leaf of the plant, or induce a gall to live in (1). There are ca. 300 000 vascular plant species (1) and it is unlikely for an individual plant to live without interactions with herbivorous arthropods. However, the intensity, frequency, and diversity of these interactions may vary with plant species and environment.

In addition to interactions between plants and herbivorous insects, plants also have interactions with carnivorous insects that consume the herbivores. Many carnivorous insects inhabit plants (e.g. 2, 3) and plant characteristics can influence carnivore behaviour and carnivore–herbivore interactions. This chapter deals with signalling between plants and herbivorous as well as carnivorous arthropods in the context of plant defence. Included are references to mites that, although not belonging to the class of insects, have very similar interactions with plants (4, 5).

2. Direct versus indirect plant defence

Plants have evolved a wide range of defences to cope with the attack of herbivorous insects and mites (1, 6, 7). Traditionally, plant characteristics that directly affect herbivores have been studied (1, 7). This so-called 'direct defence' may involve physical defences (e.g. a thick cuticle, trichomes and thorns) and chemical defences (e.g. toxins, repellents, and digestibility reducers). Defence chemicals can be found in all major classes of plant secondary metabolites: for instance, nitrogen-containing metabolites like alkaloids and glucosinolates, phenolics like phenylpropanoids and flavonoids, and terpenoids. In addition, plants may use defence-related proteins. Physical and chemical factors are often combined; examples comprise toxic, deterrent, and/or sticky compounds in glandular trichomes and deterrents in the epicuticular wax (1).

In addition to direct defences, plants can use other mechanisms to protect themselves. Herbivores have a variety of natural enemies such as predators or parasitic wasps (parasitoids) and plants may enhance the effectiveness of these enemies and employ them as 'bodyguards' (5, 8). This has been termed 'indirect defence'. This type of defence may include the provision of shelter, alternative food, and chemical information, either alone or in combinations (5, 8). For instance, many plants have so-called 'domatia', which are structures used as shelter by carnivores such as ants or predatory mites. These inhabitants provide protection to the plant by removing herbivores (9). Plants also provide floral or extrafloral nectar that carnivorous arthropods feed on and the production of these nutrient sources can be induced by herbivory (10, 11). Finally, plants may lure carnivorous arthropods with plant volatiles produced in response to herbivore attack (8, 12).

This chapter focuses on induced plant defence in response to insects and mites and the consequent interactions between plants and arthropods. The main interest is a comparison of signal transduction processes in direct and indirect defence.

3. Induction of direct defence

The inducibility of direct defence appears to be a general characteristic that has been demonstrated in more than 100 plant species in 34 families (7). Many types of direct defences can be induced in plants. This relates to both physical and chemical defences, but the induction of chemical defences has received most attention. Duration of the effect of induced defence ranges from a few hours to several years. The type and extent of defence varies with both the plant and herbivore species. Differences in induced defence can also be found within a plant: young plant parts often show a stronger induced defence than older parts.

An important aspect of herbivory is wounding of the plant and mechanical wounding can mimic the effect of herbivory in many induced direct defences. Therefore, mechanical damage has often been used to study the signal transduction pathway of defence induced by herbivory.

For many plant responses induced by insects, the underlying mechanism is unknown (7). However, the mechanisms involved in several induced responses have been especially well studied in the last decades, such as the wound-induced production of proteinase inhibitors (PIs) in solanaceous plants, especially tomato (13), and of nicotine in tobacco (14).

3.1 Proteinase inhibitors

Pioneering studies on induced direct defence were devoted to the induction of proteinase inhibitors in solanaceous plants. Herbivore feeding or mechanical wounding of potato and tomato plants result in the systemic expression of genes encoding PI proteins (15–18). PIs interfere with the digestive system of insects, retarding larval growth and development (19). This negative effect of PIs on larval growth and development has been shown for several plant–herbivore combinations (20). Elimination of PI induction can alleviate the effects on herbivores. For instance, mutant tomato plants that are impaired in a specific step in the signal transduction leading to PI-gene induction are more susceptible to feeding by *Manduca sexta* than wild-type plants (21).

PIs have been found in several plant families including the Brassicaceae, Cucurbitaceae, Fabaceae, Salicaceae, and Solanaceae. It has been estimated that each plant genome contains 100–200 different PI genes, grouped in several families and inhibiting each of the four classes of proteolytic enzymes (serine, cystein, aspartic, and metalloproteinases) (7). This may be explained by the finding that herbivores can alter the set of digestive proteinases expressed, which may (temporarily?) overcome the inhibition of their digestive capabilities by the plant (22).

Apart from being involved in direct defence, PIs may also affect indirect defence. The reduced growth rate that is caused by PIs – even when occurring temporarily – may prolong the time window during which the herbivores are exposed to their natural enemies and thus may increase mortality incurred by carnivores (23).

3.2 Plant secondary metabolites

Herbivory or mechanical wounding can also result in the induced production of low-molecular-weight secondary metabolites that may originate from all major classes of secondary metabolites (e.g. 7, 24–26). Genes encoding important enzymes in the production pathways of these metabolites can be induced by wounding or by treatment with plant hormones such as jasmonic acid. Examples are genes involved in the flavonoid biosynthesis, such as chalcone synthase (CHS) (25, 27) and phenylalanine ammonia-lyase (PAL) (28).

A well-studied system is induced nicotine production. Natural folivory or mechanical damage inflicted to tobacco induces the *de novo* biosynthesis of the alkaloid nicotine, resulting in a four- to tenfold increase in concentration in the leaves. These concentrations are high enough to be lethal for many herbivore species (14). The increased nicotine production takes place in the roots. The key regulatory enzyme in the nicotine synthesis, putrescine *N*-methyltransferase, is up-regulated at mRNA level by wounding (29). Via the xylem, nicotine is then transported to the shoots (7, 30).

The consumption of nicotine by herbivores may have a negative effect on indirect defence. For instance, mortality among *Manduca sexta* caterpillars from parasitoids and pathogens is lower when the herbivores have fed on a high-nicotine food source (31, 32).

3.3 Signal transduction

The mechanisms of induction of direct defences have been well studied for several systems. A few major signal transduction routes can be distinguished. These are centred around different plant hormones such as jasmonic acid and ethylene.

3.3.1 Jasmonic acid and other oxylipins

Jasmonic acid (JA) is a product of the lipoxygenase pathway, also called the octa-decanoid pathway (33, 34). The lipoxygenase pathway starts with the substrate linolenic acid and results in products such as JA, traumatin, and a variety of six-carbon volatile compounds such as hexanal, (E)-2-hexenal, and (Z)- and (E)-3-hexenol (33, 35). The pathway of JA biosynthesis is probably constitutively expressed (36), but can additionally be induced by, for example, wounding (17, 27), feeding by insects (37, 38), or application of microbial cell-wall components (37). Methyl jasmonate (MeJA) is a volatile derivative of JA, and may function as an airborne signal molecule (18, 39, 40). MeJA can probably be easily converted to JA in the plant (18). Therefore, MeJA is included in this section on jasmonic acid.

Jasmonic acid is a central molecule in induced direct defence against insects in many plant species (7). The importance of jasmonic acid in wound-induced defence responses has been demonstrated by the fact that:

- exogenous application of JA or MeJA induces these defence responses;
- the increase of endogenous JA after wounding correlates with the induced defence responses; and
- inhibition of the JA production pathway also inhibits the induction of the defence responses (30).

In addition, transgenic plants and mutants have been important tools to elucidate the role of jasmonic acid in signal transduction involved in direct plant defence (21, 41, 42).

Besides jasmonic acid, several other, related, oxylipins also appear to function as signalling molecules. These comprise, for example, 12-oxo phytodienoic acid (OPDA) (37) and dinor-oxo-phytodienoic acid (DN-OPDA; 43). For instance, OPDA and 10,11-dihydro-OPDA have been shown to induce the same production of secondary metabolites as JA in *Eschscholtzia california* cell suspension cultures, without the conversion to JA being necessary (37). Combinations of oxylipins may result in specific plant responses and this has resulted in the designation of 'oxylipin signatures' being important in the induction of plant defences (43).

3.3.2 Systemin

An 18-amino acid polypeptide called systemin has been identified in tomato as a potent inducer of PIs. So far, systemin homologues have only been found in members from the Solanaceae plant family, namely tomato, potato, black nightshade and bell pepper (44). Most research has been done on tomato.

Wounding induces the systemic accumulation of systemin in tomato by an increased expression of the gene encoding the precursor of systemin, prosystemin

(45). As prosystemin mRNA is systemically produced after wounding, it is not clear whether systemin is the systemic signal, inducing its own gene expression, or that another systemic signal is involved. It has been shown, however, that systemin is transported from the wound site throughout the plant within 90 minutes after wounding (46). Moreover, systemin induces prosystemin gene activity (47).

Application of systemin through the cut stems of tomato plants induced both the accumulation of PI proteins (48) and mRNA (47). In transgenic tomato plants with an antisense prosystemin gene, and thus a lowered systemin production, less PI protein accumulated after wounding. These transgenic tomato plants were more susceptible to feeding by *Manduca sexta* larvae (49). Besides the *pin* (protein *in*hibitor) genes, several other tomato genes are induced by systemin. These include genes encoding for other defensive proteins such as polyphenol oxidase (50), signal pathway-associated proteins such as lipoxygenase, proteolytic enzymes, and other proteins (47). These results show that systemin plays an important role in the signal trans-duction of wound-induced defences in tomato.

3.3.3 Ethylene

The plant hormone ethylene is produced in response to wounding (39), herbivory (51–53), and the application of systemin or JA in tomato cell suspensions (54, 55). In tomato, both genes encoding enzymes involved in ethylene production from S-adenosyl-methionine, 1-aminocyclopropane-1-carboxylate synthase (ACS) (56) and 1-aminocyclopropane-1-carboxylate oxidase (ACO) (57), are also up-regulated by wounding.

Like wounding, exogenous application of ethylene induces the production of enzymes, such as PAL, CHS, and hydroxyproline-rich glycoproteins (HPRG, involved in cell-wall strengthening) mRNA (58). However, for HPRG it was shown that different isoforms are induced by wounding compared to ethylene application (58).

3.3.4 Abscisic acid

Abscisic acid is involved in several wound-induced responses, such as induced leucine aminopeptidase (59) or PIs (60) (but see references 59 and 61). In many other wound-induced responses, ABA does not seem to play a role (e.g. 62). In several induced defences the role of abscisic acid is not clear and its role in, for example, the induction of proteinase inhibitors is under debate (59, 61). How exactly ABA is involved in the signal transduction pathways is not clear. For example, water stress promotes an increase in endogenous ABA levels with a factor of 8–10, but this does not lead to *pin2* gene expression (60). The induction of water-stress-responsive genes does not require *de novo* synthesis of proteins, whereas the induction of *pin2* by JA does. Recent data suggest that ABA is not a primary signal for *pin2* gene induction, but that it may modulate the responses to other signals (59, 61).

3.3.5 Electrical signals

Electrical signals may play a role in the systemic induction of PIs in tomato (63), but so far it remains unclear how these signals interact with chemical signals.

3.3.6 Insect damage versus mechanical damage: herbivore elicitors

As mentioned before, most research has been done on mechanical damage. Several publications report differences between herbivory- and wound-induced responses (e.g. 38, 64–67). Of course, herbivory is more than mechanical wounding and plants may use, for instance, oral secretions from herbivores to differentiate between mechanical wounding and herbivory (38, 64, 67–69).

Differences between herbivory- and wound-induced responses have been found in *pin2* gene expression in potato (69) and in nicotine accumulation in tobacco (38). Induction of the *pin2* gene in potato appeared to occur faster by feeding of *Manduca sexta* larvae than induction by artificial wounding. Similar results were obtained with the 3-hydroxy-3-methylglutaryl-coenzymeA reductase (HMGR) gene family from potato. HMGR is the first enzyme in the mevalonic-acid-derived terpenoid biosynthetic pathway. The early induction of both *pin2* and HMGR mRNA could be mimicked by application of *Manduca sexta* larvae regurgitant on artificially wounded leaves and is probably caused by an insect-derived, heat-stable elicitor from the regurgitant. These results indicate that the signalling pathways of herbivory-induced and wound-induced plant defences may be at least partially different (69). In tobacco, herbivory by *Manduca sexta* larvae or application of their regurgitant decrease the induction of nicotine production compared to mechanical damage, even though JA induction was increased (38). A similar effect was recorded for the induction of trypsin inhibitor by *Manduca sexta* in tomato (70). It seems that *Manduca sexta* larvae are able to suppress induced defences by influencing the signal pathway (38).

A component of the saliva of *Helicoverpa zea* caterpillars, the enzyme glucose oxidase, has been found to induce the salicylate pathway in soybean, tobacco, and cotton, resulting in systemic acquired resistance to *Pseudomonas syringae*. This effect was not found in response to mechanical damage (67, 71).

Although there are differences in plant responses to artificial wounding and herbivory, there are still enough indications that the signal pathways invoked by mechanical wounding or insect feeding share at least some components (Plate 4). Knowledge of the signal transduction pathway of wound-induced plant defences will help us unravel the signal transduction pathway of defences induced by herbivory.

3.3.7 Cross-talk and interactions among signal transduction pathways

The emerging picture of signal transduction in direct defence is that different signal transduction pathways interact. For instance, salicylic acid (SA), which can be induced by pathogen attack (see Chapter 10), interferes with jasmonic-acid-mediated responses involved in defence against herbivorous arthropods. SA blocks the biosynthesis of jasmonic acid and subsequent induction of gene expression (16, 17) and consequently affects the defence of plants against herbivores (72). For instance, treatment of plants with a SA mimic (benzothiodiazole-7-carbothioic acid S-methyl ester, BTH) alleviates the effect of JA treatment of tomato plants on the induction of polyphenol oxidase activity and herbivory by caterpillars of *Spodoptera exigua* (72). Vice versa, JA can inhibit the effect of salicylic acid (73) or the synthetic mimic BTH (72). For instance, JA treatment of tomato plants eliminates the induction of the *PR-4* gene and reduces the protection of BTH-treated tomato plants against *Pseudomonas syringae* pv tomato (72). In other cases, cross-induction has been reported. For instance, the spider mite *Tetranychus urticae* and the fungus *Verticillium dahliae* induced resistance against spider mites and the fungus in cotton (74). How this is related to the induction of JA and/or SA remains unclear.

JA and ethylene act in concert in the induction of *pin* gene expression in tomato (Plate 4a). These two plant hormones influence each other's level in wounded plants and are needed together for induction of PIs in tomato (54). Ethylene alone does not induce the *pin* genes in tomato (75), but inhibition of ethylene action inhibits the induction of *pin2* in tomato by wounding, systemin, JA, and oligogalacturonide fragments. These results suggest that ethylene action is downstream from JA in the wound-response pathway. However, when ethylene action is blocked, the induction of endogenous JA levels by wounding is reduced. It appears that ethylene and JA induce each other's production (54).

Signal transduction pathways can be specific for the plant species and the tissue within a plant. Tomato plants respond to cell-wall-derived oligosaccharides or chitosan with *de novo* synthesis of JA which results in the accumulation of PIs (16, 54). By contrast, in *Arabidopsis*, treatment with chitosan does not result in elevated JA levels, which is mediated by ethylene-dependent negative effects on JA effectiveness (76). This ethylene-dependent blocking of JA-mediated effects in *Arabidopsis* does not occur in systemic tissues. As a consequence, different responses are induced in local and systemic tissues (76). In addition, in nicotine induction in tobacco, ethylene and JA have opposite rather than synergistic effects (53). Thus, the specifics of the inter-action among different signal transduction pathways seem to be dependent on the plant species. Obviously, more studies are needed to obtain a better understanding of general effects and exceptions.

4. Induction of indirect defence

Research on the induction of indirect defence began in the mid-1980s (3, 77–82) and has mainly concentrated on the induction of carnivore-attracting plant volatiles. Herbivory by arthropods results in the emission of a blend of volatiles that attracts the enemies of herbivorous arthropods. This has been studied in depth for plant–spider mite–predatory mite interactions (e.g. 4) and plant–caterpillar–parasitoid interactions (e.g. 82).

The induction of plant volatiles has been recorded for more than 23 plant species from 13 families (83) and it seems that it is a common response of plants to herbivory. In all plant species investigated, the ability has been found. Among the plant families investigated are, for example, the Fabaceae, Brassicaceae, Cucurbitaceae, Poaceae, Malvaceae, Solanaceae, Rosaceae, and the Asteraceae. Two types of plant response can be distinguished (83).

- In response to herbivory the plant produces a blend that is dominated by novel compounds that are not emitted by intact or mechanically damaged plants. This type of response can be found in, for example, Lima bean (Plate 4b), cucumber, maize, and gerbera (79, 80, 84, 85).

- In response to herbivory, the plant produces a blend that is qualitatively similar to the blend emitted by intact or mechanically damaged plants. In the latter case, the emission rate from herbivore-damaged plants is much higher than from mechanically damaged or undamaged plants and it continues much longer after termination of the damage (e.g. 86). This type of response has been recorded for, for example, cabbage, cotton, tomato (Plate 4a) and potato (4, 78, 86–88).

In both types of responses, *de novo* biosynthesis of volatiles has been reported (89, 90) and thus plants invest in the production of volatiles rather than passively emitting the contents of damaged cells. Moreover, herbivore-induced plant volatiles are emitted systemically (91–94), which further supports the conclusion that the emission of volatiles is an active response rather than a passive release of cell contents.

The herbivore species that have been shown to induce plant volatiles belong to 27 species in 13 families of insects and mites. These herbivores include folivores (chewing-biting and cell-sucking species), sap-feeding insects, and species that feed in the plant such as leaf miners and stem borers (83). Plants may even induce volatiles in response to oviposition by a herbivore (95). The emitted blend varies largely among plant species in a qualitative sense. In addition, the blend also varies among plants of the same species that are damaged by different herbivores. However, this variation is much more subtle and usually relates to quantitative variation in the blend composition, i.e. the blends are composed of the same constituents but the relative contribution of different constituents to the blend varies (see reference 96 for review). Sometimes, a qualitative difference between blends of plants emitted by different herbivores has been recorded, e.g. for faba beans infested with different aphid species (97). Furthermore, the blend may be affected by herbivore instar, as was recorded for maize plants damaged by caterpillars (98).

In addition, abiotic factors may affect the emission of plant volatiles in a quantitative and qualitative sense (e.g. 99, 100), but little effort has been made to investigate the effect of abiotic conditions on the emission of herbivore-induced plant volatiles.

In conclusion, herbivory results in a change in the emission of volatiles and the composition of the emitted blend varies with biotic and abiotic factors.

4.1 Identity of herbivore-induced plant volatiles

The major volatiles emitted by plants, either constitutively or induced, belong to several classes that are produced through distinct biosynthetic pathways: e.g. fatty acid derivatives produced through the lipoxygenase pathway, terpenoids produced through the isoprenoid pathway, and phenolics produced through the shikimic acid pathway.

The fatty-acid-derived volatiles comprise C6-aldehydes, C6-alcohols and their esters such as (Z)-3-hexen-1-yl acetate. They are common plant volatiles (101) and are often referred to as green-leaf volatiles (102). They are emitted in response to artificial damage as well as herbivory (e.g. 78–80). These compounds can be perceived by parasitoid chemoreceptors (103) and some parasitoid species are attracted by these green leaf odours (104).

Terpenoids comprise the largest and most diverse chemical group in plants. They can be produced through the mevalonic acid pathway (105) or through the 1-deoxy-D-xylulose-5-phosphate pathway (106) and many of them are well known for their toxic effects on herbivores (105). Among the herbivore-induced terpenoids there are two compounds that are especially noteworthy, i.e. the homoterpenes 4,8-dimethyl-1,3(E),7-nonatriene (DMNT) and 4,8,12-trimethyl-1,3(E),7(E),11-tridecatetraene (TMTT). These terpenoids have been recorded from many plant species after the infliction of herbivory or treatment with elicitors (79, 80, 107, 108). Herbivory leads to the induction of (3S)-(E)-nerolidol synthase in Lima bean, cucumber, and corn and subsequently to the formation of DMNT (109, 110), which is a known attractant of the carnivorous mite *Phytoseiulus persimilis* (80).

Phenolics are produced through the shikimic acid pathway. Among the phenolics emitted in response to herbivory are indole and methyl salicylate. Indole is an intermediate product of tryptophan biosynthesis. It plays an important role in direct defence as a component of indole alkaloids and indole glucosinolates (e.g. 26). It has also been recorded among herbivore-induced plant volatiles, e.g. from maize, cowpea, soybean, cotton, gerbera or Lima bean (79, 85, 88, 98, 111). Methyl salicylate is the volatile methyl ester of the plant hormone salicylic acid. It has been recorded in several plant species, such as Lima bean, apple, and pear (80, 112, 113).

Nitrogen-containing compounds such as nitriles and oximes are commonly reported from herbivore-damaged plants. In crucifers these compounds can be degradation products of glucosinolates (78, 114), whereas in other plant species such as cucumber, Lima bean, or gerbera (80, 84, 85), they may be derived from amino acids (115).

4.2 Importance of herbivore-induced plant volatiles to carnivorous arthropods

For natural enemies, the most reliable information on the presence of their herbivorous victim of course originates from the herbivore itself. However, a herbivorous arthropod is only a small component of the environment, with a small biomass. Moreover, herbivores have been under natural selection to avoid being detectable by their natural enemies (116). Behavioural studies show that herbivorous arthropods or their faeces are not very attractive to carnivorous enemies (e.g. 117–120) and a chemical analysis showed that hardly any volatiles can be detected from *Spodoptera exigua* larvae and faeces (118). Plants represent a much larger biomass and detection of herbivores by their natural enemies can benefit plants (121). For example, arthropod carnivores such as predatory mites are well known to exterminate local populations of their herbivorous prey (77).

It is well established that herbivore-induced plant volatiles play an important role in the attraction of carnivores. For instance, the predatory mite *Phytoseiulus persimilis* (117), and the parasitoid wasps *Cotesia marginiventris* (118), *Cotesia glomerata* (119), *Cotesia rubecula* (120) and *Microplitis croceipes* (122) all prefer odours from the plant–host complex over those from faeces of their herbivorous host/prey. The blend of volatiles can be specific for the herbivore species that induced it. Many arthropod predators and parasitoids have been shown to discriminate between plants infested with different herbivore species and also chemical differences have been reported (see reference 96 for review). Thus, the induced plant volatiles are important cues for carnivores to locate their herbivorous victims. The discrimination among induced blends of plant odours may need to be learned or can be dependent on the physiological condition (e.g. starvation level) of the carnivore (96).

Because herbivore-induced carnivore-attractants are emitted at the site of damage as well as systemically from undamaged leaves (91–93), the odour source is much larger than the herbivore, which increases the detectability of the herbivore to its natural enemies.

4.3 Response by carnivorous arthropods and benefits to plants

Herbivore-induced plant volatiles are very important for carnivorous arthropods and the carnivores can use them during several foraging decisions. This has been shown for a large number of carnivorous arthropods, including, for example, parasitic wasps (79, 120), predatory bugs (112, 123), and predatory mites (4, 124). Parasitoids and predators can be attracted from a distance and this has been demonstrated in various setups, including olfactometers (124), windtunnels (119), semi-field setups (125, 126) and field tests (127). In addition, the volatiles can also mediate arrestment in a herbivore patch (128) or the suppression of long-range dispersal (129).

Plants can greatly benefit from the attraction of carnivorous arthropods. Carnivores such as predatory mites or predatory bugs eliminate the herbivores by predation. They can exterminate local prey populations and thus relieve the plant of their herbivores. However, in the case of parasitoids this benefit to the plant is not self-evident (121). After parasitization the herbivore is not killed but in most cases continues to feed from the plant, sometimes at a larger rate than when not being parasitized (130). The latter may be dependent on the number of parasitoid eggs deposited in the herbivore. In order to establish whether plants benefit from parasitoids, one should investigate the effect of parasitization on plant fitness, e.g. in terms of seed production. This has recently been done for *Arabidopsis thaliana* plants infested with *Pieris rapae* caterpillars and the solitary parasitoid *Cotesia rubecula*. This study showed that parasitization of the caterpillars resulted in a large reduction in fitness loss compared with plants infested with unparasitized caterpillars (121). Whether this is a general phenomenon should become clear from the investigation of other plant–herbivore–parasitoid systems.

4.4 Herbivore-induced plant volatiles and responses by herbivores

After emission of the induced volatiles, the plant is no longer in control over who exploits the information. For example, herbivores that forage for food may exploit the information to find a host plant (Plate 5). To herbivores the volatiles may represent information on:

- the presence of a host plant,
- the presence of competitors, and
- the presence of a potentially enemy-dense space

and herbivores may exploit this information. However, these are conflicting types of information and one may predict that the responses of herbivores are dependent on external and internal conditions. For instance, the information may lead to a different response in a starved herbivore than in a satiated herbivore and the response may also depend on other information, e.g. on the presence of carnivorous enemies in the environment. Indeed, different responses have been recorded for herbivores (for reviews see references 5, 131, 132). For instance, the spider mite *Tetranychus urticae* is attracted to volatiles from undamaged Lima bean plants and to volatiles from a combination of volatiles from undamaged leaves and from spider-mite-infested leaves. By contrast, when only volatiles from spider-mite-infested leaves are offered, the spider mites are repelled (133). Alternatively, herbivores may be attracted to plants infested with heterospecific herbivores because the heterospecifics provide protection from natural enemies (134).

For herbivores that feed on the volatile-emitting plant, the volatiles may also have a direct negative effect. For instance, the green-leaf volatiles have a negative effect on the rate of population increase of aphids (135). This resembles the negative effect of these compounds on bacterial proliferation (35). Thus, these induced compounds may play a role in both direct and indirect defence. Most likely other induced volatiles such as terpenes can also interfere directly with herbivore performance and thus play a dual role.

4.5 Herbivore-induced plant volatiles and effects on neighbouring plants

The information emitted by infested plants may potentially also affect downwind uninfested neighbours (Plate 5). These neighbouring plants may exploit the information to initiate defences. After all, after the upwind neighbour has been overexploited by the herbivores they may reach the downwind neighbour on wind currents just as the information did. Information transfer between herbivore-infested and uninfested plants has been investigated since the mid-1980s (reviewed in references 7, 132, 136, 137). Evidence in favour of the exploitation of the information by downwind plants is increasing and recently some exciting new data have been published (40, 138).

4.6 Signal transduction

Carnivorous arthropods can discriminate between volatiles emitted by herbivore-damaged and mechanically damaged plants (for reviews see, e.g. 83, 116). This is true both for plants that produce novel compounds in response to herbivory and for plants that produce the same compounds in response to herbivory and to mechanical damage. In the latter case the volatiles are emitted in larger amounts and especially during a longer period of time in response to herbivory than in response to wounding (e.g. 86). This has stimulated the search for herbivore elicitors that enable the plant to discriminate between wounding and herbivory. In addition, the extensive knowledge on signal transduction pathways in direct defence has been utilized to investigate their involvement in induced indirect defence.

4.6.1 Herbivore elicitors

As indicated above, mechanical damage can usually not effectively mimic herbivory in the induction of indirect defence. This can be explained by the role of herbivore elicitors. The application of oral secretions of herbivores onto mechanical damage can result in similar effects on volatile induction as herbivory itself (78, 79). The search for active components of the oral secretions has yielded two compounds: a β-glucosidase from *Pieris brassicae* caterpillars that induces volatiles in cabbage, maize, and bean plants (139, 140) and the fatty acid–amino acid conjugate N-(17-hydroxylinolenoyl)-L-glutamine, called volicitin, that induces volatiles in maize (141). In both cases, the application of the elicitor is a good mimic of herbivory, but it remains unknown what the exact effects of the elicitors are. Because glycosidically bound forms of the volatiles emitted have not been recorded, the glucosidase may release an internal elicitor. Volicitin, which has the fatty acid moiety 17-hydroxylinolenic acid, most likely activates the octadecanoid pathway (141). This pathway is known to be involved in the induction of carnivore-attracting volatiles (111, 139, 142, 143).

In some cases, mechanical damage can mimic herbivory in the induction of indirect defence. For example, caterpillar damage to cotton (*Gossypium herbaceum*) plants had the same effects on the induction of extrafloral nectar production as either mechanical damage in combination with caterpillar oral secretion or mechanical damaged plus water treatment (10). However, in general, indirect defence induced by herbivores is different from the response to mechanical damage (e.g. 5, 81, 83). An intriguing question is why herbivore elicitors appear to be more important in the induction of indirect defence than in the induction of direct defence. Is this difference an artefact, resulting from the use of different model plants? After all, the model plants of research on induced direct defence, such as potato and tomato, produce a similar odour blend in response to mechanical damage and herbivory, albeit that the effect of herbivory lasts longer (4, 86). Or is this the result of the interaction with carnivorous arthropods that base their foraging 'decisions' upon the information provided by the plant and therefore select for the emission of more specific signals in the plant?

4.6.2 Systemic elicitor

The induction of plant volatiles occurs systemically throughout the plant (91–93). An active compound has been extracted from spider-mite-infested Lima bean plants, and when applied to an uninfested plant results in attraction of carnivorous mites (94). So far, the identity of the systemic elicitor(s) has not been elucidated.

4.6.3 Jasmonic acid and octadecanoid pathway

Treatment of Lima bean, maize, and gerbera plants with JA resulted in the emission of volatiles that are also induced by herbivory (111, 139, 142, 143). It is noteworthy that treatment with JA resulted in the induction of volatiles from different bio-synthetic pathways in the same plant, such as the lipoxygenase pathway and the isoprenoid pathway. The volatile blend emitted by Lima bean after JA treatment is similar but not identical to the blend emitted by *Tetranychus urticae* infested plants. For instance, JA did not induce methyl salicylate or the homoterpene TMTT, whereas it induced two methyloximes in larger amounts than did spider-mite damage. This difference in odour blend results in differential behaviour of the carnivorous arthro-pods (Fig. 1). The carnivorous mite *Phytoseiulus persimilis* was attracted to JA-treated plants, but in a two-choice situation preferred the volatiles from spider-mite-damaged plants (111). In other plants (cucumber, tobacco, and *Arabidopsis thaliana*), MeJA induced the activity of two important enzymes in the green-leaf volatile production pathway, i.e. lipoxygenase (LOX) and hydroperoxide lyase (HPL), result-ing in the increased emission of green-leaf volatiles by plants (144). Treatment of plants with JA results in attraction of predators and parasitoids under laboratory and field conditions (111, 139, 142, 143, 145). For instance, in field-grown tomato plants,

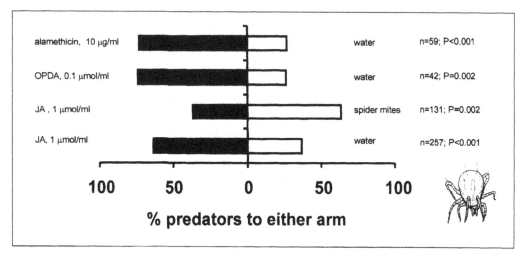

Fig. 1 Attraction of predatory mite *Phytoseiulus persimilis* to Lima bean plants infested with herbivorous spider mites (*Tetranychus urticae*) or treated with JA, OPDA, or the peptaibol alamethicin. Experiments were carried out in a Y-tube olfactometer. For details of experimental setup, see reference 111. Data based on reference 111 and M. Dicke and H. Dijkman (unpublished data).

JA treatment resulted in a larger number of parasitoids (*Hyposoter exiguae*) and in a higher parasitization percentage of *Spodoptera exigua* caterpillars (145). These results indicate an important role for JA as an endogenous signal molecule involved in induced production of volatiles. Indeed, both *Manducta sexta* herbivory and regurgitant treatment on mechanically damaged leaves resulted in an increase of endogenous JA levels in tobacco (38). Although JA is induced by herbivory, its methyl ester MeJA has never been recorded from herbivore-infested plants.

Several intermediates from the octadecanoid pathway can induce plant volatiles. For instance, OPDA induces the two homoterpenes, DMNT and TMTT, in Lima bean (146) and OPDA treatment results in attraction of the predatory mite *Phytoseiulus persimilis* (Fig. 1). Nonetheless, JA appears to be the most powerful member of the octadecanoids in terms of induction of Lima bean volatiles. A combination of octadecanoids may be responsible for the total induction pattern as recorded in response to spider-mite damage. However, no herbivory-related elicitors have been found yet that induce methyl salicylate, a constituent of spider-mite-induced Lima bean volatiles that attracts predatory mites (80).

Other compounds, such as conjugates of 1-oxo-indan-4-carboxylic acid, also appear to have a strong inducing power. Coronatin (the coronamic acid conjugate of 1-oxo-indan-4-carboxylic acid) is the phytotoxin of certain *Pseudomonas syringae* pathovars and this compound induces a similar blend in Lima bean as JA, but in addition also induces the C16 homoterpene TMTT that is not induced by JA (142). A synthetic analogue such as 1-oxo-indanoyl-isoleucine (IN-Leu) induced a similar volatile blend in Lima bean as did JA (146). Most work on the effect of octadecanoids and their analogues has been done on Lima bean plants. To what extent these results apply to other plants remains unclear. The situation seems to be different in the monocotyledon maize. For this plant the differential effects of octadecanoids do not seem to apply (146).

Whether JA plays a role in other induced indirect defences as well, such as extrafloral nectar induction, has not been investigated.

4.6.4 Salicylic acid (SA)

A remarkable compound emitted by several plant species in response to herbivory is methyl salicylate (MeSA). This compound is induced in, for example, spider-mite-infested Lima bean and tomato plants (4, 80), in psyllid-infested pear plants (112) and in Colorado potato beetle-infested potato plants (86). It has also been recorded from tobacco mosaic virus-infested tobacco plants and was suggested to be a way of disposing of salicylic acid (SA) formed in infested plants (147). In cotton, herbivory by caterpillars of *Helicoverpa zea* leads to increased levels of SA, but no effect of SA induction has been found on the performance of the caterpillars (148). In an analysis of headspace volatiles of *H. zea*-infested cotton plants, an emission of methyl salicylate has not been reported (149). The application of gaseous MeSA to Lima bean plants was reported to induce the emission of the two homoterpenes, DMNT and TMTT (150). It will be important to investigate the effect of SA production on indirect defence of plants, because SA is known to inhibit the effect of JA in direct defence. JA is important in the induction of plant volatiles and SA may inhibit this (151), but possibly the production of SA and the induction of plant volatiles by JA are spatially and/or temporally separated, which would reduce the possiblity of interference (see Section 4.6.7). If SA is induced by herbivory, its methylation and emission as MeSA is a way of avoiding the accumulation of SA in response to herbivory or possibly in response to microorganisms transmitted to the plant by the herbivore during feeding.

4.6.5 Ethylene

The induction of ethylene emission in response to herbivory or elicitors has been reported for several plants such as tobacco (53) and Lima bean (152). To our knowledge, there is no evidence that ethylene induces plant volatiles. However, the role of ethylene in induced indirect defence has received little attention to date. In tobacco, ethylene treatment did not induce the emission of $(-)$-cis-α-bergamotene or linalool (53), two compounds induced by *Manduca sexta* caterpillar feeding.

4.6.6 Other elicitors

Several other elicitors have been reported to induce plant volatiles. Among these are, for example, cellulysin and alamethicin, a mixture of peptaibols, that are produced by the fungus *Trichoderma viride* (151). The peptaibols (oligopeptides) act as ion-channel-forming compounds and are considered to mediate a very early step in the induction. In Lima bean they induce the production of a short peak of JA and of SA. The latter phytohormone is produced at a high level for a long period. It is interesting to see that administration of the peptaibol alamethicin results both in a high level of SA in the plant and in the emission of MeSA. In addition to MeSA, the only other compounds emitted are the two homoterpenes DMNT and TMTT. This indicates that the SA inhibits the octadecanoid pathway beyond OPDA (151), as this octadecanoid intermediate induces the production of the two homoterpenes (146).

It appears that the blends of volatiles induced by these elicitors affect the behaviour of carnivorous arthropods: the predatory mite *Phytoseiulus persimilis* is attracted to peptaibol-treated Lima bean plants (Fig. 1). This can be explained by the observation that two of the three volatiles emitted in response to alamethicin treatment, i.e. MeSA and the homoterpene DMNT, are known to attract this predatory mite (80). These volatiles are also induced by feeding damage of the spider mite *Tetranychus urticae*, on which this predatory mite preys.

4.6.7 Cross-talk

Research on signal transduction in induced indirect defence has a shorter history than that in induced direct defence. However, the few studies that have addressed it indicate that this will be an essential subject for future investigations. The major pathway involved in indirect defence appears to be the octadecanoid pathway. From studies on direct defence it is well known that the jasmonate and salicylate pathways interfere and MeSA is induced in several plants by herbivory (see above). Different types of herbivores have different effects on volatile induction and the relative induction of the jasmonate and salicylate pathways may play a role. In Lima bean, spider-mite feeding induces considerable emission of MeSA (80, 150), whilst caterpillar-feeding damage does not (150). Application of JA and MeSA in different combinations also resulted in differences in volatile induction in Lima bean. Treatments that included both JA and MeSA applications induced compounds that are also induced by spider-mite feeding, while application of JA exclusively resulted in the emission of volatiles that are also emitted in response to caterpillar feeding (150). The possible induction of SA by spider-mite feeding is supported by gene expression studies that show that *PR-4* gene expression is induced by spider-mite feeding and MeSA treatement, but not by caterpillar feeding or JA treatment. If herbivores, like pathogens, produce ion-channel-forming compounds that induce JA and SA in plants, then the relative timing and induced levels of these phytohormones may determine the volatile profile emitted (150, 151). This may be an important determinator of blend composition, which is known to vary with herbivore species or instar that feeds on the plant (96). In this context, it will be important to investigate the effects of pathogens on induced indirect defence against herbivores. So far, no studies are known to us that have investigated the effect of previous pathogen infestation (and thus likely induction of SA) on the ability of plants to induce volatiles that attract carnivores.

There is no evidence for an interaction of ethylene with JA-mediated volatile induction so far. In tobacco, ethylene treatment did not influence the MeJA-induced emission of $(-)$-*cis*-α-bergamotene. This contrasts with the interference of ethylene with MeJA-induced production of nicotine (53).

5. Interaction between direct and indirect defence

Plants are attacked by a wide range of herbivores and have evolved a variety of defences. The coordination of these defences seems to be a formidable task and may have several conflicts. The tailoring of defences seems to be a complex optimization problem, the outcome of which is dependent on the intensity and frequency of different types of attackers and their natural enemies.

For instance, plants are attacked by specialist and generalist herbivores. Specialist herbivores are usually well adapted to the defences of their host plant. They may even exploit secondary compounds that provide protection against generalist herbivores as token stimuli to recognize their host plant. Oviposition in *Pieris* butterflies is stimulated by glucosinolates of their cruciferous host plants and for caterpillars these compounds act as feeding stimulants (153). Specialist herbivores can also exploit secondary metabolites of their host plant in their own defence. *Manduca sexta* caterpillars derive protection from pathogens and parasitoids by the intake of nicotine from tobacco leaves (31, 32). In addition, direct defences may interfere with natural enemies of herbivores without an involvement of the herbivore. Glandular trichomes on tomato stems are a defence against herbivores, but in addition they kill the majority of predatory mites that forage on the plant by entrapment. Therefore, this direct defence is incompatible with indirect defence through predatory mites. This shows that direct defences can interfere with the effectiveness of indirect defences and one may wonder whether there is a negative correlation between investments in direct and indirect defence. In this context it is interesting to note that tomato and cabbage, which have a strong direct defence, do not emit novel volatiles in response to herbivory compared to mechanical damage. As a consequence, the information content is much lower than for plants, like Lima bean or maize, that emit novel compounds that dominate the blend in response to herbivory (132). At the individual plant level, an uncoupling of direct and indirect defence appears to be possible. In tobacco plants, damage by *Manduca sexta* caterpillars or treatment with their regurgitant results in an attenuation of induced direct defence (nicotine production) that is mediated by herbivore-induced ethylene. By contrast, there is no such effect on the induction of indirect defence (53).

6. Interaction between defences against pathogens and herbivores

Plants are also attacked by a wide range of pathogens. The defences induced against pathogens and herbivores may act synergistically (e.g. 74) or antagonistically (e.g. 72). Yet, both in direct and indirect defences against herbivores and in defences against pathogens, similar combinations of signal transduction pathways may be induced. These pathways may act antagonistically, such as in the interference between SA- and JA-mediated defences. The current knowledge suggests that, at certain points in the signal transduction pathways, switchpoints are present that influence the final outcome of the collective signal transduction pathways. With an increasing knowledge of signal transduction in induced defences against pathogens as well as those against herbivores, it will be important to unravel where these switchpoints are, where cross-talk between pathways occurs and especially how cross-talk is regulated. This will be a major challenge for the forthcoming years. It will be important to investigate different plant species and populations, especially when different populations are under different relative pressures of herbivores and pathogens.

7. Comparative analysis of signal transduction in induction of direct and indirect defence: model systems

The signal transduction pathways involved in induction of direct defence have been well studied and there is a fertile area for the study of cross-talk. The research on signal transduction in induced indirect defence is rapidly advancing and bridging the gap with the knowledge in direct defences. However, the two research fields have developed independently, mostly with different model systems. In induced direct defence the best-studied model plants are the solanaceous plants tomato and tobacco, whereas in induced indirect defence these comprise, for example, Lima bean, maize, cotton, and cabbage. There seems to be a difference in the way solanaceous and cruciferous plants employ indirect defence (few novel compounds induced) versus that of plants such as Lima bean and maize (major novel compounds induced by herbivory). To allow a better comparison of signal transduction in induced direct and indirect defence, common model systems are greatly needed. Several model systems seem to be good candidates.

Tobacco has been well studied for induced direct defence against pathogens and insects and the signal transduction pathways involved (reviewed by 7, 14, 154). Recently, these studies have also incorporated the induction of indirect defence (53). This will make tobacco a very interesting model system for a comparative analysis.

There is abundant knowledge on the induction of direct defences against pathogens and herbivores in tomato (13, 48, 72, 155, 156) and several mutants and transgenes are available (21, 49, 157). In addition, studies on the induction of indirect defence are emerging (4, 72, 145). Therefore, tomato will be an interesting model system too.

Arabidopsis thaliana has been well studied for direct defence against pathogens (158). Studies on direct defence against insects are also emerging (159, 160). In addition, there is abundant knowledge on direct defence in other crucifers (e.g. 26, 161–164) against herbivores and pathogens. Recently, the first demonstration of induced indirect defence has been made in *Arabidopsis* (121, 165). Together with the presence of various mutants and transgenes that are affected in signal transduction pathways and the near completion of the genomic analysis, *Arabidopsis* will become an important model plant for comparative studies on signal transduction in direct and indirect defence.

When concentrating on such model plants, it remains important to consider other plant species as well. The above-mentioned three plant species all have in common that they have a well-developed direct defence. The blend of volatiles induced in tomato plants by spider mites is dominated by non-novel compounds (4) and thus contrasts to the situation in plants such as Lima bean, cucumber, and maize (79, 80, 111). Therefore, it remains important to include the latter three plant species as well, to enable a comparison among plants with different types of induced indirect defence.

8. Major questions to be addressed

Developments in the research on signal transduction in induced defence against herbivorous insects have been numerous in the past decade. Direct and indirect defence induction appear to share many signal transduction pathways. The octadecanoid signal transduction pathway seems to be the major pathway involved in induced defences against insects. However, there are clear indications that other pathways also play an essential role. Recent work demonstrates that the most important signalling molecule in addition to JA is SA. This is clear from, for example, the increasing number of reports on the emission of MeSA from herbivore-damaged plants. The formation of MeSA in response to herbivory as well as the influence on JA-mediated signalling deserves further investigation.

The induction of indirect defence against herbivorous insects seems to be more specific than the induction of direct defence. In many cases there is a clearly differential response to mechanical damage versus the response to herbivory or herbivore elicitors. The compositions of volatile blends can vary substantially with the species or instar of herbivore that damages the plant and this variation can have considerable consequences for carnivore attraction and thus for defence effectiveness. What signal transduction events mediate these subtly differential responses of plants will be a rewarding subject. Most likely the involvement of signal molecules from different signal transduction pathways will be found.

A signal molecule that has not received much attention to date is ethylene. This molecule is involved in the induction of defences against pathogens or in responses to non-pathogenic microorganisms (166) and modulates the induction of nicotine in tobacco plants (53).

The rapid development in research on signal transduction in induced defence against insects enables the integration with research on induced defences against pathogens. This will be very important, because in nature plants are exposed to a gamut of attackers that include pathogens and herbivores. Investigating how plants integrate defences against all these attackers will be an important step.

In indirect defence, extensive knowledge is present on the attraction of carnivores to complete blends of volatiles emitted by herbivore-damaged or elicitor-treated plants. However, knowledge of the bioactive compounds within the blend is limited to a few studies (131). The use of elicitors that selectively induce certain blend components as well as mutants or transgenes that are modified in signal transduction pathways or in biosynthetic pathways (165) will provide excellent opportunities to elucidate which compounds are most important in attracting carnivores and how variation affects this.

With significant progress in the knowledge of mechanisms of induced plant defences, e.g. in molecular genetics, biochemistry, plant physiology, many tools will become available to investigate the function of these plant traits. Together with progress in the options of manipulating plant genotype, the ultimate step may be made, i.e. investigating the effects on plant phenotype in its interactions with biotic and abiotic components of the environment.

References

1. Schoonhoven, L.M., Jermy, T. and Van Loon, J.J.A. (1998) *Insect–Plant Biology. From Physiology to Evolution.* Chapman & Hall, London.
2. Hölldobler, B. and Wilson, E. O. (1990) *The Ants.* Harvard University Press, Cambridge MA.
3. Dicke, M. and Sabelis, M. W. (1988) How plants obtain predatory mites as bodyguards. *Neth. J. Zool.*, **38**, 148.
4. Dicke, M., Takabayashi, J., Posthumus, M. A., Schütte, C. and Krips, O. E. (1998) Plant-phytoseiid interactions mediated by prey-induced plant volatiles: variation in production of cues and variation in responses of predatory mites. *Exp. Appl. Acarol.*, **22**, 311.
5. Sabelis, M. W., van Baalen, M., Bakker, F. M., Bruin, J., Drukker, B., Egas, M., Janssen, A. R. M., Lesna, I. K., Pels, B., Van Rijn, P. and Scutareanu, P. (1999) The evolution of direct

and indirect plant defence against herbivorous arthropods. In Olff, H., Brown, V. K. and Drent, R. H. (eds), *Herbivores: Between Plants and Predators*. Blackwell Science, Oxford, p. 109.

6. Agrawal, A. A., Tuzun, S. and Bent, E. (1999) *Induced Plant Defenses Against Pathogens and Herbivores*. APS Press, St. Paul MN.

7. Karban, R. and Baldwin, I. T. (1997) *Induced Responses to Herbivory*. Chicago University Press, Chicago.

8. Dicke, M. (1999) Direct and indirect effects of plants on performance of beneficial organisms. In Ruberson, J. R. (ed.), *Handbook of Pest Management*. Marcel Dekker, New York, p. 105.

9. Grostal, P. and O'Dowd, D. J. (1994) Plants, mites and mutualism: leaf domatia and the abundance and reproduction of mites on *Viburnum tinus* (Caprifoliaceae). *Oecologia*, **97**, 308.

10. Wäckers, F. L. and Wunderlin, R. (1999) Induction of cotton extrafloral nectar production in response to herbivory does not require a herbivore-specific elicitor. *Entomol. Exp. Appl.*, **91**, 149.

11. Koptur, S. (1992) Extrafloral nectary-mediated interactions between insects and plants. In Bernays, E. A. (ed.), *Insect–Plant Interactions*. CRC Press, Boca Raton, FL, Vol. 4, p. 81.

12. Turlings, T. C. J. and Benrey, B. (1998) Effects of plant metabolites on behavior and development of parasitic wasps. *Ecoscience*, **5**, 321.

13. Ryan, C. A. (1992) The search for the proteinase inhibitor-inducing factor, PIIF. *Plant Mol. Biol.*, **19**, 123.

14. Baldwin, I. T. (1999) Inducible nicotine production in native *Nicotiana* as an example of adaptive phenotypic plasticity. *J. Chem. Ecol.*, **25**, 3.

15. Green, T. R. and Ryan, C. A. (1971) Wound-induced proteinase inhibitor in plant leaves: a possible defense mechanism against insects. *Science*, **175**, 776.

16. Doares, S. H., Narvaez-Vasquez, J., Conconi, A. and Ryan, C. A. (1995) Salicylic acid inhibits synthesis of proteinase inhibitors in tomato leaves induced by systemin and jasmonic acid. *Plant Physiol*, **108**, 1741.

17. Pena-Cortes, H., Albrecht, T., Prat, S., Weiler, E. W. and Willmitzer, L. (1993) Aspirin prevents wound-induced gene expression in tomato leaves by blocking jasmonic acid biosynthesis. *Planta*, **191**, 123.

18. Farmer, E. E. and Ryan, C. A. (1990) Interplant communication: Airborne methyl jasmonate induces synthesis of proteinase inhibitors in plant leaves. *Proc. Natl Acad. Sci. USA*, **87**, 7713.

19. Broadway, R. M., Duffey, S. S., Pearce, G. and Ryan, C. A. (1986) Plant proteinase inhibitors: a defense against herbivorous insects? *Entomol. Exp. Appl.*, **41**, 33.

20. Ryan, C. A. (1990) Protease inhibitors in plants: genes for improving defenses against insect and pathogens. *Annu. Rev. Phytopathol.*, **28**, 425.

21. Howe, G. A., Lightner, J., Browse, J. and Ryan, C. A. (1996) An octadecanoid pathway mutant (JL5) of tomato is compromised in signaling for defense against insect attack. *Plant Cell*, **8**, 2067.

22. Jongsma, M. A., Bakker, P. L., Peters, J., Bosch, D. and Stiekema, W. J. (1995) Adaptation of *Spodoptera exigua* larvae to plant proteinase inhibitors by induction of gut proteinase activity insensitive to inhibition. *Proc. Natl Acad. Sci. USA*, **92**, 8041.

23. Loader, C. and Damman, H. (1991) Nitrogen content of food plants and vulnerability of *Pieris rapae* to natural enemies. *Ecology*, **72**, 1586.

24. Berenbaum, M. R. and Zangerl, A. R. (1999) Coping with life as a menu option: inducible defences of the wild parsnip. In Tollrian, R. and Harvell, C. D. (eds), *The Ecology and Evolution of Inducible Defenses*. Princeton University Press, Princeton, NJ, p. 10.

25. Gundlach, H., Muller, M. J., Kutchan, T. M. and Zenk, M. H. (1992) Jasmonic acid is a signal transducer in elicitor-induced plant cell cultures. *Proc. Natl Acad. Sci. USA*, **89**, 2389.

26. Bodnaryk, R. P. (1992) Effects of wounding on glucosinolates in the cotyledons of oilseed rape and mustard. *Phytochemistry*, **31**, 2671.

27. Creelman, R. A., Tierney, M. L. and Mullet, J. E. (1992) Jasmonic acid/methyl jasmonate accumulate in wounded soybean hypocotyls and modulate wound gene expression. *Proc. Natl Acad. Sci. USA*, **89**, 4938.

28. Berger, S., Bell, E. and Mullet, J. E. (1996) Two methyl jasmonate-insensitive mutants show altered expression of AtVsp in response to methyl jasmonate and wounding. *Plant Physiol.*, **111**, 525.

29. Hibi, N., Higashiguchi, S., Hashimoto, T. and Yamada, Y. (1994) Gene expression in tobacco low-nicotine mutants. *Plant Cell*, **6**, 723 .

30. Baldwin, I. T., Zhang, Z. P., Diab, N., Ohnmeiss, T. E., McCloud, E. S., Lynds, G. Y. and Schmelz, E. A. (1997) Quantification, correlations and manipulations of wound-induced changes in jasmonic acid and nicotine in *Nicotiana sylvestris*. *Planta*, **201**, 397.

31. Krischik, V. A., Barbosa, P. and Reichelderfer, C. F. (1988) Three trophic level inter-actions: allelochemicals, *Manduca sexta* (L.), and *Bacillus thuringiensis* var. kurstaki Berliner. *Environ. Entomol.*, **17**, 476.

32. Barbosa, P., Saunders, J. A., Kemper, J., Trumbule, R., Olechno, J. and Martinat, P. (1986) Plant allelochemicals and insect parasitoids. Effects of nicotine on Cotesia congregata (Say) (Hymenoptera: Braconidae) and Hyposoter annulipes (Cresson) (Hymenoptera: Ichneumonidae). *J. Chem. Ecol.*, **12**, 1319.

33. Mueller, M. J. (1997) Enzymes involved in jasmonic acid biosynthesis. *Physiologia Plantarum*, **100**, 653.

34. Sembdner, G. and Parthier, B. (1993) The biochemistry and the physiological and molecular actions of jasmonates. *Annu. Rev. Plant Physiol. Plant Mol. Biol.*, **44**, 569.

35. Croft, K. P., Juttner, F. and Slusarenko, A. J. (1993) Volatile products of the lipoxygenase pathway evolved from *Phaseolus vulgaris* (L.) leaves inoculated with *Pseudomonas syringae* pv. phaseolicola. *Plant Physiol*, **101**, 13.

36. Farmer, E. E. and Ryan, C. A. (1992) Octadecanoid precursors of jasmonic acid activate the synthesis of wound-inducible proteinase inhibitors. *Plant Cell*, **4**, 129.

37. Blechert, S., Brodschelm, W., Holder, S., Kammerer, L., Kutchan, T. M., Mueller, M. J., Xia, Z. Q. and Zenk, M. H. (1995) The octadecanoic pathway: signal molecules for the regulation of secondary pathways. *Proc. Natl Acad. Sci. USA*, **92**, 4099.

38. McCloud, E. S. and Baldwin, I. T. (1997) Herbivory and caterpillar regurgitants amplify the wound-induced increases in jasmonic acid but not nicotine in *Nicotiana sylvestris*. *Planta*, **203**, 430.

39. Enyedi, A. J., Yalpani, N., Silverman, P. and I., R. (1992) Signal molecules in systemic plant resistance to pathogens and pests. *Cell*, **70**, 879.

40. Karban, R., Baldwin, I. T., Baxter, K. J., Laue, G. and Felton, G. W. (2000) Communication between plants: induced resistance in wild tobacco plants following clipping of neigh-boring sagebrush. *Oecologia*, **125**, 66.

41. McConn, M., Creelman, R. A., Bell, E., Mullet, J. E. and Browse, J. (1997) Jasmonate is essential for insect defense in Arabidopsis. *Proc. Natl Acad. Sci. USA*, **94**, 5473.

42. Bell, E., Creelman, R. A. and Mullet, J. E. (1995) A chloroplast lipoxygenase is required for wound-induced jasmonic acid accumulation in *Arabidopsis. Proc. Natl Acad. Sci. USA*, **92**, 8675.

43. Weber, H., Vick, B. A. and Farmer, E. E. (1997) Dinor-oxo-phytodienoic acid: a new hexadecanoid signal in the jasmonate family. *Proc. Natl Acad. Sci. USA*, **94**, 10473.

44. Constabel, C. P., Yip, L. and Ryan, C. A. (1998) Prosystemin from potato, black night-shade, and bell pepper: primary structure and biological activity of predicted systemin polypeptides. *Plant Mol. Biol.*, **36**, 55.

45. McGurl, B., Pearce, G., Orozco-Cardenas, M. and Ryan, C. A. (1992) Structure, expression, and antisense inhibition of the systemin precursor gene. *Science*, **255**, 1570.

46. Narvaez-Vasquez, J., Pearce, G., Orozco-Cardenas, M. L., Franceschi, V. R. and Ryan, C. A. (1995) Autoradiographic and biochemical evidence for the systemic translocation of systemin in tomato plants. *Planta*, **195**, 593.

47. Bergey, D. R., Howe, G. A. and Ryan, C. A. (1996) Polypeptide signaling for plant defensive genes exhibits analogies to defense signaling in animals. *Proc. Natl Acad. Sci. USA*, **93**, 12053.

48. Pearce, G., Strydom, D., Johnson, S. and Ryan, C. A. (1991) A Polypeptide from tomato leaves induces wound-inducible proteinase inhibitor proteins. *Science*, **253**, 895.

49. Orozco-Cardenas, M., McGurl, B. and Ryan, C. A. (1993) Expression of an antisense prosystemin gene in tomato plants reduces resistance toward *Manduca sexta* larvae. *Proc. Natl Acad. Sci. USA*, **90**, 8273.

50. Constabel, C. P., Bergey, D. R. and Ryan, C. A. (1995) Systemin activates synthesis of wound-inducible tomato leaf polyphenol oxidase via the octadecanoid defense signaling pathway. *Proc. Natl Acad. Sci. USA*, **92**, 407.

51. Martin, W. R., Morgan, P. W., Sterling, W. L. and Meola, R. W. (1988) Stimulation of ethylene production in cotton by salivary enzymes of the cotton fleahopper (Heteroptera: Miridae) *Environ. Entomol.*, **17**, 930.

52. Rieske, L. K. and Raffa, K. F. (1995) Ethylene emission by a deciduous tree, *Tilia americana*, in response to feeding by introduced basswood thrips, *Thrips calcaratus. J. Chem. Ecol.*, **21**, 187.

53. Kahl, J., Siemens, D. H., Aerts, R. J., Gäbler, R., Kühnemann, F., Preston, C. A. and Baldwin, I. T. (2000) Herbivore-induced ethylene suppresses a direct defense but not a putative indirect defense against an adapted herbivore. *Planta*, **210**, 336.

54. O'Donnell, P. J., Calvert, C., Atzorn, R., Wasternack, C., Leyser, H. M. O. and Bowles, D. J. (1996) Ethylene as a signal mediating the wound response of tomato plants. *Science*, **274**, 1914.

55. Felix, G. and Boller, T. (1995) Systemin induces rapid ion fluxes and ethylene bio-synthesis in *Lycopersicon peruvianum* cells. *Plant J.*, **7**, 381.

56. Yip, W. K., Moore, T. and Yang, S. F. (1992) Differential accumulation of transcripts for four tomato 1-aminocyclopropane-1-carboxylate synthase homologs under various conditions. *Proc. Natl Acad. Sci. USA*, **89**, 2475.

57. Barry, C. S., Blume, B., Bouzayen, M., Cooper, W., Hamilton, A. J. and Grierson, D. (1996) Differential expression of the 1-aminocyclopropane-1-carboxylate oxidase gene family of tomato. *Plant J.*, **9**, 525 .

58. Ecker, J. R. and Davis, R. W. (1987) Plant defense genes are regulated by ethylene. *Proc. Natl Acad. Sci. USA*, **84**, 5202.

59. Chao, W. S., Gu, Y.-Q., Pautot, V., Bray, E. A. and Walling, L. L. (1999) Leucine amino-peptidase RMAs, proteins and activities increase in response to water deficit, salinity

and the wound signals systemin, methyl jasmonate and abscisic acid. *Plant Physiol.*, **120**, 979.

60. Pena-Cortes, H., Sanchez-Serrano, J. J., Mertens, R., Willmitzer, L. and Prat, S. (1989) Abscisic acid is involved in the wound-induced expression of the proteinase inhibitor II gene in potato and tomato. *Proc. Natl Acad. Sci. USA*, **86**, 9851.

61. Birkenmeier, G. F. and Ryan, C. A. (1998) Wound signaling in tomato plants – evidence that ABA is not a primary signal for defense gene activation. *Plant Physiol*, **117**, 687.

62. Laudert, D. and Weiler, E. W. (1998) Allene oxide synthase: a major control point in *Arabdopsis thaliana* octadecanoid signaling. *Plant J.*, **15**, 675.

63. Wildon, D. C., Thain, J. F., Minchin, P. E. H., Gubb, I. R., Reilly, A. J., Skipper, Y. D., Doherty, H. M., O'Donell, P. J. and Bowles, D. J. (1992) Electrical signalling and systemic proteinase inhibitor induction in the wounded plant. *Nature*, **360**, 62.

64. Hartley, S. E. and Lawton, J. H. (1991) Biochemical aspects and significance of the rapidly induced accumulation of phenolics in birch foliage. In Tallamy, D. W. and Raupp, M. J. (eds), *Phytochemical Induction by Herbivores*. John Wiley & Sons, New York, p. 105.

65. Stout, M. J., Workman, J. and Duffey, S. S. (1994) Differential induction of tomato foliar proteins by arthropod herbivores. *J. Chem. Ecol.*, **20**, 2575.

66. Baldwin, I. T. (1990) Herbivory simulations in ecological research. *Trends Ecol. Evol.*, **5**, 91.

67. Felton, G. W. and Eichenseer, H. (1999) Herbivore saliva and its effect on plant defense against herbivores and pathogens. In Agrawal, A. A., Tuzun, S. and Bent, E. (eds), *Induced Plant Defenses Against Pathogens and Herbivores. Biochemistry, Ecology and Agriculture.* APS Press, St. Paul, MN, p. 19.

68. Lin, H., Kogan, M. and Fischer, D. (1990) Induced resistance in soybean to the Mexican bean beetle (Coleptera: Coccinellidae): comparison of inducing factors. *Environ. Entomol.*, **19**, 1852.

69. Korth, K. L. and Dixon, R. A. (1997) Evidence for chewing insect-specific molecular events distinct from a general wound response in leaves. *Plant Physiol.*, **115**, 1299.

70. Jongsma, M. A., Bakker, P. L., Visser, B. and Stiekema, W. J. (1994) Trypsin inhibitor activity in mature tobacco and tomato plants is mainly induced locally in response to insect attack, wounding and virus infection. *Planta*, **195**, 29.

71. Eichenseer, H., Mathews, M. C., Bi, J. L., Murphy, B. and Felton, G. W. (1999) Salivary glucose oxidase: multifunctional roles for *Helicoverpa zea*? *Arch. Insect Biochem. Physiol.*, **42**, 99.

72. Thaler, J. S., Fidantsef, A. L., Duffey, S. S. and Bostock, R. M. (1999) Trade-offs in plant defense against pathogens and herbivores: A field demonstration of chemical elicitors of induced resistance. *J. Chem. Ecol.*, **25**, 1597.

73. Sano, H. and Ohashi, Y. (1995) Involvement of small GTP-binding proteins in defense signal-transduction pathways of higher plants. *Proc. Natl Acad. Sci. USA*, **92**, 4138.

74. Karban, R., Adamchak, R. and Schnathorst, W. C. (1987) Induced resistance and inter-specific competition between spider mites and a vascular wilt fungus. *Science*, **235**, 678.

75. Ryan, C. A. (1974) Assay and biochemical properties of the proteinase inhibitor-inducing factor, a wound hormone. *Plant Physiol.*, **54**, 328.

76. Rojo, E., Leon, J. and Sanchez-Serrano, J. J. (1999) Cross-talk between wound signalling pathways determines local versus systemic gene expression in *Arabidopsis thaliana*. *Plant J.*, **20**, 135.

77. Sabelis, M. W. and Dicke, M. (1985) Long-range dispersal and searching behaviour. In Helle, W. and Sabelis, M. W. (eds), *Spider Mites: Their Biology, Natural Enemies and Control. World Crop Pests 1A*. Elsevier, Amsterdam, p. 141.

78. Mattiacci, L., Dicke, M. and Posthumus, M. A. (1994) Induction of parasitoid attracting synomone in brussels sprouts plants by feeding of *Pieris brassicae* larvae: role of mechanical damage and herbivore elicitor. *J. Chem. Ecol.*, **20**, 2229.

79. Turlings, T. C. J., Tumlinson, J. H. and Lewis, W. J. (1990) Exploitation of herbivore-induced plant odors by host-seeking parasitic wasps. *Science*, **250**, 1251.

80. Dicke, M., van Beek, T. A., Posthumus, M. A., Ben Dom, N., van Bokhoven, H. and de Groot, A. E. (1990) Isolation and identification of volatile kairomone that affects acarine predator-prey interactions. Involvement of host plant in its production. *J. Chem. Ecol.*, **16**, 381.

81. Takabayashi, J. and Dicke, M. (1996) Plant-carnivore mutualism through herbivore-induced carnivore attractants. *Trends Plant Sci.*, **1**, 109.

82. Turlings, T. C. J., Loughrin, J. H., McCall, P. J., Rose, U. S. R., Lewis, W. J. and Tumlinson, J. H. (1995) How caterpillar-damaged plants protect themselves by attracting parasitic wasps. *Proc. Natl Acad. Sci. USA*, **92**, 4169.

83. Dicke, M. (1999) Evolution of induced indirect defence of plants. In Tollrian, R. and Harvell, C. D. (eds), *The Ecology and Evolution of Inducible Defenses*. Princeton University Press, Princeton, NJ, p. 62.

84. Takabayashi, J., Dicke, M., Takahashi, S., Posthumus, M. A. and van Beek, T. A. (1994) Leaf age affects composition of herbivore-induced synomones and attraction of predatory mites. *J. Chem. Ecol.*, **20**, 373.

85. Krips, O. E., Willems, P. E. L., Gols, R., Posthumus, M. A. and Dicke, M. (1999) The response of *Phytoseiulus persimilis* to spider-mite induced volatiles from gerbera: influence of starvation and experience. *J. Chem. Ecol.*, **25**, 2623.

86. Bolter, C. J., Dicke, M., van Loon, J. J. A., Visser, J. H. and Posthumus, M. A. (1997) Attraction of Colorado potato beetle to herbivore damaged plants during herbivory and after its termination. *J. Chem. Ecol.*, **23**, 1003.

87. Agelopoulos, N. G. and Keller, M. A. (1994) Plant-natural enemy association in the tritrophic system, *Cotesia rubecula-Pieris rapae*-Brassicaceae (Crucifera): III. Collection and identification of plant and frass volatiles. *J. Chem. Ecol.*, **20**, 1955.

88. McCall, P. J., Turlings, T. C. J., Loughrin, J., Proveaux, A. T. and Tumlinson, J. H. (1994) Herbivore-induced volatile emissions from cotton (*Gossypium hirsutum* L.) seedlings. *J. Chem. Ecol.*, **20**, 3039.

89. Donath, J. and Boland, W. (1994) Biosynthesis of acyclic homoterpenes in higher plants parallels steroid hormone metabolism. *J. Plant Physiol.*, **143**, 473.

90. Pare, P. W. and Tumlinson, J. H. (1997) De Novo biosynthesis of volatiles induced by insect herbivory in cotton plants. *Plant Physiol*, **114**, 1161.

91. Dicke, M., Sabelis, M. W., Takabayashi, J., Bruin, J. and Posthumus, M. A. (1990) Plant strategies of manipulating predator-prey interactions through allelochemicals: prospects for application in pest control. *J. Chem. Ecol.*, **16**, 3091.

92. Turlings, T. C. J. and Tumlinson, J. H. (1992) Systemic release of chemical signals by herbivore-injured corn. *Proc. Natl Acad. Sci. USA*, **89**, 8399.

93. Röse, U. S. R., Manukian, A., Heath, R. R. and Tumlinson, J. H. (1996) Volatile semiochemicals released from undamaged cotton leaves – a systemic response of living plants to caterpillar damage. *Plant Physiol.*, **111**, 487.

94. Dicke, M., Baarlen, P.v., Wessels, R. and Dijkman, H. (1993) Herbivory induces systemic production of plant volatiles that attract predators of the herbivore: extraction of endogenous elicitor. *J. Chem. Ecol.*, **19**, 581.

95. Meiners, T. and Hilker, M. (2000) Induction of plant synomones by oviposition of a phytophagous insect. *J. Chem. Ecol.*, **26**, 221.

96. Dicke, M. (1999) Are herbivore-induced plant volatiles reliable indicators of herbivore identity to foraging carnivorous arthropods? *Entomol. Exp. Appl.*, **92**, 131.
97. Du, Y., Poppy, G. M., Powell, W., Pickett, J. A., Wadhams, L. J. and Woodcock, C. M. (1998) Identification of semiochemicals released during aphid feeding that attract parasitoid *Aphidius ervi*. *J. Chem. Ecol.*, **24**, 1355.
98. Takabayashi, J., Takahashi, S., Dicke, M. and Posthumus, M. A. (1995) Developmental stage of herbivore *Pseudaletia separata* affects production of herbivore-induced synomone by corn plants. *J. Chem. Ecol.*, **21**, 273.
99. Blaakmeer, A., Geervliet, J. B. F., van Loon, J. J. A., Posthumus, M. A., van Beek, T. A. and de Groot, A. E. (1994) Comparative headspace analysis of cabbage plants damaged by two species of *Pieris* caterpillars: consequences for in-flight host location by *Cotesia* parasitoids. *Entomol. Exp. Appl.*, **73**, 175.
100. Takabayashi, J., Dicke, M. and Posthumus, M. A. (1994) Volatile herbivore-induced terpenoids in plant-mite interactions: variation caused by biotic and abiotic factors. *J. Chem. Ecol.*, **20**, 1329.
101. Hatanaka, A., Kajiwara, T. and Sekiya, J. (1987) Biosynthetic pathway for C6-aldehydes formation from linolenic acid in green leaves. *Chem. Phys. Lipids*, **44**, 341.
102. Visser, J. H. and Avé, D. A. (1978) General green leaf volatiles in the olfactory orientation of the Colorado beetle, *Leptinotarsa decemlineata*. *Entomol. Exp. Appl.*, **24**, 538.
103. van Loon, J. J. A. and Dicke, M. (2001) Sensory ecology of arthropods utilizing plant infochemicals. In Barth, F. G. and Schmid, A. (eds), *Sensory Ecology*. Springer Verlag, Heidelberg, p. 253.
104. Whitman, D. W. and Eller, F. J. (1992) Orientation of *Microplitis croceipes* (Hymenoptera: Braconidae) to green leaf volatiles: Dose-response curves. *J. Chem. Ecol.*, **18**, 1743.
105. Gershenzon, J. and Croteau, R. (1991) Terpenoids. In Rosenthal, G. A. and Berenbaum, M. R. (eds), *Herbivores: Their Interactions with Secondary Plant Metabolites*. Academic Press, New York. Vol. 1, p. 165.
106. Lichtenthaler, H. K. (1999) The 1-deoxy-D-xylulose-5-phosphate pathway of isoprenoid biosynthesis in plants. *Annu. Rev. Plant Physiol. Plant Mol. Biol.*, **50**, 47.
107. Boland, W., Feng, Z., Donath, J. and Gäbler, A. (1992) Are acyclic C11 and C16 homoterpenes plant volatiles indicating herbivory? *Naturwissenschaften*, **79**, 368.
108. Dicke, M. (1994) Local and systemic production of volatile herbivore-induced terpenoids: Their role in plant-carnivore mutualism. *J. Plant Physiol.*, **143**, 465.
109. Bouwmeester, H. J., Verstappen, F., Posthumus, M. A. and Dicke, M. (1999) Spider-mite induced (3S)-(E)-nerolidol synthase activity in cucumber and Lima bean. The first dedicated step in acyclic C11-homoterpene biosynthesis. *Plant Physiol.*, **121**, 173.
110. Degenhardt, J. and Gershenzon, J. (2000) Demonstration and characterization of (E)-nerolidol synthase from maize: a herbivore-inducible terpene synthase participating in (3E)-4,8-dimethyl-1,3,7-nonatriene biosynthesis. *Planta*, **210**, 815.
111. Dicke, M., Gols, R., Ludeking, D. and Posthumus, M. A. (1999) Jasmonic acid and herbivory differentially induce carnivore-attracting plant volatiles in lima bean plants. *J. Chem. Ecol.*, **25**, 1907.
112. Scutareanu, P., Drukker, B., Bruin, J., Posthumus, M. A. and Sabelis, M. W. (1997) Volatiles from *Psylla*-infested pear trees and their possible involvement in attraction of anthocorid predators. *J. Chem. Ecol.*, **23**, 2241.
113. Takabayashi, J., Dicke, M. and Posthumus, M. A. (1991) Variation in composition of predator-attracting allelochemicals emitted by herbivore-infested plants: relative influence of plant and herbivore. *Chemoecology*, **2**, 1.

114. Geervliet, J. B. F., Posthumus, M. A., Vet, L. E. M. and Dicke, M. (1997) Comparative analysis of headspace volatiles from different caterpillar-infested and uninfested food plants of *Pieris* species. *J. Chem. Ecol.*, **23**, 2935.

115. Kaiser, R. A. J. (1993) On the scent of orchids. In Teranishi, R., Buttery, R. G. and Sugisawa, H. (eds), *Bioactive volatile compounds from plants.* American Chemical Society, Washington, DC, ACS Symposium Series 525, p. 240.

116. Vet, L. E. M. and Dicke, M. (1992) Ecology of infochemical use by natural enemies in a tritrophic context. *Annu. Rev. Entomol.*, **37**, 141.

117. Sabelis, M. W., Afman, B. P. and Slim, P. J. (1984) Location of distant spider mite colonies by *Phytoseiulus persimilis*: localization and extraction of a kairomone. *Acarology VI*, **1**, 431.

118. Turlings, T. C. J., Tumlinson, J. H., Eller, F. J. and Lewis, W. J. (1991) Larval-damaged plants: source of volatile synomones that guide the parasitoid *Cotesia marginiventris* to the micro-habitat of its hosts. *Entomol. Exp. Appl.*, **58**, 75.

119. Steinberg, S., Dicke, M. and Vet, L. E. M. (1993) Relative importance of infochemicals from first and second trophic level in long-range host location by the larval parasitoid *Cotesia glomerata*. *J. Chem. Ecol.*, **19**, 47.

120. Geervliet, J. B. F., Vet, L. E. M. and Dicke, M. (1994) Volatiles from damaged plants as major cues in long-range host-searching by the specialist parasitoid *Cotesia rubecula*. *Entomol. Exp. Appl.*, **73**, 289.

121. van Loon, J. J. A., de Boer, J. G. and Dicke, M. (2000) Parasitoid-plant mutualism: parasitoid attack of herbivore increases plant reproduction. *Entomol. Exp. Appl.*, **97**, 219.

122. McCall, P. J., Turlings, T. C. J., Lewis, W. J. and Tumlinson, J. H. (1993) Role of plant volatiles in host location by the specialist parasitoid *Microplitis croceipes* Cresson (Braconidae: Hymenoptera). *J. Insect Behav.*, **6**, 625.

123. van Loon, J. J. A., de Vos, E. W. and Dicke, M. (2000) Orientation behaviour of the predatory hemipteran *Perillus bioculatus* to plant and prey odours. *Entomol. Exp. Appl.*, **96**, 51.

124. Sabelis, M. W. and van de Baan, H. E. (1983) Location of distant spider mite colonies by phytoseiid predators: demonstration of specific kairomones emitted by *Tetranychus urticae* and *Panonychus ulmi. Entomol. Exp. Appl.*, **33**, 303.

125. Wiskerke, J. S. C. and Vet, L. E. M. (1994) Foraging for solitarily and gregariously feeding caterpillars: a comparison of two related parasitoid species. *J. Insect. Beh.*, **7**, 585.

126. Janssen, A. (1999) Plants with spider-mite prey attract more predatory mites than clean plants under greenhouse conditions. *Entomol. Exp. Appl.*, **90**, 191.

127. de Moraes, C. M., Lewis, W. J., Paré, P. W., Alborn, H. T. and Tumlinson, J. H. (1998) Herbivore-infested plants selectively attract parasitoids. *Nature*, **393**, 570.

128. Sabelis, M. W., Vermaat, J. E. and Groeneveld, A. (1984) Arrestment responses of the predatory mite, *Phytoseiulus persimilis*, to steep odour gradients of a kairomone. *Physiol. Entomol.*, **9**, 437.

129. Sabelis, M. W. and Afman, B. P. (1994) Synomone-induced suppression of take-off in the phytoseiid mite *Phytoseiulus persimilis* Athias-Henriot. *Exp. Appl. Acarol.*, **18**, 711.

130. Harvey, J. (2000) Dynamic effects of parasitism by an endoparasitoid wasp on the development of two host species: implications for host quality and parasitoid fitness. *Ecol. Entomol.*, **25**, 267.

131. Dicke, M. and van Loon, J. J. A. (2000) Multitrophic effects of herbivore-induced plant volatiles in an evolutionary context. *Entomol. Exp. Appl.*, **97**, 237.

132. Dicke, M. and Vet, L. E. M. (1999) Plant-carnivore interactions: evolutionary and ecological consequences for plant, herbivore and carnivore. In Olff, H., Brown, V. K. and Drent, R. H. (eds), *Herbivores: Between Plants and Predators.* Blackwell Science, Oxford, p. 483.

133. Dicke, M. (1986) Volatile spider-mite pheromone and host-plant kairomone, involved in spaced-out gregariousness in the spider mite *Tetranychus urticae*. *Physiol. Entomol.*, **11**, 251.

134. Shiojiri, K., Takabayashi, J., Yano, S. and Takafuji, A. (2000) Flight response of parasitoids toward plant-herbivore complexes: A comparative study of two parasitoid-herbivore systems on cabbage plants. *Appl. Entomol. Zool.*, **35**, 87.

135. Hildebrand, D. F., Brown, G. C., Jackson, D. M. and Hamilton-Kemp, T. R. (1993) Effects of some leaf-emitted volatile compounds on aphid population increase. *J. Chem. Ecol.*, **19**, 1875.

136. Bruin, J., Sabelis, M. W. and Dicke, M. (1995) Do plants tap SOS signals from their infested neighbours? *Trends Ecol. Evol.*, **10**, 167.

137. Shonle, I. and Bergelson, J. (1995) Interplant communication revisited. *Ecology*, **76**, 2660.

138. Arimura, G., Ozawa, R., Shimoda, T., Nishioka, T., Boland, W. and Takabayashi, J. (2000) Herbivory-induced volatiles elicit defence genes in lima bean leaves. *Nature*, **406**, 512.

139. Hopke, J., Donath, J., Blechert, S. and Boland, W. (1994) Herbivore-induced volatiles: the emission of acyclic homoterpenes from leaves of *Phaseolus lunatus* and *Zea mays* can be triggered by a β-glucosidase and jasmonic acid. *FEBS Lett.*, **352**, 146.

140. Mattiacci, L., Dicke, M. and Posthumus, M. A. (1995) beta-Glucosidase: an elicitor of herbivore-induced plant odor that attracts host-searching parasitic wasps. *Proc. Natl Acad. Sci. USA*, **92**, 2036.

141. Alborn, T., Turlings, T. C. J., Jones, T. H., Steinhagen, G., Loughrin, J. H. and Tumlinson, J. H. (1997) An elicitor of plant volatiles from beet armyworm oral secretion. *Science*, **276**, 945.

142. Boland, W., Hopke, J., Donath, J., Nueske, J. and Bublitz, F. (1995) Jasmonic acid and coronatin induce odor production in plants. *Angew Chem. Int. Ed. Engl.*, **34**, 1600.

143. Gols, R., Posthumus, M. A. and Dicke, M. (1999) Jasmonic acid induces the production of gerbera volatiles that attract the biological control agent *Phytoseiulus persimilis*. *Ent. Exp Appl*, **93**, 77.

144. Avdiushko, S., Croft, K. P. C., Brown, G. C., Jackson, D. M., Hamiltonkemp, T. R. and Hildebrand, D. (1995) Effect of volatile methyl jasmonate on the oxylipin pathway in tobacco, cucumber, and arabidopsis. *Plant Physiol.*, **109**, 1227.

145. Thaler, J. S. (1999) Jasmonate-inducible plant defenses cause increased parasitism of herbivores. *Nature*, **399**, 686.

146. Boland, W., Koch, T., Krumm, T., Piel, J. and Jux, A. (1999) Induced biosynthesis of insect semiochemicals in plants. In Chadwick, D. J. and Goode, J. (eds), *Insect–Plant Interactions and Induced Plant Defence*. Novartis Foundation Symposium 223. Wiley, Chicester, p. 110.

147. Shulaev, V., Silverman, P. and Raskin, I. (1997) Airborne signalling by methyl salicylate in plant pathogen resistance. *Nature*, **385**, 718.

148. Bi, J. L., Murphy, J. B. and Felton, G. W. (1997) Does salicylic acid act as a signal for induced resistance in cotton to *Helicoverpa zea*? *J. Chem. Ecol.*, **23**, 1805.

149. Turlings, T. C. J., Wäckers, F. L., Vet, L. E. M., Lewis, W. J. and Tumlinson, J. H. (1993) Learning of host-finding cues by hymenopterous parasitoids. In Papaj, D. R. and Lewis, A. C. (eds), *Insect Learning: Ecological and Evolutionary Perspectives*. Chapman & Hall, New York, p. 51.

150. Ozawa, R., Arimura, G., Takabayashi, J., Shimoda, T. and Nishioka, T. (2000) Involvement of jasmonate- and salicylate-related signaling pathway for the production of specific herbivore-induced volatiles in plants. *Plant Cell Physiol.*, **41**, 391.

151. Engelberth, J., Koch, T., Kühnemann, F. and Boland, W. (2000) Channel-forming peptaibols are a novel class of potent elicitors of plant secondary metabolism and tendril coiling. *Angew. Chem. Intl. Ed.*, **39**, 1860.

152. Piel, J., Atzorn, R., Gabler, R., Kuhnemann, F. and Boland, W. (1997) Cellulysin from the plant parasitic fungus *Trichoderma viride* elicits volatile biosynthesis in higher plants via the octadecanoid signalling cascade. *FEBS Lett.*, **416**, 143.

153. Chew, F. S. and Renwick, J. A. A. (1995) Host plant choice in *Pieris* butterflies. In Carde, R. T. and Bell, W. J. (eds), *Chemical Ecology of Insects*. Chapman & Hall, New York, Vol. 2, p. 214.

154. Baldwin, I. T. and Preston, C. A. (1999) The eco-physiological complexity of plant responses to insect herbivores. *Planta*, **208**, 137.

155. Stout, M. J., Fidantsef, A. L., Duffey, S. S. and Bostock, R. M. (1999) Signal interactions in pathogen and insect attack: systemic plant-mediated interactions between pathogens and herbivores of the tomato, *Lycopersicon esculentum*. *Physiol. Mol. Plant Pathol.*, **54**, 115.

156. Jones, D. A., Thomas, C. M., Hammond-Kosack, K. E., Balint-Kurti, P. J. and Jones, J. D. G. (1994) Isolation of the tomato Cf-9 gene for resistance to *Cladosporium fulvum* by transposon tagging. *Science*, **266**, 789.

157. McGurl, B., Orozco-Cardenas, M., Pearce, G. and Ryan, C. A. (1994) Overexpression of the prosystemin gene in transgenic tomato plants generates a systemic signal that constitutively induces proteinase inhibitor synthesis. *Proc. Natl Acad. Sci. USA*, **91**, 9799.

158. Dangl, J. L. (1993) The emergence of *Arabidopsis thaliana* as a model for plant-pathogen interactions. *Adv. Plant Path.*, **10**, 127.

159. Grant-Petersson, J. and Renwick, J. A. A. (1996) Effects of ultraviolet-B exposure of *Arabidopsis thaliana* on herbivory by two crucifer-feeding insects (Lepidoptera). *Environ. Entomol.*, **25**, 135.

160. Mauricio, R. and Rausher, M. D. (1997) Experimental manipulation of putative selective agents provides evidence for the role of natural enemies in the evolution of plant defense. *Evolution*, **5**, 1435.

161. Avdiushko, S. A., Brown, G. C., Dahlman, D. L. and Hildebrand, D. F. (1997) Methyl jasmonate exposure induces insect resistance in cabbage and tobacco. *Environ. Entomol.*, **26**, 642.

162. Palaniswamy, P. and Lamb, R. J. (1993) Wound-induced antixenotic resistance to flea beetles, *Phyllotreta cruciferae* (Goeze) (Coleoptera: Chrysomelidae) in crucifers. *Can. Entomol.*, **125**, 903.

163. Agrawal, A. A. (1998) Induced responses to herbivory and increased plant performance. *Science*, **279**, 1201.

164. Shapiro, A. M. and DeVay, J. E. (1987) Hypersensitivity reaction of *Brassica nigra* L. (Cruciferae) kills eggs of *Pieris* butterflies (Lepidoptera: Pieridae). *Oecologia*, **71**, 631.

165. van Poecke, R. M. P., Posthumus, M. A. and Dicke, M. (2001) Herbivore-induced volatile production by *Arabidopsis thaliana* leads to attraction of the parasitoid *Cotesia rubecula*: chemical, behavioral and gene-expression analysis. *J. Chem. Ecol.*, **27**, 1911.

166. Pieterse, C. M. J., van Wees, S. C. M., van Pelt, J. A., Knoester, M., Laan, R., Gerrits, H., Weisbeek, P. J. and van Loon, L. C. (1998) A novel signaling pathway controlling induced systemic resistance in *Arabidopsis*. *Plant Cell*, **10**, 1571.

Index

Printed in the United States
By Bookmasters